PERGAMON INTERNATIONAL LIBRARY
of Science, Technology, Engineering and Social Studies
The 1000-volume original paperback library in aid of education,
industrial training and the enjoyment of leisure
Publisher: Robert Maxwell, M.C.

INTERNATIONAL SERIES ON
MATERIALS SCIENCE AND TECHNOLOGY
VOLUME 22 - EDITOR: W. S. OWEN, D.ENG., PH.D.

Dimensional Instability
AN INTRODUCTION

THE PERGAMON TEXTBOOK
INSPECTION COPY SERVICE

An inspection copy of any book published in the Pergamon International Library will gladly be sent to academic staff without obligation for their consideration for course adoption or recommendation. Copies may be retained for a period of 60 days from receipt and returned if not suitable. When a particular title is adopted or recommended for adoption for class use and the recommendation results in a sale of 12 or more copies, the inspection copy may be retained with our compliments. The Publishers will be pleased to receive suggestions for revised editions and new titles to be published in this important International Library.

Dimensional Instability

AN INTRODUCTION

BY

CHARLES W. MARSCHALL

AND

ROBERT E. MARINGER

Battelle, Columbus Laboratories, Columbus, Ohio, U.S.A.

PERGAMON PRESS

OXFORD · NEW YORK · TORONTO · SYDNEY · PARIS · FRANKFURT

U.K.	Pergamon Press Ltd., Headington Hill Hall, Oxford OX3 0BW, England
U.S.A.	Pergamon Press Inc., Maxwell House, Fairview Park, Elmsford, New York 10523, U.S.A.
CANADA	Pergamon of Canada Ltd., 75 The East Mall, Toronto, Ontario, Canada
AUSTRALIA	Pergamon Press (Aust.) Pty. Ltd., 19a Boundary Street, Rushcutters Bay, N.S.W. 2011, Australia
FRANCE	Pergamon Press SARL, 24 Rue des Ecoles, 75240 Paris, Cedex 05, France
WEST GERMANY	Pergamon Press GmbH, 6242 Kronberg-Taunus, Pferdstrasse 1, West Germany

First edition 1977

Library of Congress Cataloguing in Publication Data
Marschall, Charles Walter.
Dimensional instability.
(International series in materials science & technology; 22)
1. Strength of materials. 2. Stability.
I. Maringer, Robert Edward, joint author. II. Title.
TA407.M34 1977   620.1'123   76-53726
ISBN 0-08-021305-7

Printed in Great Britain by Pitman Press, Bath

Contents

Preface

The need for a book on dimensional instability became evident to us as a result of numerous conversations with designers, materials specialists, and manufacturing engineers. They are being asked to design and build components that will exhibit dimensional stability similar to, or better than, that required of high-quality gage blocks, i.e., unit dimensional changes no greater than about one part per million over long periods of time. Unlike gage blocks, however, the components often have to carry load and operate in a fluctuating environment. Furthermore, for various reasons it is not advisable to build components from gage block materials. Hence the designers and materials engineers find themselves in the position of trying to design with materials for which little or no dimensional stability information is available. To add to the predicament, little is available concerning general principles of dimensional stability that might provide direction in solving specific engineering problems.

This book was written in response to that need. It is based on the premise that engineering materials exhibit perfect stability and perfect elasticity only to a first approximation--as requirements for constancy of shape and dimensions grow increasingly demanding, a point is reached beyond which dimensional changes in service must be expected and attention directed to keeping these changes within tolerable levels. As an introductory book, it deals primarily with general principles of how dimensional instability can occur and offers general suggestions on how it can be minimized. Through an understanding of these principles, the reader should be better prepared to anticipate and to analyze specific dimensional instability problems and to plan a course of action that will ultimately produce a satisfactory solution.

In addition to dealing with general principles, the book also describes in some detail methods commonly employed to investigate dimensional instability in laboratory test coupons. It is through the use of such tests that the effects of material variables and service conditions on dimensional instability are evaluated and applied to practical situations.

The book's final two chapters deal with special materials that frequently find application in precision design. In certain situations, a component's performance may be impaired by the dimensional changes resulting from even relatively small temperature fluctuations. Use of materials that exhibit very low thermal expansivity can lead to greatly improved performance in such situations. Materials of this type are described in Chapter 9. Situations also arise in which it would be advantageous if the elastic moduli of a component remained invariant with changing temperature. Materials that exhibit near-zero thermoelastic coefficients are discussed in Chapter 10.

All authors, we suspect, discover at some point that they are unable to cover their subject to the depth and breadth that they had hoped at the outset. We are no exceptions. Our outlook doubtless would have been broadened by more opportunities for discussion with knowledgeable persons of various nationalities actively working in this area of technology. Time and budget considerations limited such discussions, however, leaving us with the realization that the final result of our efforts is biased toward our own views and is based largely on work that has been conducted in the United States. In

addition, more attention is devoted to metals than to ceramics and other
nonmetals because of our greater familiarity with this type of material.

We express our appreciation to the Metals and Ceramics Information
Center of Battelle's Columbus Laboratories for financial assistance in the
planning, data gathering, organizing, and initial writing phases of this
project and to Battelle Institute for financial assistance in the major
portion of manuscript preparation. The cooperation of the personnel at the
Battelle Seattle Research Center, in particular the late Mrs. Bernice Ives,
where much of the manuscript preparation was accomplished, is gratefully
acknowledged. We acknowledge also the assistance of our colleague Mr. P.R.
Held, both in gathering reference material for this book and in carefully
carrying out experiments to investigate microyield and microcreep behavi-
our. Finally, we thank various companies and individuals for granting us
permission to use unpublished data and the many publishers and authors who
permitted us to use copyright material.

Chapter 1.

Introduction

The ability to produce a reliable precision device such as a gyroscope or a telescope mirror depends upon the capability of the designer and upon the designer's understanding of material behavior, including the effects of processing. In meeting the exacting requirements of the device, particularly with respect to long-term reliability, the designer must recognize two aspects of the problem. First, the critical components of the device must be made to meet initial property and dimensional tolerances, and second, these properties and dimensions must be maintained throughout the life of the device in its service environment.

The first of these is the most obvious. A precision device must be made of the appropriate materials, with the right properties and the right dimensions, or it will never function in the first place. For example, a gage block must meet rigid specifications for hardness, flatness, parallelism and dimensions before it is ever put up for sale. A mirror must exhibit specified values of reflectivity, flatness or curvature before it is ever mounted into the optical device for which it is intended. A ball or race must be of certain dimensions and hardness before it is ever assembled as part of a complete bearing. Once the device is assembled and put into operation, its proper function depends upon the stability of its component properties and dimensions. It is this latter aspect of the problem with which this book is concerned.

Many engineers first become aware of dimensional instability through unfortunate experiences--a gyro drifts, a valve sticks, a mirror warps, or a standard changes--all after apparently acceptable processing and handling procedures have been followed. Understandably, such experiences can lead to a definition that voices the associated frustrations: dimensional instability is an unacceptable change in dimensions or shape in an environment where such changes would not normally be expected. This definition is clearly inadequate because all materials are unstable to some degree, and certain types of instability should be expected in most real applications. The problem really narrows down to keeping the instabilities within acceptable limits.

1.1 TYPES OF DIMENSIONAL INSTABILITY

It is helpful to define dimensional instability in a more general sense than implied above. In this book, dimensional instability is considered to be:

(a) A distortion or dimensional change occurring as a function of time in a fixed environment.
(b) A distortion or dimensional change, measured under a fixed environment, after exposure to a variable environment.
(c) A distortion or dimensional change, measured under a fixed environment, and depending upon the environmental path used to reach the fixed environment.

Examples of these three types of instability are pertinent.

1

Gage blocks have been observed to change dimensions over the years, even under carefully controlled conditions in the metrology laboratory. Such behavior illustrates dimensional change with time under a fixed environment. The magnitude of the changes may be on the order of ten parts per million (ppm) per year, but usually is much less. Some data to illustrate this are given in Fig. 1.1.

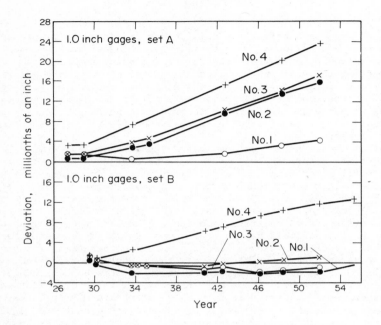

Fig. 1.1.   Example of length change with time under fixed environment for two sets of nominally similar 1-inch steel gage blocks (Emerson 1957).

One might take similar gage blocks, which normally are relatively stable in the metrology laboratory, and subject them to thermal cycling, say from -100° to +150° C (-150° to +300° F). The blocks may tend to grow or to shrink, depending upon their processing history, by several ppm per cycle. The relatively large irreversible dimensional changes that can be accumulated in response to environmental changes are illustrative of the second type of instability. Some metals show a significant dimensional change after each thermal cycle, as can be seen for uranium in Fig. 1.2. Uranium, however, is a special case, since its thermal expansion behavior is anisotropic.

Another important example of the second type of instability can be observed be loading a gage block in compression to a relatively high stress, taking care not to damage the measuring surfaces. Measurement of the block after removal of the load will probably reveal that the length has decreased, depending on the magnitude of the stress and processing history. If the length of the block is then monitored as a function of time, a gradual length increase may be observed, illustrative of the first type of instability described earlier.

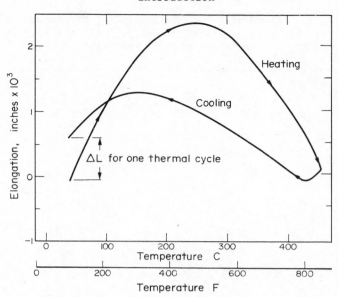

Fig. 1.2. Example of length change, measured under fixed environment, after exposure to variable environment; dilatometer curve for $300^{\circ}C$ rolled, alpha annealed uranium (3-in specimen).

Finally, the third type of instability can be illustrated by observing the elastic load-versus-deflection behavior of a gage block. If the sensitivity of the measurement instrumentation were great enough, it would be observed that the "elastic" displacement on gradual removal of the load did not exactly retrace the displacement observed on application of the load. This "hysteresis" indicates that the dimensions corresponding to some given load differ, depending upon whether the specific load is arrived at by increasing or by decreasing the load from some previous value. The possible magnitude of such hysteresis, symptomatic of the third type of instability, is illustrated in Fig. 1.3, where the data are for prestrained molybdenum crystals.

It might seem at first glance that dimensional instability is a relatively modern problem. A little reflection will show that this is not the case. Perhaps one of the most unstable and yet most useful of natural products is wood. The tendency of wood to warp, swell, or crack in response to its environment, especially the water content of the environment, is well known. A good deal of research has been conducted on the stabilization of wood. Modern kiln-drying and impregnation techniques have helped to alleviate the problem. The same may be said about shrinkage in clothing. Research has led to the "alleviation" of the problem. Note, however, that these instabilities have not been entirely eliminated. Most wood still warps a little, and most cotton fabrics still exhibit some shrinkage, but modern practice has reduced the instability to acceptable amounts.

Instability problems similar to those described above for natural products can be observed in many man-made products. Castings warp, rolled and drawn metal products twist and bend, epoxy resins shrink and crack, lacquers and enamels shrink and craze. Manufacturers and users learn to live with and

deal with all of these instabilities. Engineers try to process materials in
such a way that the instabilities are not sufficiently great to impair their
function, even though some instability still remains.

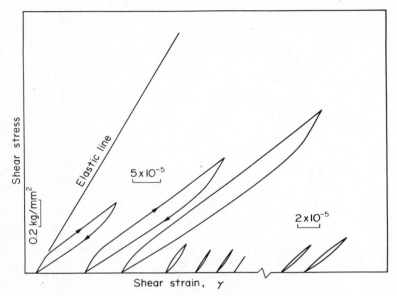

Fig. 1.3.   Example of length change, measured under
             fixed environment, and depending upon load-
             ing path; single crystal molybdenum specimen
             prestrained to a shear strain of 3×10^{-3} in
             compression (Meakin, 1967; copyright National
             Research Council of Canada, used by permis-
             sion).

    Two converging factors in modern technology require that the problems of
dimensional instability be reconsidered. The demand for instrumentation with
greater accuracy and reliability is continually increasing while, at the same
time, there is an ever-increasing demand for miniaturization.

    It is easy to rationalize the need for greater dimensional stability where
greater accuracy and reliability are required. Specifications for dimensions
are tighter, and performance depends upon maintenance of specified dimensions.
The warpage of an astronomical mirror by $\lambda/10$ (one-tenth of a wavelength of
light) is not too serious if the usage envisioned requires a surface accurate
to $\lambda/4$. The same change would be completely unacceptable in a mirror whose
function demanded a surface accurate to $\lambda/100$.

    To understand why miniaturization aggravates problems of dimensional insta-
bility requires a further appreciation of measurement problems and of the
mechanisms of instability. Let it suffice to say here that residual stresses
in surface layers can be a prime cause of dimensional instability. As parts
become smaller, the ratio of the amount of material in the surface layers to
that in the volume becomes larger, and the effects of these residual stresses
are magnified. A useful example can be found in the performance of Elgiloy,
a widely used spring material. When used in thicknesses of about 2 mm (0.080 in),
spring performance is good, as expected. When the material is reduced in
thickness to 1/8 mm (0.005 in) or less by grinding, burnishing, or lapping,

the spring performance is markedly impaired, and significant hysteresis in the stress – strain response is observed. This behavior is a result of the alteration of the properties of the alloy near the surface by the process of reducing the specimen size. In the thin sample, virtually the whole of the cross-section is made up of altered material, while in the thick sample, the altered material comprises only a small fraction of the total specimen.

Although this book deals almost exclusively with dimensional changes, it must be noted that dimensional, mechanical, and physical properties are often so interrelated that it is not to be expected that one property will change independently of another. The behavior of a specimen of iron containing a little carbon in solution can be used to illustrate this interdependence. When the carbon precipitates from solution in the form of carbide particles, the dimensions of the specimen decrease, the yield strength, hardness, and elastic modulus increase, and the electrical resistivity decreases. Often, these individual property changes combine to affect the performance of a device. For example, if carbon precipitates during the use of an iron or steel tuning fork, the dimensions will decrease and the elastic modulus will increase. Each of these will cause the resonant frequency of the tuning fork to increase. Thus the effect of the precipitation of a small amount of carbon is multiplied. On the other hand, in a steel gage block, dimensional instability might be made up of a gradual tempering of the martensite (dimensional decrease) and of a gradual transformation of retained austenite to martensite (dimensional increase). In this case, assuming equal magnitudes for the two effects, the gage blocks might be dimensionally stable, even while some structural instabilities are occurring.

All materials are unstable to some degree, including metals and ceramics as well as the organic materials. In the real world of engineering, there is no such thing as a perfectly stable material or structure, nor is there anything with a perfectly stable resistivity, nor is there anything with a perfectly constant density--*ad infinitum*. The degree of acceptability of instability of any property is dictated by the function of that property in its application. A unit dimensional change of one part per 100 000 is trivial in a meter stick, and wouldn't cause much of a problem in a standard hand micrometer, but it would be troublesome in a gage block, and it could destroy the value of a telescopic mirror.

It is the objective of this book to draw the designer's attention to some of the many parameters that control the behavior of materials used in the manufacture of highly precise devices. It will soon become apparent to the reader that all of the answers are not contained here. Time and again, a specific example of an unusual behavior will be followed by a statement that the behavior is not understood, thus precluding accurate predictions of the extent to which that particular anomaly will occur in a different material under different conditions.

To keep the scope within reasonable bounds, the discussion centers on small dimensional changes. Materials such as wood and plastic, that exhibit gross dimensional changes, are not discussed. For practical purposes, dimensional changes greater than about 0.05% (500 ppm) are considered gross.

1.2 STANDARDIZATION OF UNITS

In discussing dimensional instability, the units employed often vary, depending on the investigator and on his nationality. This sometimes leads to misunderstanding and misinterpretation. Fortunately, real progress is being made as a result of international efforts to standardize units. Current standards applicable to the discussion of dimensional instability are reviewed briefly here.

Metric Practice

In most parts of the world where they still survive, the customary U.S. and British units, such as the pound and inch, are gradually being replaced by those of a modernized metric system known as Le Système International d'Unités. This system is abbreviated as SI in all languages. SI is outlined in detail in ASTM Metric Practice Guide (ASTM, 1966).

The SI system consists of six base units, along with several supplementary and derived units. These are given, with their appropriate symbols, in Tables 1.1, 1.2, and 1.3. The symbols and prefix terminology for multiples of these units are shown in Table 1.4.

Table 1.1   Base Units (Copyright ASTM, 1966; used by permission)

Quantity	Unit	SI symbol
length	meter	m
mass	kilogram	kg
time	second	s
electric current	ampere	A
thermodynamic temperature	degree Kelvin	°K
luminous intensity	candela	cd

Table 1.2   Supplementary Units (Copyright ASTM, 1966; used by permission)

plane angle	radian	rad
solid angle	steradian	sr

Table 1.3   Derived Units (Copyright ASTM, 1966; used by permission)

Quantity	Unit	SI symbol	Formula
acceleration	meter per second squared	m/s^2	----
angular acceleration	radian per second squared	rad/s^2	---
angular velocity	radian per second	rad/s	---
area	square meter	m^2	---
capacitance	farad	F	A.s/V
density	kilogram per cubic meter	kg/m^3	---
electric capacitance	farad	F	A.s/V

Table 1.3. (cont.)

Quantity	Unit	SI symbol	Formula
electric charge	coulomb	C	A.s
electric field strength	volt per meter	V/m	---
electric resistance	ohm	Ω	V/A
electromotive force	volt	V	W/A
energy	joule	J	N.m
force	newton	N	$kg.m/s^2$
frequency	hertz	Hz	s^{-1}
illumination	lux	lx	lm/m^2
inductance	henry	H	V.s/A
kinematic viscosity	square meter per second	m^2/s	---
luminance	candela per square meter	cd/m^2	---
luminous flux	lumen	lm	cd.sr
magnetic field strength	ampere per meter	A/m	---
magnetic flux	weber	Wb	V.s
magnetic flux density	tesla	T	Wb/m^2
magnetomotive force	ampere	A	---
potential difference	volt	V	W/A
power	watt	W	J/s
pressure	newton per square meter	N/m^2	---
quantity of heat	joule	J	N.m
stress	newton per square meter	N/m^2	---
velocity	meter per second	m/s	---
viscosity	newton-second per square meter	$N.s/m^2$	---
voltage	volt	V	W/A
volume	cubic meter	m^3	---
work	joule	J	N.m

Table 1.4 Multiple and Submultiple Units
(Copyright ASTM, 1966; used by permission)

Multiplication factors	Prefix	SI symbol
1 000 000 000 000 = 10^{12}	tera	T
1 000 000 000 = 10^{9}	giga	G
1 000 000 = 10^{6}	mega	M
1 000 = 10^{3}	kilo	k
100 = 10^{2}	hecto	h
10 = 10^{1}	deka	da
0.1 = 10^{-1}	deci	d
0.01 = 10^{-2}	centi	c
0.001 = 10^{-3}	milli	m
0.000 001 = 10^{-6}	micro	μ
0.000 000 001 = 10^{-9}	nano	n
0.000 000 000 001 = 10^{-12}	pico	p
0.000 000 000 000 001 = 10^{-15}	femto	f
0.000 000 000 000 000 001 = 10^{-18}	atto	a

Four of the six base units are of direct concern to dimensional stability; these are defined below:

(1) Meter--In 1960 the 11th General Conference on Weights and Measures (CGPM) adopted the meter as the length of exactly 1 650 763.73 wavelengths of the radiation in vacuum corresponding to the unperturbed transition between the levels $2p_{10}$ and $5d_5$ of the atom of krypton-86, the orange-red line. The transition corresponds to a wavelength (krypton-86 orange-red) of 6 057.802 x 10^{-10}m or 605.7802 nm.

(2) Kilogram--The 3rd CGPM (1901) adopted the kilogram as the mass of a particular cylinder of platinum - iridium alloy called the International Prototype Kilogram which is preserved by the International Bureau of Weights and Measures in a vault at Sevres, France.

(3) Second--The 11th CGPM (1960) adopted the ephemeris second, based on the annual motion of the Earth around the Sun, as the standard unit of time. The 12th CGPM (1964) recommended the adoption of a new atomic frequency standard to define the second and empowered the International Committee on Weights and Measures to designate atomic or molecular standards of frequency for temporary use. The standard chosen is the time of transition between the hyperfine levels $F = 4$, $M = 0$, and $F = 3$, $M = 0$ of the fundamental state $^2S_{1/2}$ of the cesium-133 atom unperturbed by external fields. The value of 9 192 631 770 hertz was assigned to the frequency of this transition.

(4) Degree Kelvin--The 10th CGPM (1954) adopted the thermodynamic Kelvin degree as the unit of temperature determined by the Carnot cycle with the triple-point temperature of water defined as exactly 273.16K.

Dimensional Practice

The specification of dimensions in practice has had a long and colorful history, with standards ranging all the way from the king's bodily dimensions to wavelengths of light. The king's dimensions presented obvious problems from the standpoint of specifying standards. The problem that the current user of information has is more directly concerned with modern data, where the standards have been more or less accepted--only the terminology varied. The current user has to be able to translate the often weird and wonderful units of the past into a language which is understood by today's technology.

A. *Length*

Despite the use of metric measurements in most parts of the world, the English-speaking countries, and particularly the United States, have for many years clung to British length units. Although these countries are officially committed to changing over to the SI system, the fact remains that most published information more than a few years old from engineers and scientists of these countries employs British units. Accordingly, in the field of precision devices and dimensional stability, the most common unit of length in English-language publications has been the microinch--one-millionth of an inch, often abbreviated 1 μin. Gage blocks are often calibrated as something like 2.000 000 ± 0.000 003 in. The equivalent in the metric system is the micron--one-millionth of a meter, often abbreviated as μm or simply as μ. The micron is, therefore, 39.37 μin. Because of the importance and common usage of the microinch and because there is no common SI unit of approximately the same size, it may be some time before it is replaced by metric units in American science and industry.

For conversion, the inch is defined as 25.4 mm exactly. Thus, through the definition of the meter, the inch and the microinch are traceable to the standard of krypton-86 light.

B. *Flatness*

In optics, especially those involving mirrors and interferometry, dimensional changes are often cited in terms of wavelengths of light. A specific wavelength that is used sometimes is that of the red cadmium line in air at 760 mm pressure and at 15°C. This wavelength is 6 438.469 6 angstrom units (10^{-10} m) or approximately 25.34 μin. Since, in interferometry, two diffraction lines represent one wave, the line spacings represent 0.3219 μm (12.67 μin). A diffraction line deviating from straightness by half the distance to its neighboring line would then indicate that the surface being observed deviated from flatness by 1/4 wave (λ/4) or by some 0.161 μm (6.3 μin). Optical flats are often specified as being flat to a tenth-wave (λ/10) or a twentieth-wave (λ/20). Curved surface mirrors can also be said to be accurate to some fraction of a wave.

When measurements are expressed in microinches and when flatness is expressed as fractions of wavelengths of light, it is convenient to specify flatness in terms of the green radiation of mercury-198, cadmium, or helium. The values for these light sources are very close to 20 μin for the wavelength and 10 μin for the fringe spacing, as tabulated overleaf. This permits convenient fractionalizing.

Light source	Wavelength (μin)	Fringe spacing (μin)
Mercury-198	20.15	10.08
Helium	19.75	9.87
Cadmium	20.02	10.01

C. *Angularity*

Angular measurements are frequently specified in the precision industries, and the terms most often used are seconds or minutes of arc. An arc-second is 1/3600 of a degree. An arc-second is therefore the equivalent of a displacement, at a distance of 1 meter, of

$$1 \times \left(\frac{1}{3600}\right)\left(\frac{1}{360}\right) \times 2\pi = 4.85 \times 10^{-6} \text{ meters}$$

or approximately 5 μm. Metric practice recommends the use of the radian. An arc-second then becomes 4.848 137 x 10^{-6} radians, and, in terms of displacements at unit distances, is the equivalent of an angle of about 5 microradians, a displacement at 1 inch of 5 μin, or a displacement at 1 meter of 5 μm.

D. *Stress and Pressure*

It has been customary for many years to use pounds (lb) or kilograms (kg) over a unit of area as a measure of stress or pressure. Such usage fails to differentiate between mass and force units. This is sometimes corrected by using pound-force (lbf) or kilogram-force (kgf). In the SI system of units, this problem is avoided by the use of the Newton as a unit of force. The Newton (N) is equal to a kilogram-meter per second per second, stemming directly from the first law of motion ($F = ma$). Thus, the SI unit for stress or pressure is Newton per meter-squared (N/m^2). It is also acceptable in the SI system to express stress in Pascals (abbreviated Pa), where 1 Pa = 1 N/m^2. Conversion to N/m^2 from other units can be accomplished as follows:

1 lbf/in^2 = 1 psi = 6 894.757 N/m^2
1 ksi = 1 000 psi = 6.894 757 MN/m^2 (meganewtons per meter-squared)
1 dyne/cm^2 = 0.100 000 0 N/m^2
1 kg/mm^2 (or 1 kgf/mm^2) = 9.806 650 MN/m^2

In this book, stress is expressed in both SI and British units.

E. *Strain*

If an object changes dimensions, e.g., as a result of the application of a load or of a change in temperature, the change can be referred to either in terms of the actual dimension of the change, or as a ratio of the increment of change to one of the original dimensions. The latter method is almost universally used and results in a dimensionless parameter known as strain. Thus, if a meter stick elongates by one-millionth of a meter, the strain is 10^{-6}. Although dimensionless, strain is sometimes reported in the literature with units, such as 10^{-6}in/in, or 1 μin/in, or 1 μm/m; all of these are obviously equivalent to 10^{-6}. In some publications, a strain of 10^{-6} is referred to as one "microstrain", abbreviated $\mu\epsilon$.

It is also common to refer to larger strains in terms of percent. Thus, a strain of 0.002 can be called 0.2% strain. Percentage is not often used in precision work because of the cumbersome number of zeroes, e.g., a strain of

10^{-6} equals a strain of 0.000 1%.

Brown (1968) has somewhat arbitrarily divided the scale of strain measurement into millimicro-, micro-, and macrostrain, as shown in Fig. 1.4.

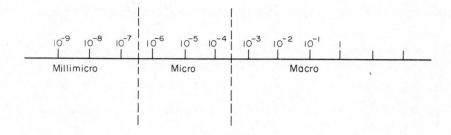

Fig. 1.4.   Scale of strain sensitivity as suggested by Brown (Brown, 1968; copyright Wiley, used by permission).

This is a useful division, and can facilitate discussions of precision designs. Strains in excess of about 5×10^{-4} would be called macrostrains, and related terms such as macroproperties, or macromechanical properties, or macroplasticity, would imply properties derived from macrostrain measurements. Strains below about 5×10^{-4} would then be called microstrains and related terms such as micromechanical properties, or microplasticity would imply properties derived from microstrain measurements. The terms millimicrostrain or millimicroplasticity are rather cumbersome, and might well be avoided.

1.3   CHAPTER SUMMARY

The discussion in this chapter emphasizes that all materials are susceptible to dimensional instability of several types and that some instability is to be expected in most real applications. The designer must be aware of those instabilities that can impair performance of a particular component and take steps to ensure that they are kept within acceptable limits.

The units employed in discussing dimensional instability have historically been diverse. Attempts to standardize these units are discussed briefly.

REFERENCES

ASTM. 1966. *ASTM Metric Practice Guide,* American Society for Testing and Materials, Philadelphia.

Brown, N. 1968. Observations of microplasticity, *Microplasticity,* C. J. McMahon, Jr., editor, Interscience, New York, 45.

Emerson, W. B. 1957. Secular length changes of gage blocks during twenty-five years, *Metrology of Gage Blocks,* U.S. Nat. Bur. Stand. Circular 581, 71.

Meakin, J. D. 1967. Microstrain behavior of body-centered cubic metals, *Can. J. Phys.* 45, 1121.

Chapter 2.

Material Properties Pertinent to Precision Design

In selecting materials for a specific precision device, the designer must
consider many of the same material characteristics that are important in
conventional designs, including fabricability, density, corrosion resistance,
electrical and thermal conductivity, and cost. However, other properties
assume equal or greater importance if the designer is to be able to predict
accurately the effects of stress and temperature on component dimensions.
Important properties include:

> Elastic properties
>> modulus values
>> practical limit of elasticity
>> time-dependent elasticity
>> thermoelastic coefficient
>
> Microplastic properties
>> microyield behavior
>> microcreep behavior
>
> Thermal expansion coefficient

Elastic properties, plastic properties, and thermal expansion are important
also in conventional design, and handbooks have been prepared from which the
designer can obtain various "constants" that describe material behavior. For
example, a handbook might show that a certain steel has a Young's modulus
value of 210 GN/m^2(30 x 10^6 psi) and a proportional limit of 350 MN/m^2 (50
ksi). With access to these published values, it is relatively easy for a
designer to accept these numbers as material constants and to assume that
at stresses below 350 MN/m^2, strain will be linearly proportional to
stress; i.e. the material will follow Hooke's Law.

In this chapter, the concepts of conventional design are reexamined
relative to the requirements of precision design. The validity of using
"laws" and "constants" to describe material behavior is examined in Section
2.2, followed by consideration of time and rate effects on mechanical behavior
in Section 2.3.

Because nonelastic strains can be of such great significance in precision
design, Section 2.4 is devoted to an elementary discussion of crystalline
imperfections known as dislocations and their role in microplastic deformation.
Although knowledge of dislocation concepts is not essential to the designer,
these concepts can help to reveal why and how real materials deviate so readily
from the perfectly Hookean behavior that is often assumed in conventional
design.

Following the discussion of elementary dislocation concepts, some of the
terminology surrounding the measurement of nonelastic strain is reviewed in
Section 2.5. Various terms used in subsequent chapters are defined.

2.1 EXAMINATION OF CONVENTIONAL DESIGN ASSUMPTIONS

For conventional design, the deformation of metals in response to an
applied stress is often viewed very simply. Below a certain level of stress,

the metal is assumed to deform elastically and to obey Hooke's Law; i.e., strain, ϵ, is linearly proportional to stress, σ, as indicated by line OA in Fig. 2.1. In simple tension or compression, this is expressed as $\sigma = E\epsilon$, where the constant of proportionality, E, is commonly called Young's modulus.

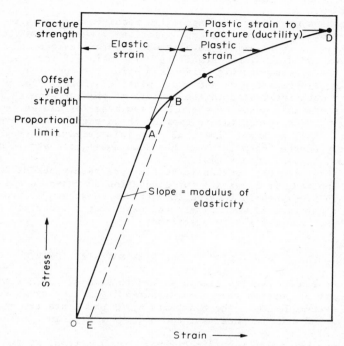

Fig. 2.1.   Schematic stress – strain diagram to illustrate assumed material behavior for conventional design.

If the loading of a test specimen should be discontinued below point A and the load removed, the stress and strain will retrace line OA back to the origin. Above the proportional limit, whose reported value depends on the sensitivity of the measurement, both elastic and plastic strains occur as indicated by curve segment AB. To avoid the problem of deciding precisely where the stress – strain curve deviates from a straight line, it is common to measure an offset yield strength (point B) at some small arbitrary amount of plastic strain. In order for the term yield strength to be meaningful, the amount of plastic strain must be specified. If not specified, the term yield strength is often understood to imply a plastic strain of 0.2% (2 x 10-3). Some metals, including iron and steel in certain conditions, yield abruptly, i.e., they undergo a large amount of plastic strain with little or no increase in the applied stress. Sometimes the stress may actually decrease with increasing strain. In those instances where a stress drop is observed, it is common to specify an upper and lower yield stress.

    If, after reaching point B, the load is gradually removed, the stress and strain now follow line BE. When the stress reaches zero, the elastic strain disappears and the plastic strain, OE, remains. Reloading is now assumed to follow line EBC. In other words, a small amount of plastic strain raises the

proportional limit from point A to point B. Similarly, the offset yield
strength is increased. This phenomenon is referred to as strain hardening.

Continuing the test beyond point C continues to produce both elastic and
plastic strain, culminating in fracture, or some other criterion of failure,
at point D. In tension tests, it is common to report the amount of plastic
strain at fracture.

For many, perhaps most, engineering design requirements, this model of
mechanical behavior is a reasonable approximation of the way real materials
behave. Nonetheless, it is almost totally useless for designs calling for
extreme precision and dimensional stability. The main difficulties arise in
the portion of the stress – strain curve from point O to point B, i.e., the
region comprising elastic strains and small plastic strains. The precision
designer is very much interested in the details of the curve in this region
and needs parameters that describe where Hookean behavior ceases and nonelastic
deformation begins. As investigators examine these regions with ever more
sensitive tools, it is apparent that real materials rarely, if ever,
exhibit perfectly Hookean behavior and that some nonelastic strain occurs at
stresses that may be vanishingly small.

In spite of the fact that nonelastic strains are rarely absent completely
when stress is applied, it is useful for discussion purposes to assume that
stress and strain are linearly related through the constants of proportionality referred to as elastic moduli. In the next section, certain precautions
in employing these "constants" are described.

2.2 USE OF "CONSTANTS" TO DESCRIBE MATERIAL BEHAVIOR

The elastic constants used to relate stress to strain, while simple in
concept, exemplify properties that become increasingly complex as requirements
for greater accuracy develop. The constants E, G, and ν are the ones given in
design handbooks and are the ones most commonly generated in the laboratory
for design use.

Young's modulus, E, relates unidirectional normal stress, σ, and strain in
the same direction, ϵ:

$$\sigma = E\epsilon.$$

It is usually measured in a simple tension or compression test. Similarly,
the shear modulus or rigidity modulus, G, relates shear stress, τ, and shear
strain, γ:

$$\tau = G\gamma.$$

The Poisson ratio, ν, is the ratio of lateral strain to longitudinal strain
under the action of a normal stress applied in the longitudinal direction.
For an isotropic material, ν can be expressed as a function of E and G:

$$\nu = \frac{E}{2G} - 1.$$

The reader is referred elsewhere [see McClintock and Argon (1966), for
example] for a detailed discussion of the generalized stress – strain relationships involving more complex stress states and anisotropic elastic constants.

When dealing with crystalline materials, it must be recognized that certain
dangers exist in trying to describe the elastic behavior of a specific material
in terms of single values of the above constants. These dangers are most
obvious for single crystals, because the elastic constants are strong functions
of the crystalline direction in which they are measured. Maximum and minimum
values of E and G and the corresponding crystallographic direction are shown

in Table 2.1 for metals in single crystal form. The variation in E for iron
is also sketched in Fig. 2.2. Note that in a crystal of alpha-iron (body-

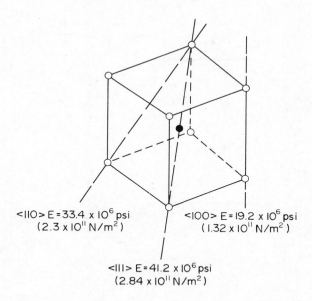

<110> E = 33.4 x 10^6 psi
(2.3 x 10^{11} N/m^2)

<100> E = 19.2 x 10^6 psi
(1.32 x 10^{11} N/m^2)

<111> E = 41.2 x 10^6 psi
(2.84 x 10^{11} N/m^2)

Fig. 2.2.   Anisotropy of Young's modulus in alpha-iron.

centered-cubic structure), the E values in various crystallographic
directions differ by a fraction greater than two. Of course, the variation
will not be as great in a polycrystalline aggregate. In fact, if the
aggregate is a truly random arrangement of a large number of individual crystals,
the elastic constants of the aggregate will be independent of direction.
Since truly random aggregates are uncommon, some anisotropy will persist.
Thus, in precision design, one should *expect* the elastic "constants" to have
values that depend on the direction of measurement. The extent of the
directional dependence will depend strongly on the processing history of the
test specimen.

An example will illustrate the extent to which an early step in processing
can affect Young's modulus. Two specimens of square cross-section measuring
6 x 6 x 25 mm (1/4 x 1/4 x 1 in) were cut from a zone-melted ingot of high-
purity iron. One was cut with its long dimension normal and one with its
long dimension parallel to the direction of zone-melting. The specimens were
drawn to 1.5 mm (0.060 in) diameter wire, and heat treated at 850°C (455°F)
for recrystallization. Young's modulus for the specimens cut normal to the
zone-pass direction was about 180 GN/m^2 (26.0 x 10^6 psi), while for those cut
parallel the modulus was about 228 GN/m^2 (33.0 x 10^6 psi).

Poisson's ratio also will be a unique number for a given material only if
the material is isotropic. Poisson showed theoretically that ν should be
about 0.25 for isotropic materials, and actual measurements show a value of
about 0.25 to 0.33 for most nominally isotropic materials. Where significant
anisotropy exists, ν may differ appreciably from these values.

Table 2.1 Maximum and Minimum values for Young's Modulus (E)
and Shear Modulus (G) of Metal Crystals

(After Schmid and Boas, 1950)

Metal	E_{max}			E_{min}			G_{max}			G_{min}		
	GN/m²	ksi	Direction	GN/m²	ksi	Direction	GN/m²	ksi	Direction	GN/m²	ksi	Direction
Al	75.4	10.9		62.7	9.1		28.4	4.1		24.5	3.5	
Cu	190	27.5		66.6	9.7		75.4	10.9		30.4	4.4	
Ag	115	16.6	[111]	43.1	6.2	[100]	43.6	6.3	[100]	19.3	2.8	[111]
Au	112	16.2		41.1	6.0		40.2	5.8		17.6	2.6	
α-Fe	284	41.2		132	19.2		116	16.8		59.8	8.7	
W	392	56.8		392	56.8		152	22.0		152	22.0	
Mg	50.3	7.30	0[a]	42.8	6.20	53.3	18.0	2.61	44.5	16.7	2.43	90
Zn	124	17.92	70.2	34.9	5.06	0	48.7	7.05	90	27.2	3.95	41.8
Cd	81.3	11.8	90	28.2	4.08	0	24.6	3.56	90	18.0	2.61	30
β-Sn	84.7	12.3	[001]	26.2	3.8	[110]	17.8	2.68	45.7[b]	10.4	1.50	[100]

[a] Angle with the hexagonal axis.

[b] Angle with the tetragonal axis in prism plane type II.

These examples hardly represent the exception. They more accurately represent the rule. Published moduli for engineering materials are almost invariably approximations. These variations must be taken into account in precision applications.

There are factors other than directionality that can affect the elastic properties of a material, one of the most obvious being temperature. This factor can be critical in such an application as a tuning fork. Ideally, the frequency of the fork would be independent of temperature. Since Young's modulus of the material largely controls the resonant frequency of the fork, the temperature dependence of the modulus can have an important effect on fork behavior.

Normally, the experimentally determined modulus values of a metal decrease as the temperature increases, as shown schematically in Fig. 2.3. At the low end of the temperature range, the relationship is approximately linear.

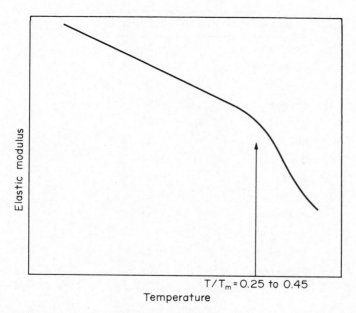

Fig. 2.3.   Schematic representation of the "normal" temperature dependence of the elastic modulus.

As temperature is increased further, a point is reached beyond which the modulus values decrease with temperature at a higher rate. This is indicative of the inclusion of some nonelastic strain in the modulus measurement. The breakover point will depend on the method of measuring the modulus--dynamic methods will indicate higher breakover temperatures than will static methods because they permit less time for nonelastic strain to occur.

The effect of temperature on Young's modulus is often expressed in terms of a thermoelastic coefficient, g, defined as

$$g = \frac{1}{E}\frac{dE}{dT}.$$

Some representative values of g are given in Table 2.2. As is evident from these values, thermoelastic coefficients for most materials are negative in sign, indicating that the modulus decreases as the temperature increases.

Table 2.2 Thermoelastic Coefficients (g)

for various Pure Metals near 20°C

(Koster, 1948)

Metal	$g \times 10^6$ per degree C
Aluminum	−430
Manganese	−312
Silver	−416
Copper	−296
Gold	−240
Cobalt	−245
Beryllium	−140
Iron	−260
Palladium	−105
Platinum	−133
Tantalum	−133
Molybdenum	−128
Tungsten	−101

Such behavior is by no means universal. A particularly striking example of an exceptional behavior, and of the sensitivity of the modulus behavior to heat treatment, is given in Fig. 2.4. The material is a copper – manganese alloy containing about 0.8 weight % vanadium. The data were obtained on a torsional pendulum, where the square of the pendulum frequency is proportional to the shear modulus of the material.

In many such cases of apparently anomalous behavior, g is strongly sensitive to processing variables. Careful attention must be paid to processing of materials like Ni-Span-C (an iron – nickel – chromium – titanium alloy), which is one of a family of alloys in which it is possible to obtain $g = 0$. The effects on g of both prior deformation and heat treatment temperature for Ni-Span-C are shown in Fig. 2.5. Additional discussion of materials with near-zero thermoelastic coefficients appears in Chapter 10.

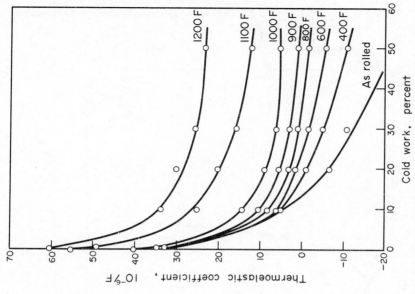

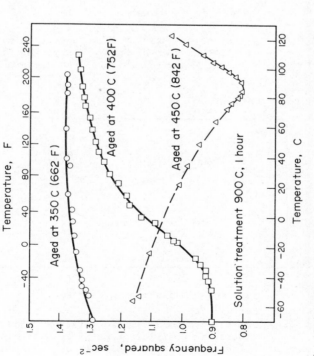

Fig. 2.5. Effect of cold work and 5-hr heat treatment at temperature shown on thermoelastic coefficient of Ni-Span-C (copyright International Nickel Co., 1963; used by permission).

Fig. 2.4. Effect of aging treatment on modulus – temperature relationship of specimens of Cu-Mn-V alloy (data courtesy of American Potash and Chemical Corporation).

In ferromagnetic materials, not only temperature but magnetic fields can influence both the elastic modulus values and the thermoelastic coefficient. This effect is shown in Figs. 2.6 and 2.7.

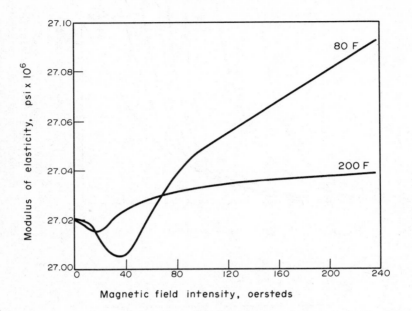

Fig. 2.6.   Effect of magnetic field intensity on modulus of elasticity of Ni–Span–C (copyright International Nickel Co., 1963; used by permission).

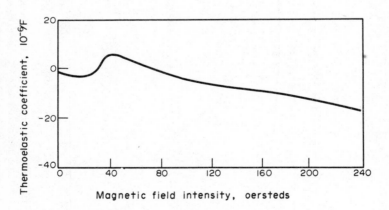

Fig. 2.7.   Effect of magnetic field intensity on thermoelastic coefficient of Ni–Span–C (copyright International Nickel Co., 1963; used by permission).

Yet another property that is closely related to elastic properties and which is often specified as a "constant" in handbooks is the thermal expansion coefficient, α, defined as

$$\alpha = \frac{1}{L}\frac{dL}{dT} \; ,$$

where L is the length of a sample and T is temperature. In fact, α is not a constant, even for a pure single crystal, as the data in Fig. 2.8 clearly show. Since data books often are made for designers with less exacting problems, thermal expansion data may be expressed in them as

$$\bar{\alpha} = \frac{L_1 - L_0}{L_0(T_1 - T_0)} \; ,$$

where L_0 is the length at temperature T_0 and L_1 is the length at temperature T_1. This expression provides a rough approximation of the true thermal expansion coefficient, but its use can lead to serious errors in a device requiring precise compensation for thermal expansion.

A further difficulty in precision design can be avoided by recognizing that many polycrystalline materials are anisotropic in their thermal expansion.

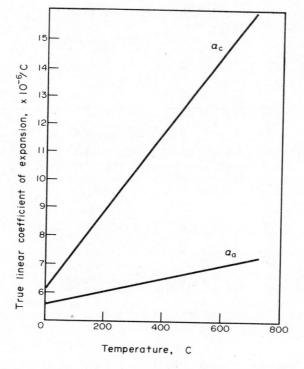

Fig. 2.8.   Effect of temperature on the linear coefficients of expansion along the c and a axes of
a single crystal of alpha-zirconium (Johnson
and Honeycombe, 1962; copyright Elsevier
Sequoia, used by permission).

In polycrystalline aggregates, in which form most engineering metals are used, the orientation of the individual crystals is seldom random and some of the anisotropy of the type shown in Fig. 2.8 for a single crystal persists. Uranium provides an extreme example of this (see Fig. 1.2).

Rosenfield and Averbach (1956) have shown that the thermal expansion coefficient is sensitive to both tensile stress and strain. A typical set of measurements on 1020 steel is shown in Fig. 2.9. Tensile stress appears to increase the thermal expansion coefficient of 1020, 1040, and 1060 steels, but to decrease it in some other materials, such as regular and free-cut Invar.

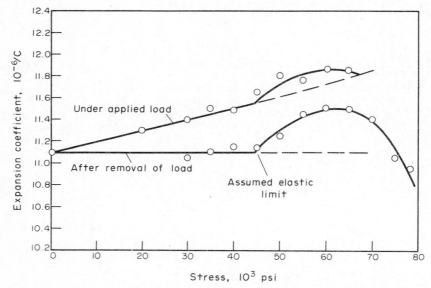

Fig. 2.9.   Effect of stress on the expansion coefficient
of a 1020 steel (Rosenfield and Averbach,
1956; copyright American Institute of
Physics, used by permission).

2.3   TIME AND RATE EFFECTS ON MECHANICAL BEHAVIOR

A second difficulty in treating metals as if they behave as shown in Fig. 2.1 pertains to time and rate effects. It is commonly assumed that most structural materials used at, or near, room temperature are relatively insensitive to time and rate effects. Although this assumption is adequate for many conventional applications, it may be totally inadequate for precision design. The generalized picture of time and rate effects presented below deals primarily with behavior in the "elastic" and small plastic region (*OAB* of Fig. 2.1).

Application of a Square-wave of Stress

When a tensile stress is applied instantaneously to a rod or wire specimen, held constant for some time, and then removed, the specimen can respond in several ways, as shown schematically in Fig. 2.10(a). The simplest response is purely elastic as in curve *E*. The specimen suddenly lengthens when the load is applied, retains this length as long as the load is maintained, and returns immediately to its original length when the load is removed. The load is simply changing the spacing between atoms. Metals always show some elastic response to load. In addition, they may exhibit one or more other types of response, the detection of which depends on the sensitivity of the experimental apparatus.

A second type of mechanical response is time-dependent elastic behavior. The term anelastic will be used to denote such behavior. Three different curves (A_r, A_i, and A_s) are used in Fig. 2.10(a) to denote rapid, intermediate, and slow anelastic response times, respectively. The reason for this distinction will become clear in the next section. In each instance, no anelastic strain is observed at the instant of load application. Thereafter, anelastic strain increases gradually toward a saturation level until the load is suddenly removed. At this point, the anelastic strain gradually returns toward zero. Combined effects of anelastic strain (curve A_i) and elastic strain are shown in curve $E + A_i$ of Fig. 2.10(b).

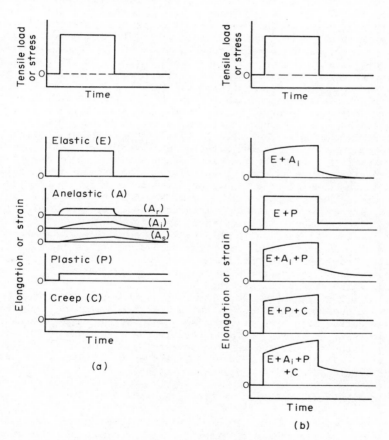

Fig. 2.10.   Schematic illustration of types of response
              to an instantaneously applied and removed
              tensile stress: (a) individual types of
              response, and (b) combined responses.

It is important to point out that various meanings have been attached to the term anelastic. The meaning adopted here conforms to a definition advanced by Nowick and Berry (1972). They indicated that anelastic behavior differs from elastic behavior only in that anelastic response is not instantaneous. To define anelasticity, they employ three postulates: (1) for every stress, there is a unique equilibrium value of strain, and vice versa, (2) the equilibrium response is achieved only after the passage of sufficient time, and (3) the stress – strain relationship is linear. Zener (1948), on the other hand, used the term to denote "that property of solids in virtue of which stress and strain are not single-valued functions of one another in that low stress range in which no permanent set occurs and in which the relation of stress to strain is still linear". Dieter (1961) states that a "nonelastic body is said to behave anelastically when the stress and strain are not single valued functions of each other *and* the internal friction is independent of amplitude". He has also used the term anelasticity to describe the disappearance of a small amount of plastic strain with time after removal of load. Lubahn (1961) states simply that "anelastic strain is slowly recoverable strain".

A third type of response that may take place is plastic (curve P). As shown in Fig. 2.10(a), a lengthening occurs virtually instantaneously upon application of the load and remains after removal of the load. In crystalline materials, this type of response is due almost entirely to the movement of dislocations, as discussed in Section 2.4. The combined elastic and plastic response is shown in curve $E + P$ of Fig. 2.10(b).

Time-dependent plastic elongation (creep) is illustrated in curve C. Little or no plastic strain is observed at the instant of load appplication. However, as time progresses, elongation gradually appears, at an ever decreasing rate, until the load is removed. Thereafter, the elongation remains constant. When creep occurs at an ever decreasing rate, as pictured here, it is commonly called primary, first stage, or transient creep. As was noted for plastic strain, creep also occurs as a result of dislocation motion in crystalline materials. In the case of creep, however, dislocations become blocked by internal barriers. They can move only as time and temperature act to relax some of these barriers or to activate new dislocation sources. Combined elastic, plastic, and creep response is illustrated in curve $E + P + C$.

Although not shown specifically in Fig. 2.10, some recovery of both plastic and creep strain may occur with time after unloading. This can usually be differentiated from anelastic effects in that only partial recovery is normally observed and the duration of the recovery period may far surpass the period during which stress was applied.

Curve $E + P + C + A_i$ in Fig. 2.10(b) illustrates the general response of a material to a sudden applied load when each of the foregoing types of deformation occurs. Each type of response should be anticipated in a real material. However, depending on the material, the magnitude of the stress, the duration of the stress, and the temperature, one or more of the types of response may be negligible or beyond the limits of measurement. The elastic response, as mentioned earlier, is always present.

Application and Removal of Stress at Finite Rates

Although the curves of Fig. 2.10 are useful in illustrating the general nature of a material's response to stress, it is uncommon to study mechanical behavior by rectangular load pulses of this type.* Especially in examining

---

*Creep tests are one exception to this; here, loads are normally applied abruptly, but without shock, and held constant for long periods of time.

microstrain behavior, load – unload procedures are frequently used at moderate
strain rates. Loads are applied to a specimen at a nominally constant rate
until some preset level is reached, at which point the specimen is unloaded
at the same, or in some cases higher, rate. It is useful to derive the
load – unload stress – strain curves that can result from the
various types of mechanical response described in foregoing paragraphs.

Figure 2.11(a) illustrates schematically the various types of strain response to a load – unload cycle. As in Fig. 2.10(a), each response is identified by an appropriate letter. Of course, the elastic response is always
present; the others may or may not occur. By plotting stress values versus

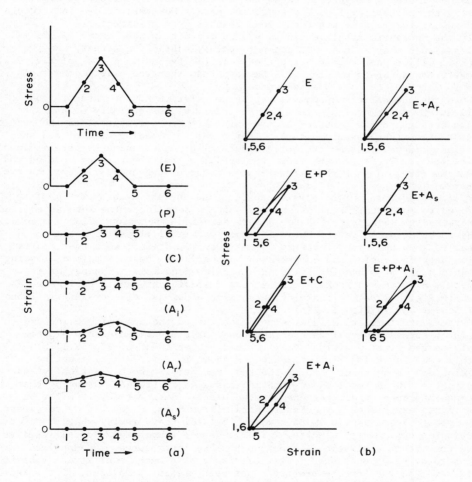

Fig. 2.11.   Schematic illustration of types of response
to a load – unload sequence at a finite
loading rate: (a) individual types of
response, and (b) stress – strain curves
for various combined types of response,
constucted from information in (a).

strain values at any particular time during the load – unload test, it is
possible to construct stress – strain curves. These are shown in Fig. 2.11(b)
for various combinations of strain response.

In Fig. 2.11(b), curve E illustrates perfectly elastic behavior while $E + P$ illustrates the addition of a small amount of plastic flow. Curve $E + A_i$
illustrates the development of a hysteresis loop. It develops because, as
shown in curve A_i of Fig. 2.11(a), the anelastic strain lags slightly behind
the stress. At time 2, the anelastic strain is just beginning to take on a
positive value, even though the stress has already reached half its peak value.
Beyond time 3, when the stress has begun to drop, the anelastic strain continues to increase slightly because the stress is still relatively high.
Shortly thereafter it, too, begins to decrease. However, it does not return
to zero until some time after the load reaches zero at time 5.

If the anelastic strain exhibits a very rapid response, curve A_r of Fig.
2.11(a), such that it does not lag perceptibly behind the stress, the
material will appear to exhibit perfect elasticity--curve $E + A_r$ of Fig. 2.11(b).
However, the apparent modulus will be less than the true elastic modulus.
This is referred to as a "relaxed modulus".

When the anelastic response is very slow relative to the rate of load
application and removal, curve A_s of Fig. 2.11(a), essentially no anelastic
strain occurs and the stress – strain curve $E + A_s$ is identical to the perfectly
elastic curve (unrelaxed modulus). If the rate of loading were slowed substantially, however, some anelastic strain might occur and a hysteresis loop
again appear. It is evident from this analysis that the rate of loading and
unloading can influence the stress – strain curve when anelastic effects are
present.

A simple arrangement of elastic springs and anelastic dashpots is often
employed to model this type of stress – strain behavior. The interested reader
is referred to Kennedy (1963) for a more detailed discussion of such models.

By procedures similar to those followed in Fig. 2.11, it is a simple
matter to show the type of stress – strain curve obtained when the load is
cycled between tension and compression, either at a constant rate of loading
and unloading, or sinusoidally as in a torsion pendulum. If behavior is
perfectly elastic and Hookean, the stress – strain curve will be linear, as
shown in Fig. 2.12(a). If anelastic strains occur as well, hysteresis loops
will appear, as in Fig. 2.12(b).

Certain problems arise in attempting to extend the analysis to include the
effect of stress reversal on specimens that have experienced some plastic
strain on the first positive cycle of load--curve $E + P$ of Fig. 2.11(b).
In this case, a stress – strain curve as shown in Fig. 2.13(a) might be
anticipated. From point O to A, the specimen deforms both elastically and
plastically. When the load is removed, curve AB is traced. Assuming that the
specimen has been strain-hardened by the plastic deformation, it might be
expected to experience only elastic strain from point B to C. However, most
real metals do not behave in this way. Instead, they display what is referred
to as a *Bauschinger effect*--Fig. 2.13(b). When a compressive load is applied
after some tensile plastic deformation has been introduced, the specimen
begins to display nonelastic (plastic) strain at a much lower stress than it
would have had the tensile prestrain not been present. In fact, if the tensile
prestrain is relatively large, the specimen may begin to show appreciable deviation from elastic response even before the tensile load is completely removed. This is illustrated in Fig. 2.14 with some data obtained on 70-30 brass.
This also illustrates a point made earlier--plastic strain may be partially
recoverable upon removal of load.

The Bauschinger effect emphasizes the striking difference between elastic
and anelastic behavior on the one hand, and plastic behavior on the other.

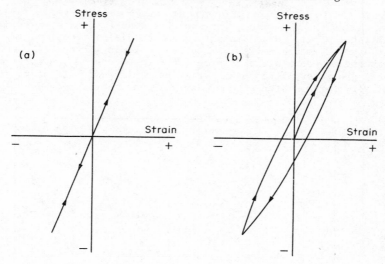

Fig. 2.12.   Schematic illustration of stress – strain
curves obtained when load is cycled between
tension and compression: (a) elastic strain
only, and (b) elastic plus anelastic strain.

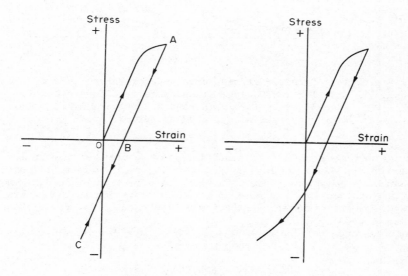

Fig. 2.13.   Schematic illustration of stress – strain
curves to show compressive loading after
plastic deformation in tension: (a) compressive elastic strain only, and (b) Bauschinger
effect.

In the case of the former, an applied stress causes relatively short-range effects,
e.g., atoms change their spacing, interstitial impurity atoms may jump to
an adjacent lattice site, or a pinned segment of dislocation line may bow.
When the stress is removed, it is a relatively simple matter for these occurr-
ences to reverse. When plastic deformation occurs, however, even if only a
small amount, the material has been permanently altered. Dislocations have
moved relatively large distances and many will have been stopped by obstacles.

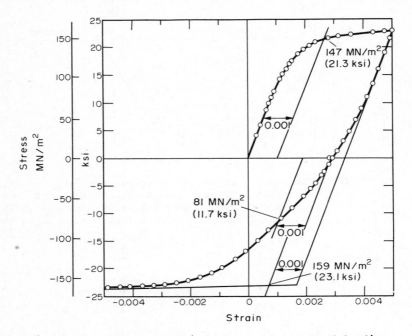

Fig. 2.14.   Stress – strain curve under reversed loading
              for 70-30 alpha brass; data from Corten and
              Elsesser (1952) (Lubahn and Felgar, 1961;
              copyright Wiley, used by permission).

In addition, many more "mobile" dislocations may be present than in the unstrained material. When the load is removed, the dislocations will not return
to their original positions. However, when the direction of loading is
reversed, the dislocations can move appreciable distances in the reverse
direction at low stresses because there are likely to be far fewer obstacles
to their motion.
     Dislocations and their role in plastic strain are discussed in Section 2.4.

2.4 DISLOCATIONS AND PLASTIC STRAIN

     In crystalline materials, many mechanical behavior characteristics stem
from imperfections known as dislocations. Because of their importance, some
dislocation concepts and terminology are discussed briefly below. Readers
familiar with dislocation theory can proceed to Section 2.5 without loss of
text continuity.

Dislocations are crystalline imperfections of a particular type that exist in all metals and crystalline ceramics. Their existence was first postulated in the 1930's, but conclusive proof of their existence did not come until many years later with the development of transmission electron microscopy and etch pitting techniques.

Upon first inspection, an imperfection such as a dislocation would seem to be a relatively innocuous feature of a crystal. However, discovery of the dislocation has had a greater impact on man's understanding of deformation mechanisms in crystals than any other discovery of recent times. It has long been known that plastic deformation in metals takes place by a shear displacement of one plane of atoms over another along certain preferred crystalline planes. However, computation of the theoretical stress that should be required to shear one crystalline plane over another, based on known interatomic forces, gives values of approximately $G/10$, where G is the shear modulus. In other words, stresses less than about $G/10$ should produce essentially elastic behavior. For iron, where $G = 80$ GN/m^2(12 x 10^6 psi), this would mean that the yield strength in shear should be about 8 GN/m^2 (1.2 x 10^6 psi). Experimentally determined values of yield strength of iron and other metals are more like $G/100$ to $G/10\ 000$. Clearly, this is a large discrepancy between theory and observation. Dislocation concepts can help to explain these discrepancies.

Examination of the simplest type of dislocation, called an edge dislocation, will reveal why dislocations have had such a profound effect on deformation theory. Figure 2.15(a) depicts a perfect crystal and Fig. 2.16(a) a crystal

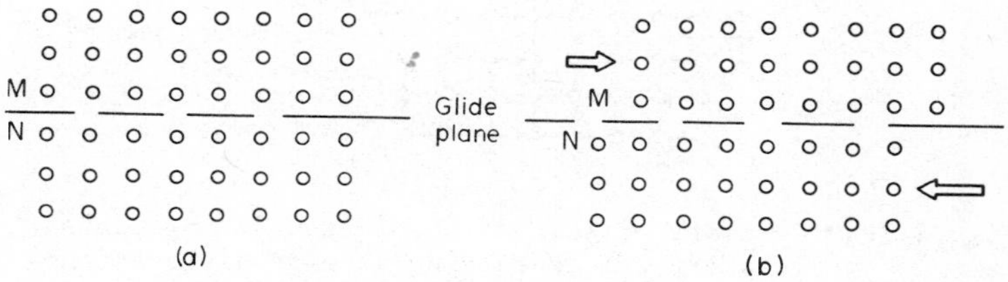

Fig. 2.15. Schematic illustration of atom arrangements in a perfect crystal: (a) before, and (b) after shear displacement.

containing an edge dislocation. Assume that similar rows of atoms are located behind those shown, i.e., that the crystal extends into the page. It is evident that the dislocation is simply an extra plane of atoms in the upper half of the crystal. Although the origin of this extra half-plane of atoms is uncertain--a growth defect, perhaps--it will suffice to say here that such defects occur commonly, even in well annealed crystals.

When a shear stress is applied to the perfect crystal, Fig. 2.15(a), each atom on plane M must simultaneously stretch and break its bond with its neighbor on plane N if a permanent shear displacement, i.e., plastic strain, is to occur [Fig. 2.15(b)]. This requires an extremely large shear stress. In the crystal containing the dislocation, on the other hand, a relatively small shear stress is required to obtain a permanent shear displacement, because only a few bonds need to be arranged at any one time. For example, row A is initially in an equilibrium position between rows X and Y[Fig. 2.16 (a)]. A small shear stress will now bring row A nearer to row Y than is row B. Consequently, row B now becomes the extra half plane of atoms or, in other words, the dislocation has moved one atom spacing [Fig. 2.16(b)]. It will be an equally simple matter for it to move entirely out of the crystal, leaving the upper half of the crystal permanently sheared relative to the lower half [Fig. 2.16(c)].

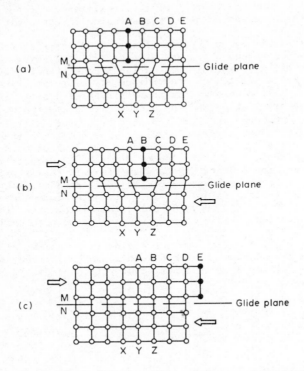

Fig. 2.16.   Schematic illustration of atom arrangements in a crystal containing an edge dislocation: (a) before shear displacement, (b) after dislocation has moved one lattice space, and (c) after dislocation has moved out of crystal.

Analogies have been made between shearing one plane of atoms across another in a metal crystal and shifting a rug, spread out on a floor, to a new position. If the rug covers a large area and an attempt is made to move it by gripping one edge and pulling, a very large force will be required. It is a simple matter to reduce the force required to move the rug by creating a wrinkle along one edge, as shown in Fig. 2.17. A relatively small force applied to this wrinkle will cause it to move from left to right. What this is doing, in effect, is moving a small part of the rug at a time. When the wrinkle is swept out of the carpet, the entire rug will have been successfully moved. This accomplishment will be the result of applying a small force over a large distance rather than a large force over a small distance.

Since the postulation of dislocations, the problem in deformation theory now becomes completely reversed. Where it formerly was difficult to explain why the yield strength of metals is not extremely high, 10 to 1000 times that observed experimentally, it now becomes difficult to explain why the yield strength is not extremely low, a fraction of that observed experimentally, since dislocations theoretically can move under very small stresses.

The fact is, as the behavior of metals is examined on an ever finer scale, that many metals do exhibit evidence of plastic deformation, i.e., irreversible movement of dislocations, at stresses that approach zero. Tinder and Washburn (1964), for example, have reported permanent strains of about 10^{-8} in annealed

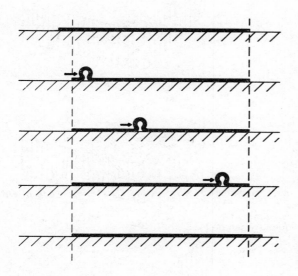

Fig. 2.17.   Schematic illustration of a rug being moved by sweeping a wrinkle from one side to the other, to show analogy with dislocation role in shear of metal crystals.

pure copper at a shear stress of about 30×10^3 N/m^2 (4.3 psi). In zinc single crystals, Tinder and Trzil (1973) detected plastic deformation at a shear stress of only 200 N/m^2 (0.03 psi)!

The picture of dislocations presented here is greatly oversimplified and non-quantitative. Many quantitative and detailed treatments are available elsewhere.* To assist the reader who may be unfamiliar with dislocations to better understand subjects covered in this book, certain general observations concerning dislocations are listed below.

(a) The original source of dislocations in a crystal is not entirely clear. Nonetheless, they are nearly always present in real crystals.
(b) Plastic deformation increases the number of dislocations present; annealing causes the number to decrease but does not remove them entirely.
(c) Most dislocations are not simple edge dislocations. They normally contain both edge and screw components. Such a dislocation line is pictured in Fig. 2.18.

---

*See, for example, Cottrell (1964), Friedel (1964), Weertman and Weertman (1964), Nabarro (1967), Hirth and Lothe (1968), and Rosenfield *et al.* (1968).

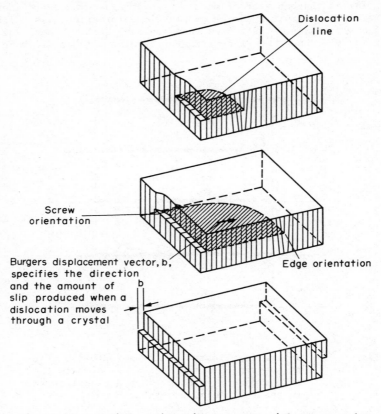

Fig. 2.18. A dislocation line segment with screw and
edge components (Carnahan, 1964).

(d) The number of dislocations present in a material is termed dislocation
     density and is expressed as the length of dislocation line per unit
     volume (cm/cm^3). In Fig. 2.16, the dislocation line extends into the
     page. This is sometimes expressed also as the number of dislocation
     lines that intersect a unit area of surface (cm^{-2}). In well-annealed
     crystals, densities of 10^4 to 10^6 are common. Cold working can
     increase this to 10^{10} to 10^{12} cm/cm^3.*

(e) In an otherwise ideal crystal, dislocation motion is impeded only by
     a lattice friction stress, sometimes called the Peierls stress or the
     Peierls - Nabarro stress. In real crystals, there are generally other
     obstacles to dislocation motion; these include foreign atoms, lattice
     vacancies, dispersed or precipitated particles, grain boundaries, and
     other dislocations. By deliberately introducing appropriate obstacles,
     it is possible to increase greatly the stress required to produce

---

*It is interesting to express this range of dislocation densities in another
way. At 10^4 cm/cm^3, each cubic centimeter contains about 100 m (over 300 ft)
of dislocation line. At 10^{12} cm/cm^3, the same volume of material contains
10 million km (over 6 million miles) of dislocation line!

large-scale motion of the dislocations, i.e., the yield strength is raised. Methods commonly employed to accomplish this include alloying to produce solid solution hardening or precipitation hardening, refinement of grain size, and cold working.

(f) Dislocations may be immobilized to varying degrees as a result of "locking" by foreign impurity atoms. Locking occurs when atoms whose size deviates from that of the host are attracted to the region of abnormal atom spacing near the dislocation.

(g) Plastic deformation arising from dislocation motion is not limited to that produced by sweeping out of the crystal those dislocations present originally. Under certain circumstances, moving dislocations can generate new dislocations in almost limitless quantities.

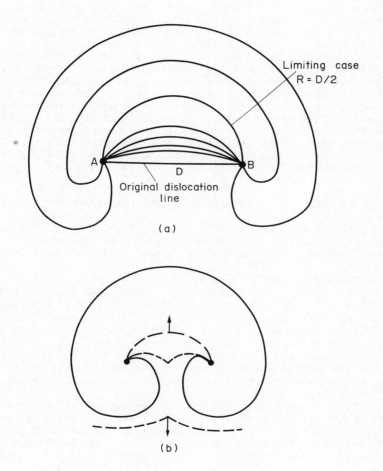

Fig. 2.19.   The generation of dislocations by a double-pinned dislocation segment (Kennedy, 1963; copyright Wiley, used by permission).

(h)   A particular dislocation line may be mobile in certain regions and
      immobile in others, i.e., it may be "pinned" at various points. Between
      the pinning points, the dislocation can be thought of as possessing
      properties much like those of a stretched rubber band. Thus, under
      an imposed shear stress, it may bow between pinning points and return
      toward its unbowed position upon removal of the stress. As will be
      discussed later, this can have important implications in mechanical
      hysteresis.

(i)   One commonly described mechanism of dislocation multiplication depends
      on a Frank – Read source, named after the investigators who first
      postulated it. This requires a pinned length of dislocation line. As
      the stress acting on this line segment is increased, the dislocation
      will first bow out, as shown in Fig. 2.19(a). The loop will then
      continue to expand under stress and will loop back on each side, until
      the two sides join. This results in the formation of a closed loop
      which continues to expand, leaving a dislocation between A and B as
      before, which can repeat the cycle. The results of the process are
      shown schematically in Fig. 2.19(b).

(j)   Another dislocation multiplication mechanism involves grain boundaries
      and grain boundary ledges (Price and Hirth, 1972). Many investigators
      believe that grain boundary sources of dislocations are extremely
      important in the microstrain region of metal deformation.

(k)   The rate at which plastic flow occurs depends on both the density
      of mobile dislocations and their velocity. Employing the principles
      of "dislocation dynamics", it is possible to compute stress – strain
      curves that agree closely with those observed experimentally. Stein
      (1968) has applied dislocation dynamics to the microplastic strain
      region of molybdenum with a fair degree of success.

2.5 TERMINOLOGY APPLICABLE TO NON–HOOKEAN BEHAVIOR

In describing the mechanical behavior of materials in the nominally elastic
region and the very beginning of the plastic region, various terms and definitions have been employed by different investigators. Some of these are reviewed briefly here.

According to Timoshenko (1940), a "body is perfectly elastic, if it recovers
its original shape completely after unloading". This is an interesting concept
from a mathematical point of view, but it leads to considerable complication
when applied to real materials. The data shown in Fig. 2.20 would be considered
indicative of "perfectly elastic" behavior according to Timoshenko's definition,
for, on removal of the load, the residual strain observed is zero. In this
case, the "elastic limit" would be reported as greater than 400 MN/m^2 (60 ksi),
but a designer would be hard pressed to design an acceptable altimeter, for
example, based on such spring properties. The material, an Al-Cu-Ni alloy,
is clearly non-Hookean.

A somewhat different approach is involved in the concept of the *proportional
limit,* defined as the stress at which the stress – strain curve deviates from
a straight-line, or Hookean, relationship. Again, unless the sensitivity of
the equipment used to measure the strain is specified, the data have little
meaning. In fact, with instrumentation of sufficient sensitivity, it can be
shown that the proportional limit approaches zero for all materials.

A more rational approach to the problem of defining meaningful parameters
appears in the work of Hughel (1960), who measured the "precision elastic limit".
For measurement of this parameter, load is applied to a specimen, then lowered
to zero, and the residual strain is measured. The procedure is repeated,

with the load being increased incrementally, until the residual plastic strain is equal to 10^{-6}. The technique is the same as that commonly used to measure the elastic limit, but now the *precision elastic limit* (PEL) is defined as an offset of a given amount.

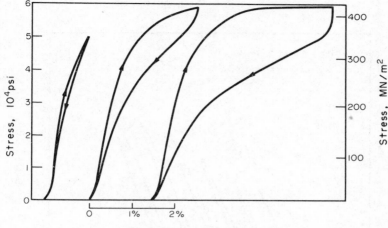

Fig. 2.20.   Surface stress versus surface strain curves
in four-point bending for an Al-Cu-Ni alloy
(Rachinger, 1958; copyright Institute of
Physics, used by permission).

One major objection to this term is one of semantics. It seems a poor use of terms to define an "elastic" property in terms of a given amount of plastic deformation. The term *microyield strength* (MYS) is favored over PEL in this respect. Care still should be taken in interpreting tabulated data, however, for other definitions of MYS have been put forward. Bonfield and Li (1966), for example, refer to the *microscopic yield strength* (MYS) as the stress associated with a residual strain of 2×10^{-6}. Kyle *et al.* (1966) refer to a *microinch offset yield stress* (MOYS), which is again the stress to cause a permanent plastic strain of 10^{-6}.

In this book, the term σ_y, followed by an amount of plastic strain in parentheses, is used to denote offset yield strengths. For example, $\sigma_y(10^{-6})$ means the stress required to produce a permanent strain of 10^{-6}. Similarly, the 0.2% offset yield strength is denoted as $\sigma_y(2 \times 10^{-3})$. This terminology accomplishes two things: (1) it makes it clear that the property being reported is an offset yield strength, rather than an elastic limit, and (2) it states specifically the amount of offset, or plastic, strain.

Brown (1968) uses somewhat different terminology. A typical set of data from his experiments is shown in Fig. 2.21. Generally, the initial portion of the stress – strain curve is linear. The first indication of a nonlinear behavior is the formation of a closed loop upon unloading. The stress magnitude at which the closed loop first forms, σ_E, Brown calls the *elastic limit*. Note that this definition differs fundamentally from that previously offered. As the amplitude of the load cycle is increased, the area of the loop increases until it no longer closes. The stress where the loop first does not close, σ_A, is called the *anelastic limit*. When the strain at the open end of the loop equals 1×10^{-6}, σ_A is equal to the MOYS.

Brown's definition of elastic limit is intuitively more attractive than the classical definition. The difficultly with Timoshenko's definition, where it is necessary to think of a material such as that exemplified by Fig. 2.20 as being elastic, is avoided. However, it must be remembered that both σ_E and σ_A defined by Brown depend upon the sensitivity of the equipment, and that if this is not specified, the data lose their meaning.

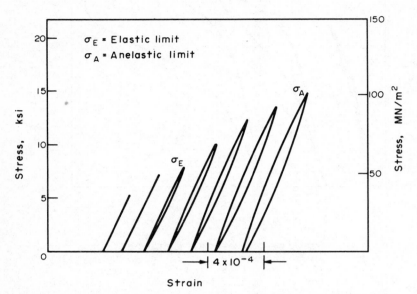

Fig. 2.21.   Typical set of load – unload stress – strain
curves to define σ_E and σ_A (Brown, 1968;
copyright Wiley, used by permission).

Yet another term was introduced by Plenard (1968), who defines a "limit of accommodation". She accepts the earlier definition of the elastic limit as the highest stress at which, on removal of the stress, the specimen will return to its original dimensions--Fig. 2.22 (a). When this stress is exceeded, above Brown's "anelastic limit", plastic deformation (ϵ_r) remains--Fig.2.22(b). However, if the specimen is stress cycled to stresses slightly above this limit, subsequent loops will tend to close (as in Fig. 2.23) until some cyclic stress, called the *limit of accommodation* (σ_L) is reached, above which the stress – strain loops remain open, and the residual strain increases with each succeeding cycle. This behavior is shown in Fig. 2.24.

All of the micromechanical properties discussed so far have been related to the behavior of a material subjected to forces for a short period of time. If a specimen now is subjected to sufficiently high stress for an extended period of time, strain will continue with time. This phenomenon is known as creep or, in the microstrain range, *microcreep*. Hughel (1960), in his measurements of precision elastic limit, loaded a specimen to the same stress level three successive times. If no difference in residual strain accompanied the successive loadings, the stress level would then be increased. If the residual strain increased after successive loadings, he had exceeded what he called the *microcreep limit*, or MCL. In a sense, this MCL is related to the limit of accommodation (σ_L) as depicted in Figs. 2.23 and 2.24. However, it has been shown by Maringer and Imgram (1968) for a number of materials that significant

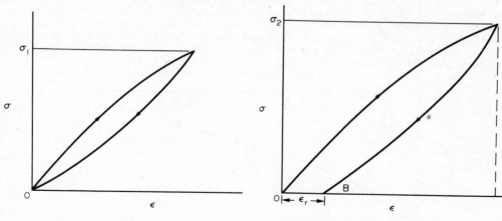

(a) σ_1 is less than the elastic
    limit–no residual deformation

(b) σ_2 is greater than the elastic
    limit–residual deformation = OB

Fig. 2.22. Schematic diagram of elastic and plastic be-
            havior showing Plenard's terminology
            (Plenard, 1968; copyright *Revue de
            Metallurgie*, used by permission).

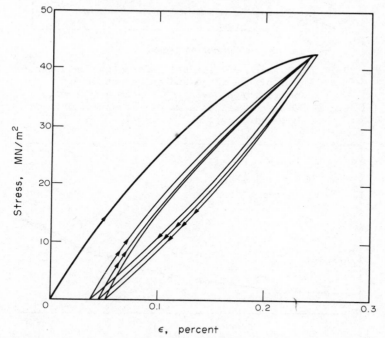

Fig. 2.23. The evolution of hysteresis loops in the
            course of repeated cycling of a Mg–Zr alloy
            (Plenard, 1968; copyright *Revue de
            Metallurgie*, used by permission).

microcreep can occur at stresses well below $\sigma_y(10^{-6})$. The microcreep of
5456–H34 aluminum is shown in Fig. 2.25. The fact that no creep was detected
until well over 1 h had passed under load did not preclude a rather significant
amount of creep at longer times. It seems doubtful that any such material
property as a microcreep limit exists.

It should be apparent from the discussion above that the terminology

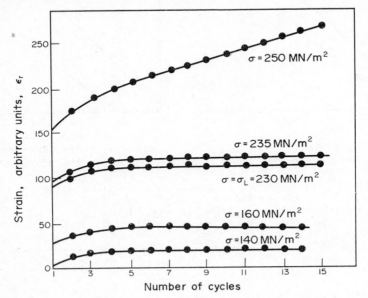

Fig. 2.24. Variation of residual deformation as a func-
            tion of the range and number of stress
            cycles for a cast iron specimen (Plenard,
            1968; copyright *Revue de Metallurgie*, used
            by permission).

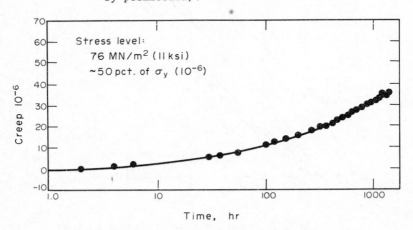

Fig. 2.25. Microcreep behavior of 5456–H34 aluminum
            (Maringer and Imgram, 1968; copyright
            American Society of Mechanical Engineers,
            used by permission).

describing micromechanical proerties is in a state of evolution. Agreement on the meaning of the terms in general use is nonexistent. Furthermore, most of the parameters described are quite arbitrary in nature. There is no real physical significance to a residual strain of 10^{-6} as a definition of PEL or MYS. Strains of 10^{-5} or 10^{-7} could be used almost as well. Strains near 10^{-6} are chosen because they are convenient to use and not excessively difficult to measure. In most cases, it is desirable to have the full stress - strain or strain - time curve, along with the experimental conditions of temperature, strain-rate, etc., in order to have a better "feel" for the material behavior.

2.6 CHAPTER SUMMARY

In precision design, it is necessary to reexamine many of the conventional design assumptions pertaining to material behavior. This applies particularly to the early portions of the stress - strain curve, including elastic, anelastic and microplastic strain. Terms such as elastic limit and proportional limit, while useful for discussion purposes, should perhaps be discarded because they create the impression that material behavior can be described with a few simple models and equations. It might be better if Hooke's Law were renamed Hooke's rule-of-thumb and taught as such.

Various aspects of material properties for precision design are considered in this chapter. Dangers in using "constants" to describe material behavior are emphasized for elastic moduli. Because most real crystalline materials are not truly random arrangements of many individual crystals, anisotropy of the elastic moduli is to be expected. The moduli are also a function of temperature and, in certain cases, of external fields.

Because the strain response of a material under stress is not always instantaneous, time and rate effects on mechanical behavior are described in general terms. It is shown that anelastic strain can lead to the formation of hysteresis loops in the stress - strain curve and, depending on the strain rate, apparent reduction of the elastic moduli. Elastic strain and anelastic strain are shown to be completely recoverable, while plastic strain is only partially recoverable.

Dislocation concepts are introduced to help clarify why crystalline materials often exhibit unexpected behavior, such as plastic deformation at very small stresses, partial recovery of plastic strain, strain hardening, and the Bauschinger effect.

Terminology surrounding the area of elastic, anelastic, and microplastic strains is still evolving. The importance of specifying the sensitivity of the strain sensing equipment in reporting data is emphasized. Terms such as "precision elastic limit", "anelastic limit", and "microyield strength" have meaning only when defined in terms of strain.

REFERENCES

American Potash and Chemical Corporation. 1971. Unpublished data.

Bonfield, W. and Li, C. H. 1966. The microstrain characteristics of beryllium, *Beryllium Technology*, Vol. 1, Gordon & Breach, New York, 539.

Brown, N. 1968. Observations of microplasticity, *Microplasticity*, C. J. McMahon, Jr., editor, Wiley, New York, 45.

Carnahan, R. D. 1964. Microplasticity, Aerospace Corp. Report No. TDR-269 (4240-10)-14, prepared for U.S. Air Force Systems Command.

Corten, H. T. and Elsesser, T. M.   1952. The effect of slightly elevated
    temperature treatment upon microscopic and submicroscopic residual stresses
    induced by small inelastic strains in metals, *Trans. ASME*, 74, 1297.
Cottrell, A. H.   1964. *Theory of Crystal Dislocations*, Gordon & Breach, New
    York.
Dieter, G. E., Jr.   1961. *Mechanical Metallurgy*, McGraw-Hill, New York.
Friedel, J.   1964. *Dislocations*, Pergamon, Oxford.
Hirth, J. P. and Lothe, J.   1968. *Theory of Dislocations*, McGraw-Hill, New
    York.
Hughel, J.   1960. An investigation of the precision mechanical properties of
    several alloys of beryllium, General Motors Report No. MR-120.
International Nickel Company.   1963. Engineering properties of Ni-Span-C,
    Alloy 902, Tech. Bull. T-31, Huntington Alloy Products Div., The Inter-
    national Nickel Co., Huntington, W. Virginia.
Johnson, R. H. and Honeycombe, R. W. K.   1962. Some microstructural observations
    during the thermal cycling of zirconium, *J. Less-Common Metals*, 4, 226.
Kennedy, A. J.   1963. *Processes of Creep and Fatigue in Metals*, Wiley, New
    York.
Koster, W.   1948. The temperature dependence of the elastic modulus of pure
    metals, *Z. Metallkunde*, 39, 1.
Kyle, P. E. , Papirno, R., Tang, C. N. and Becker, H.   1966. Photomechanical
    investigation of structural behavior of gyroscope components, Allied
    Research Associates, Final Report on NASA Contract 8-11294.
Lubahn, J. D. and Felgar, R. P.   1961. *Plasticity and Creep of Metals*, Wiley,
    New York.
Maringer, R. E. and Imgram, A. G.   1968. Effects of some processing variables
    on dimensional stability, *Trans. ASME*, Series F, 90, 846.
McClintock, F. A. and Argon, A. S.   1966. *Mechanical Behavior of Materials*,
    Addison-Wesley, Reading, Mass.
Nabarro, F. R. N.   1967. *Theory of Crystal Dislocations*, Oxford University
    Press(Clarendon), London.
Nowick, A. S. and Berry, B. S.   1972. *Anelastic Relaxation in Crystalline
    Solids*, Academic Press, New York.
Plenard, E.   1968. A new mechanical characteristic: the limit of accommodation,
    Rev. Met. 65, 845.
Price, C. W. and Hirth, J. P.   1972. A mechanism for the generation of screw
    dislocations from grain boundary ledges, *Mater. Sci. Eng.* 9, 15.
Rachinger, W. A.   1958. A super-elastic single crystal calibration bar,
    Brit. J. Appl. Phys. 9, 250.
Rosenfield, A. R. and Averbach, B. L.   1956. Effect of stress on the expansion
    coefficient, *J. Appl. Phys.* 27, 154.
Rosenfield, A. R., Hahn, G. T., Bement, A. L., Jr. and Jaffee, R. I.   1968.
    Dislocation Dynamics, McGraw-Hill, New York.
Schmid, E. and Boas, W.   1950. *Plasticity of Crystals*, Hughes, London.
Stein, D. F.   1968. A dislocations-dynamics treatment of microstrain, *Micro-
    plasticity*, C. J. McMahon, Jr., editor, Wiley, New York, 141.
Timoshenko, S.   1940. *Strength of Materials*, Part 1, Van Nostrand, New York.
Tinder, R. F. and Trzil, J. P.   1973. Millimicroplastic burst phenomena in
    zinc monocrystals, *Acta Met.* 21, 975.
Tinder, R. F. and Washburn, J.   1964. The initiation of plastic flow in
    copper, *Acta Met.* 12, 129.
Weertman, J. and Weertman, J. R.   1964. *Elementary Dislocation Theory*, Mac-
    millan, New York.
Zener, C.   1948. *Elasticity and Anelasticity of Metals*, University of
    Chicago Press, Chicago.

Chapter 3.

Sources of Dimensional Instability

At the beginning of Chapter 1 examples of three types of dimensional instability were given. Each of these might be viewed as a deviation from perfectly Hookean behavior. For example, when no stress is applied to a body, Hooke's Law states that the strain should be exactly zero. Thus, no matter how long the period of observation, Hooke's Law predicts no change of dimensions with time, measured at constant temperature, in the absence of an applied stress. Similarly, if a constant stress is applied for a long time, the strain should be predictable and constant. If stress is varied, the strain at any time should be a function only of the stress.

These are precisely the assumptions made in conventional design of components that operate near room temperature, modified only by a recognition that stresses must be kept below certain levels to avoid plastic deformation or fracture. In precision design, these assumptions must be reexamined with an intensity that depends upon the particular application. A body can, in fact, undergo gradual dimensional changes in the presence of zero or small constant applied stress. The strain in a body does, in fact, depend not only on the stress but on the path employed in reaching that stress--and so on.

In this chapter various sources of dimensional instability or, in the sense noted above, deviations from perfectly Hookean behavior, are outlined briefly to set the stage for more detailed discussions in Chapters 4 through 7. Sources include applied stress, alteration of internal stress, and microstructural changes. Neither thermal expansion nor change in elastic modulus with temperature are included as sources of dimensional instability. since they do not fall within the definition employed here. Nonetheless, because of their importance in precision design, each is treated in detail in Chapters 9 and 10, respectively.

3.1 APPLIED STRESS

From a dimensional stability standpoint, applied stress can produce various important types of response, as described in Chapter 2. A material may exhibit anelastic strain, mechanical hysteresis, microplastic strain, or microcreep, in addition to the ever-present elastic strain. Each type of response is reviewed briefly in subsequent sections of this chapter. A more detailed discussion of anelasticity and mechanical hysteresis appear in Chapter 4 and of microplastic strain and microcreep in Chapter 5.

Elastic Strain

True elastic strain results only from changes in atom spacing that accompany application of a stress. No atom rearrangement or change in atom neighbors is involved; hence, elastic response is virtually instantaneous and completely reversible. Because it is instantaneous and completely reversible, elastic strain is not considered to be a manifestation of dimensional instability. Only when significant nonelastic strains occur, in addition to elastic strains, is the term dimensional instability used.

There is at least one apparent exception to the instantaneity characteristic. When any material is suddenly strained elastically, it will experience a small temperature change--decrease on extension, increase on compression. During the period that the strain is imposed, the material will gain or lose heat from its surroundings; i.e., its temperature will gradually change. Because of thermal expansion, it will gradually change dimensions, giving the erroneous impression that the elastic strain is not instantaneous (see Fig. 5.9, for example).

Anelastic Strain

In addition to changes in atom spacing (elastic strains) that occur when a stress is applied to a material, other changes may also take place on an atomic level. For example, impurity atoms located in interstitial positions, such as carbon in iron, may move from one type of lattice position to another and then return to their original position when the stress is removed. This will produce a small amount of anelastic strain. The time dependency of such strain is a function of the particular mechanism giving rise to the strain, as opposed to the instantaneous nature of true elastic strain. Like elastic strain, anelastic strain is completely reversible.

Other mechanisms that have been identified as sources of anelastic strain in metals and alloys include:

- movement of segments of dislocation lines; dislocations are discussed in Section 2.4 ;
- rearrangement of solute atoms into an ordered array, as opposed to a random arrangement;
- relaxation of grain boundaries;
- movement of magnetic domain boundaries in ferromagnetic materials.

When anelastic strains occur at relatively low rates, components subjected to stress for long periods of time can be expected to exhibit gradually increasing strain with time. If the stress is subsequently removed, the strain will gradually disappear, approaching zero after long times. Consequently, components made from materials susceptible to low rate anelastic strain are potentially unstable dimensionally, both during sustained loading and after unloading.

Mechanical Hysteresis

When anelastic strains accompany elastic strains, mechanical hysteresis may be observed; i.e., the stress - strain curve on loading may differ from that on unloading. This can have serious consequences in precision springs and precision instruments where even minor deviations from Hooke's Law are viewed unfavorably. The formation of hysteresis loops will depend both on the imposed strain rate and the time dependency of the anelastic strain. For any particular anelastic strain mechanism, there will be a certain strain rate or, more correctly, load - unload frequency, which will produce a maximum mechanical hysteresis effect. Mechanical hysteresis is manifested by a hysteresis loop and by energy loss in each cycle.

Microplastic Strain

It was noted earlier that anelastic strain can result from the movement of segments of dislocation lines. It was noted also that anelastic strain is completely reversible; hence, the dislocation motion responsible for anelastic strain must also be reversible. However, under appropriately high applied stresses, the motion of dislocations ceases to be reversible; i.e., they do not return to their original positions when the stress is removed. It is at this point that plastic strain is first observed.

There is a fundamental difference between this type of strain and those mentioned earlier. Elastic and anelastic strains produce only very short-range changes in the internal structure. Given sufficient time after stress removal, the original structure will be restored. Plastic strain, however, produces changes in the internal structure such that the original structure cannot be restored simply by removing the stress. In crystalline materials, most of these changes are the result of movement of dislocations over relatively long distances, creation of new dislocations, and interactions of dislocations with various internal barriers.

As was noted also for anelastic strain, plastic strain does not occur instantaneously. Thus, the plastic strain response to an applied stress depends on loading rate.

The implications of microplastic strain relative to dimensional stability are easily appreciated. If such strain occurs, the dimensions of the component will have been changed permanently.

Microcreep

It is evident from the foregoing discussions that dislocations tend to move under the action of an applied stress to produce either anelastic or plastic strain. Generally, however, there are internal barriers that oppose their motion. Only if these barriers can be overcome will there be plastic deformation. When stresses are imposed for only short periods of time, the primary means for overcoming these barriers is through an increase in the applied stress. If the stresses are imposed for long periods, however, it may be possible, through the combined effects of time and temperature, to gradually relax the barriers or to activate additional dislocations, thereby permitting some dislocation motion. The resulting time-dependent strain is frequently referred to as microcreep.

Microcreep can be very damaging in precision design. Components subjected to stress for long time periods may experience gradually increasing strain and permanent changes in dimensions. Furthermore, if microcreep occurs, it is likely that the material can also undergo stress relaxation at the same temperature, with additional consequences, as discussed in Section 3.2.

3.2 ALTERATION OF INTERNAL STRESS

In Section 3.1, the role of applied stresses in dimensional instability was discussed briefly. It is possible also for internal stresses, i.e., stresses that exist in a body in the absence of external forces, to contribute to dimensional instability. The net forces and moments in the body must, of course, equal zero.

Two types of internal stress are involved. One is termed long-range and the other short-range. Long-range internal stresses are those commonly associated with fabrication processes, such as welding, forming, machining, and uneven cooling. They are often referred to simply as residual stresses. For example, a long cylindrical bar quenched from elevated temperatures, so that the surface cools much faster than the interior, will exhibit compressive residual stresses over its entire surface to a considerable depth, while the interior of the bar will exhibit tensile residual stresses. The stress pattern spreads over distances that are large relative to the microstructural features of the material.

Short-range internal stresses are associated with microstructural features of the material. For example, in a two-phase alloy consisting of finely dispersed particles of an intermetallic compound in a metal matrix, slow cooling from an elevated temperature will give rise to internal stresses at the particle – matrix interfaces because of thermal expansion differences. Short-range internal stresses can also develop between adjacent grains in a polycrystalline material as a result of temperature changes, particularly in materials that exhibit large anisotropy of thermal expansion coefficients.

Yet another source of short-range internal stresses is moving dislocations associated with metals undergoing plastic deformation. As the moving dislocations approach internal barriers, back stresses are developed opposing further motion. When the stress producing the plastic strain is removed, the dislocations tend to reverse their direction under the influence of the back stress. However, there will be other obstacles that will limit the amount of this reversal. Thus, a balanced system of short-range internal stresses will develop as a result of plastic strain.

Since they are confined to extremely small volumes and vary appreciably over short distances, these short-range internal stresses are difficult to detect or measure. It should not be assumed, however, that their magnitudes are necessarily low. Depending on the material and particular circumstances, they can be large enough to cause plastic deformation or, in brittle materials, microcracks.

Internal stresses, *per se*, are not necessarily detrimental. In fact, they are frequently introduced intentionally to improve certain types of behavior. For example, to reduce the likelihood of tensile fracture or fatigue failure, it is common to treat metals so that they have a high compressive residual stress at the surface. Even where extreme dimensional stability is required, internal stresses are not necessarily harmful. However, if internal stresses exist and if, during service, these internal stresses undergo alteration for any reason, then it is a virtual certainty that there will be an accompanying change in shape and dimensions. In the next sections, consideration is given to ways in which such alteration of internal stresses might arise. Each is discussed in greater detail in Chapter 6.

Stress Relaxation

In the present context, stress relaxation means simply a reduction in the magnitude of internal stresses existing in a body. It results as the elastic strains that give rise to these stresses are gradually converted to permanent or plastic strains through the combined effect of time and temperature. This is similar to microcreep, under a constant stress, discussed in Section 3.1, except that each increment of additional plastic strain that occurs in stress relaxation further reduces the stress, making further relaxation continually more difficult.

Stress relaxation, or stress relief, is often applied purposely to a dimensionally critical component to reduce long-range internal stresses to a level such that they will not relax further in service. Such treatments usually involve holding for some time at an elevated temperature, where plastic strain occurs more readily.

It is common to assume that most structural metals and alloys exhibit neither creep nor stress relaxation near room temperature. Such an assumption is dangerous in applications where shape and dimensions are critical.

Surface Removal

When the surface of a body containing long-range internal stresses is removed, for example, by machining or chemical polishing, the shape and dimensions will change, often by substantial amounts. In fact, one method for measuring the magnitude and distribution of residual stresses in a body is to note the dimensional changes that occur as material is gradually removed from the surface.

This observation has practical implications in precision applications. In precision finishing operations, for example, it is virtually impossible to achieve the exact desired shape if residual stresses are present, because of warpage that occurs each time additional material is removed. There are also implications for dimensionally critical finished parts in service. If they contain long-range internal stresses, they should be protected against any type of surface alteration, such as corrosion, abrasion, or wear.

Thermal Cycling

Often, in specifying treatments for dimensionally critical components, a thermal cycling treatment is included as the final step. The intention in specifying this treatment is to reduce any long-range residual stresses that may exist in the finished component so that they will be less likely to relax further in service. In certain materials, such as hardened steels, it may also be employed to produce a more stable microstructure.

The rationale behind thermal cycling can be traced back to the discussion of short-range internal stresses at the beginning of Section 3.2. Any temperature change will produce short-range stresses at interfaces where thermal expansion differences exist.* If these stresses, when added to the long-range internal stresses present, become great enough to cause localized plastic flow, then the long-range internal stresses should be diminished, that is, some of the elastic strain will have been changed to plastic strain. To be most effective, the temperature is cycled both above and below the service temperature to induce short-range stresses first of one sign and then the other.

The effectiveness of thermal cycling in relaxing long-range internal stresses depends on the nature of the material. In materials where there is little thermal expansion anisotropy among individual grains and no second-phase particles, little effect is anticipated.

Mechanical Exercising

Alteration of internal stresses by mechanical exercising--vibration or repeated gentle impacts, for example--is cloaked with a certain amount of mystery. Little real understanding exists. Historically, it was observed that metal parts could be stabilized by transporting them over cobblestone streets. More recently, there are reports that forced vibrations to nominally small stress amplitudes produces improved stability, presumably as a result of relaxation of long-range internal stresses. On this basis, some manufacturers of dimensionally critical components routinely employ vibration cycling as a precautionary measure.

From a very simple viewpoint, it might be assumed that addition of any stress to a body already containing high internal stresses will cause plastic deformation in certain localized regions. As noted earlier, this is equiva-

---

*More correctly, short-range stresses already exist at these interfaces. Changing the temperature causes the short-range stresses to change in proportion to the temperature change and to the difference in expansion coefficients. To a first approximation, this effect is reversible as the temperature changes are reversed.

lent to stating that the internal stress pattern will be altered. Reversing the sign of the added stress will simply increase the likelihood of reducing the peak stresses, regardless of their sign. Reversing the direction of the stress may also allow dislocations that are blocked from motion in one direction to move in the opposite direction. This may, in fact, be a part of the explanation of apparently successful mechanical exercising operations.

Viewed in this way, one stress cycle at a low rate would appear to be as effective as many cycles at a higher rate. However, there are some indications that both the number of cycles and the frequency have an effect on the process.

Debate continues regarding the ability of mechanical vibration to significantly alter long-range internal stresses. From a practical viewpoint, any dimensionally critical structure that will experience vibration in service probably should be exposed to similar vibration of equal or slightly greater severity prior to placing it in service, simply as a type of proof test.

3.3 MICROSTRUCTURAL CHANGES

Any change in the internal structure of a material will be accompanied by a change in the specific volume and, hence, a change in external dimensions. To minimize the likelihood that such microstructural changes will occur in service, leading to dimensional instability, materials are normally treated in such a way as to produce a relatively stable microstructure.

Definition of what constitutes a stable microstructure involves practical considerations. In the strictest thermodynamic sense, it is doubtful that any commercial alloys ever achieve what might be termed an equilibrium microstructure. Thus, there will always be a driving force tending to change the existing structure toward the equilibrium structure. However, unless the temperature is changed substantially, such changes are generally small in magnitude and, with certain exceptions, occur very slowly near room temperature.

In spite of their generally small magnitude and slow rate, these subtle microstructural changes can lead to significant dimensional changes in dimensionally critical structures over long periods of time. Several types of microstructure change are described in subsequent sections. Each is discussed in more detail in Chapter 7.

Changes in Vacancy Concentration

All crystalline materials contain imperfections known as vacancies. A vacancy is simply an unoccupied atom-site in the crystal lattice. At ordinary temperatures, the equilibrium concentration of vacancies is extremely small but at temperatures approaching the melting point the concentration can become significant. If, on cooling to room temperature, the vacancies are trapped, the volume of the crystal will be slightly greater than it would be ordinarily. There will now exist a tendency for the number of vacancies to gradually decrease toward their equilibrium concentration. Should this, in fact, occur, the volume and external dimensions of the crystal will grow smaller with time.

Phase Changes

Phase changes in solid metals refer to changes in crystal structure. For example, in the hardening of steel the face-centered-cubic (fcc) crystal structure that exists at high temperatures is changed to a body-centered-cubic (bcc) structure at lower temperatures. During the course of the phase change, the specific volume of the steel increases significantly because atom packing in bcc structures is less efficient than in fcc structures.

In an operational part in service, one would not anticipate that the component would undergo a total phase change. However, it is entirely possible that some small percentage of a certain phase may gradually transform to a more stable phase, with a concomitant change in dimensions. The phase change described above for steel will help illustrate this. Even though the fcc structure is not stable at room temperature, it is often observed that a small percentage of this structure persists for long periods of time at room temperature after cooling from high temperatures. Over periods of weeks, months, and years, a portion of this retained fcc phase may gradually change to the more stable bcc structure, resulting in a gradual increase in dimensions.

Precipitation and Re-solution

In many alloys, the solid-solubility of the alloying element in the host lattice depends on temperature. Thus, at elevated temperatures, much or all of the alloying element is held in solution; at low temperatures, there will be a tendency for the alloying element to be rejected from solution in some form. For example, Fig.3.1 shows schematically a portion of the equilibrium

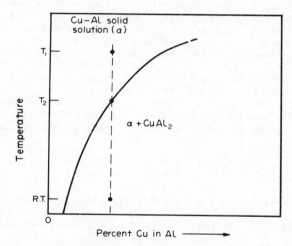

Fig. 3.1.   Schematic equilibrium phase diagram for aluminum-rich portion of Al-Cu system.

phase diagram for alloys formed by adding copper to aluminum.

At temperature T_1, given sufficient time, all the Cu will dissolve in the Al. Below T_2, a two-phase structure will exist--a solid solution phase (α) and an intermetallic compound ($CuAl_2$). However, if the alloy is cooled rapidly from T_1 to room temperature, the copper will essentially remain trapped in solution in the aluminum. The explanation for this is that, in order for $CuAl_2$ to form, copper atoms must diffuse through the crystal lattice to nucleating sites. This takes some time at any temperature, and at low temperatures is extremely slow. Thus, in a thermodynamic sense, the structure resulting from rapid cooling is not stable--there exists a driving force tending to cause $CuAl_2$ to be rejected from the solution. Should this occur, there will be an accompanying change of dimensions. From a practical standpoint, the important question is not whether the first structure is stable, which it clearly is not, but whether it will display any significant instability during the service life of a component made from this alloy.

Still referring to Fig. 3.1, a simple treatment can be employed that will greatly enhance the microstructural stability, and hence, the dimensional stability. If the alloy had been rapidly cooled to room temperature to retain the copper in solution, reheating to a temperature just below T_2 will greatly accelerate the precipitation of $CuAl_2$, so that in a relatively short time the reaction will be virtually completed. Now prolonged exposure at room temperature will be far less likely to result in additional precipitation of $CuAl_2$. It should be noted, however, that some tendency for additional precipitation will still exist, because Cu is less soluble in Al at room temperature than it is just below T_2.

The type of treatment just described is commonly called precipitation hardening. Not only does the precipitation treatment enhance stability, but it also increases the conventional yield and tensile strengths. It also increases the resistance to microplastic flow, though it is not yet certain that the treatments that provide the best conventional yield strengths also produce the highest microyield strengths.

It follows from the previous argument that if precipitation of a second phase produces a change in dimensions, then redissolving or re-solution will likewise produce a change in dimensions, of opposite sign. There will be a tendency for this to occur whenever the temperature is raised above that where an equilibrium two-phase structure existed previously.

An alternate treatment to the precipitation hardening schedule described earlier would be simply to cool from T_1 to room temperature very slowly. This would permit the $CuAl_2$ to begin precipitating at T_2 and would probably produce a more stable microstructure at room temperature than would rapid cooling and reheating. Such a treatment is not normally employed, however, because the associated conventional strength properties are generally inferior. Little is known about the effect of such treatments on microplastic properties.

Atom Rearrangement and Ordering

In principle, the considerations involved in discussing atom rearrangement and ordering are similar to those in the previous section for precipitation and re-solution. Instead of forming precipitates of a second phase, the alloying atoms may simply tend to occupy specific positions, as opposed to random positions, relative to like or unlike atoms in the host lattice. Sometimes this ordering may be associated with an applied stress or magnetic field. In any event, small dimensional changes will accompany such atomic rearrangements.

Irradiation Effects

Neutron irradiation can produce sizable changes in the dimensions of a body. These changes can be of two types--growth, which is anisotropic and arises through the creation of lattice defects, and swelling, which is an isotropic volume change that results from accumulation of fission products. Furthermore, irradiation can promote microstructural changes such as transformation of retained austenite in hardened steels and can accelerate relaxation of internal stresses, each of which will be accompanied by dimensional changes. Additional discussion of irradiation effects appears in Sections 6.4 and 7.5.

3.4 MAGNETIC AND ELECTRICAL FIELDS

Certain materials will experience reversible dimensional changes when they
are subjected to external fields. Specifically, ferromagnetic materials
exhibit *magnetostriction* when subjected to magnetic fields, while electric
fields give rise to *electrostriction* in dielectric materials. Both are
discussed further in Section 7.6. Ferromagnetic effects also play important
roles in the behavior of metallic materials that exhibit near-zero thermal
expansion coefficients and near-zero thermoelastic coefficients, as described
in Chapters 9 and 10.

3.5 CHAPTER SUMMARY

Various potential sources of dimensional instability are outlined in this
chapter. Some manifest themselves only under the application of stress to a
component, giving rise to strains that can differ significantly from those predicted by Hooke's Law. Included are anelastic, microplastic and microcreep
strains.

Other sources of dimensional instability require no applied stress. Internal stresses that develop for a variety of reasons and which are difficult to
avoid completely will produce dimension and shape changes whenever they are
altered during service. Similarly, microstructural changes that occur gradually within materials as they attempt to achieve an equilibrium structure will
be accompanied by changes in volume. Irradiation also can produce dimensional
changes by altering the material through creation of lattice defects and, in
fissionable materials, introduction of fission products.

In the special case of ferromagnetic materials and dielectric materials,
dimensions are influenced by magnetic fields and electrical fields, respectively.

The greater the dimensional stability requirements of a particular application,
the more important it becomes that the designer be aware of the role of each
of the potential sources of instability. Unfortunately, as permissible dimensional changes in service approach 1 part in 10^6, or less, it becomes increasingly difficult for the designer to find appropriate data for most engineering
materials. In such cases, the designer will be forced to devise test programs
in cooperation with materials specialists to ensure that dimensional instability can be held within acceptable limits.

Chapter 4.

Applied Stress Effects: Anelasticity and Mechanical Hysteresis

In conventional design, it is common to assume that the response of a material
to an applied stress is either elastic or a combination of elastic and plastic.
From discussions in earlier chapters, however, it is apparent that there is
another type of response--anelastic--that can be important in precision design
because it produces non-Hookean behavior at stress levels where essentially
Hookean behavior is normally anticipated.

Anelastic strain closely resembles elastic strain and, for this reason, the
two are often difficult to distinguish. The main distinguishing features are:

- elastic strain occurs virtually instantaneously and results simply from
  reversible changes in atom spacing under the action of an applied stress;
- anelastic strain is time dependent and results from reversible, relatively
  short-range changes in atom arrangements; the time dependency is governed
  by the particular anelastic strain mechanisms that are operative.

Anelastic strains are the source of mechanical hysteresis loops, discussed
in Section 2.3.2, and of internal friction and damping. The latter effects
are often studied at very low stresses and with exquisite sensitivity in torsion pendulums or in vibrating bars, for example, to gain improved understanding
of the various mechanisms by which anelastic strain occurs. Detailed treatments of internal friction and damping are outside the scope of this discussion.
The interested reader is referred to a book by Nowick and Berry (1972).

In this chapter, attention is directed primarily toward mechanical hysteresis
as observed in stress - strain curves obtained in load - unload tests at moderate rates and employing strain sensitivities of the order of 10^{-5} to 10^{-7}.
Many of the materials that have been studied in this manner are of little practical interest to the designer. They include various pure metals, such as iron,
copper, zinc, beryllium, and magnesium, often in the form of single crystals,
investigated for the purpose of obtaining a better understanding of deformation
mechanisms. Although the materials themselves may be of little practical
interest, results obtained from them can provide valuable clues as to the type
of behavior to expect in structural metals and alloys. Consequently, a portion of the discussion in this chapter makes reference to some of the fundamental studies on pure metals.

4.1 ELASTIC LIMIT AND ANELASTIC LIMIT

In much of the literature dealing with the fundamentals of microstrain
behavior of materials, it is common to define an elastic limit and an anelastic
limit, as shown in Fig. 2.21. When a load - unload technique is employed, the
stress at which closed hysteresis loops first appear is called the elastic
limit, σ_E (or τ_E for shear or torsion). The anelastic limit, σ_A, is the stress
above which the hysteresis loops fail to close upon unloading. It is evident
that both σ_E and σ_A will be a function of strain measurement sensitivity.

In spite of the wide usage and acceptance of this terminology, the origina-
tors (Brown and Ekvall, 1962) pointed out that this type of behavior is typical

only of materials that have previously been plastically strained some amount.
During initial loading of annealed specimens, closed loops are not observed,
but once a specimen has been given a permanent strain, it will begin to show
hysteresis loops at or below the maximum stress level reached during the prestraining (Roberts and Brown, 1960). Other investigators (Brown and Lukens,
1961; Roberts and Brown, 1962; Roberts and Hartman, 1964; Tinder, 1965; Lukas
and Klesnil, 1965; Rutherford and Swain, 1966; and Bilello and Metzger, 1968)
have reported similar findings. There are exceptions, however. Bonfield and
Li (1965) found no evidence of plastic strain occurring prior to the formation
of closed hysteresis loops in beryllium. Likewise, Rutherford and Swain (1966)
observed closed hysteresis loops in Al_2O_3 with no evidence of permanent strain.
Some of these observations are summarized in Table 4.1.

With the exception of beryllium, the materials listed in Table 4.1 are
of relatively high purity. The list also contains a ceramic, Al_2O_3. Only the
ceramic and QMV beryllium appear to behave abnormally; that is, they exhibit
closed hysteresis loops without detectable prior plastic strain. Since the
results reported for QMV beryllium were obtained with less strain sensitivity
than any of the others and since results for S200 beryllium follow the general
behavior pattern, it appears reasonable to generalize that plastic deformation
does, in fact, precede the appearance of closed hysteresis loops, at least in
annealed metals. Ceramics have received too little attention to be included
in this generalization.

If plastic strain does indeed precede the appearance of closed hysteresis
loops, it might be asked whether there is a limiting stress below which plastic
strain does not occur in an annealed material. Some investigators feel that
the evidence points toward some plastic strain, however small, at any stress
in an annealed material. Others have evidence that a threshold stress exists
below which no plastic deformation will occur. This question has not been
answered satisfactorily because of practical limitations on sensitivity of
strain measurement. Further discussion appears in Section 5.1.

4.2 HYSTERESIS LOOPS FOLLOWING PLASTIC PRESTRAIN

The fact that hysteresis loops begin to appear in load – unload stress –
strain curves after some small amount of plastic strain has occurred, and
continue to be present even after relatively large plastic strains, suggests
that plastic strain alters the material in some way to cause it to respond
anelastically, as well as elastically, to an applied stress. This has been the
subject of much discussion. Helgeland (1967), McMahon (1968), and Lukas and
Klesnil (1965) give detailed derivations of the conditions under which closed
hysteresis loops originate in load – unload cycles. Each reference emphasizes
the importance of an internal "back stress", developed during the course of
plastic strain, that acts in a direction opposite to that of the external
stress. For example, if a small tensile plastic strain is imposed, there will
remain, after unloading, an internal stress acting to shorten or compress the
specimen. This internal stress presumably results from dislocations piling up
against barriers, creating a back stress. When the external load is removed,
the back stress will cause the dislocations to move in the reverse direction
until a balance is achieved between the back stress and the friction stress
of the lattice, which opposes dislocation movement in any direction. Subsequent load – unload cycles in tension to the same, or lower, stress level,
will produce closed hysteresis loops. However, a compressive load – unload
cycle following plastic strain in tension *will not* show a closed hysteresis
loop, presumably because the sign of the internal stress is opposite that
required. Only after some compressive plastic strain has been introduced will

Table 4.1 Summary of Observations of Load–Unload Hysteresis
Loops by Several Investigators

Material	Form	Type of stressing	Approximate strain sensitivity	Plastic strain detected before hysteresis loop?	Reference
Zinc (99.994% pure)	Single crystals	Tension	10^{-6}	Yes	Roberts and Brown (1960)
Magnesium (high purity)	Single crystals	Tension	10^{-6}	Yes	Roberts and Hartman (1964)
QMV beryllium	Polycrystalline Hot-pressed Annealed	Tension Tension	2×10^{-6} 2×10^{-6}	No No	Bonfield and Li (1965)
Zinc (99.995% pure)	Single crystals	Torsion	5×10^{-9}	Yes	Tinder (1965)
Iron (99.95% pure)	Polycrystalline	Tension and compression	5×10^{-7}	Yes	Lukas and Klesnil (1965)
S200 beryllium	Polycrystalline	Compression	2×10^{-7}	Yes	Rutherford and Swain (1966)
Al_2O_3 (hot-pressed)	Polycrystalline	Compression	2×10^{-7}	No	Rutherford and Swain (1966)
High density Al_2O_3 (Lucalox)	Polycrystalline	Compression	2×10^{-7}	No	Rutherford and Swain (1966)
Copper (99.999% pure)	Polycrystalline	Tension	5×10^{-7}	Yes	Bilello and Metzger (1969)

a closed loop be observed in compressive load – unload cycles. Lukas and Klesnil (1965) have demonstrated this in polycrystalline iron specimens.

Others, including Roberts and Brown (1960), Roberts and Hartman (1964), and Bonfield and Li (1965), ascribe closed hysteresis loops to the bowing of pinned dislocation line segments (see discussion in Section 2.4). Resistance to this bowing comes from a frictional stress made up of intrinsic lattice friction and various other components, including impurities, intersecting dislocations, etc. Such bowing would presumably be possible in nonprestrained materials as well as prestrained, unless the prestraining is necessary to create sufficient length of pinned dislocation lines to make the hysteresis loops discernible with present detection techniques.

Magnitude of Hysteresis Effects

Hysteresis loops that arise from internal back stresses on dislocations or from dislocation bowing can have an appreciable magnitude. In some instances, the anelastic strain can actually exceed the elastic strain. Examples of such large effects are shown in Figs. 4.1 and 4.2, for single crystals of

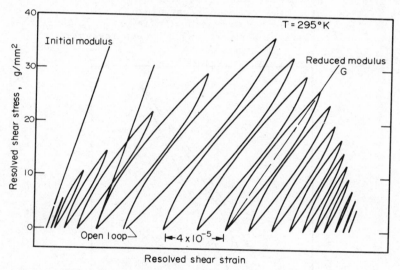

Fig. 4.1.   Hysteresis loops observed for a single crystal of magnesium, prestrained 6.6×10^{-3} (Roberts and Hartman, 1964; copyright AIME, used by permission).

magnesium and beryllium, respectively. In the examples shown, the apparent modulus at maximum load is less than half the initial elastic modulus. A somewhat smaller effect is shown in Fig. 4.3 for polycrystalline zirconium, where the apparent modulus after prestraining is about 17% less than the initial modulus.

The magnitude of the hysteresis effect depends both on the amount of the prestrain and on the magnitude of the stress cycle. Increasing either of these variables generally increases the magnitude of the hysteresis effect. It appears also to be a much larger effect in single crystals than in polycrystalline materials.

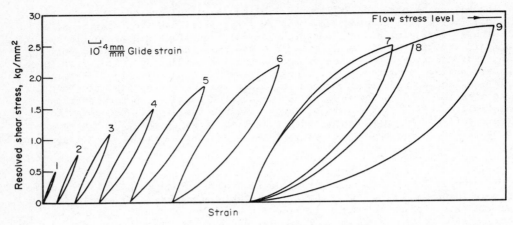

Fig. 4.2.   Hysteresis loops observed for a single crystal of beryllium, prestrained 4×10^{-1}
(shear) in compression (Lawley *et al.*, 1966;
copyright Pergamon, used by permission).

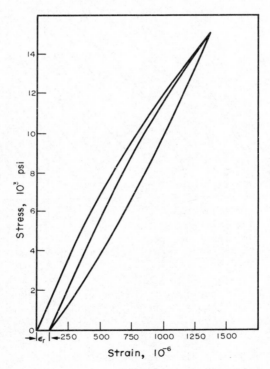

Fig. 4.3.   Hysteresis loops observed for polycrystalline
zirconium, prestrained 0.65% by rolling at
$77^{\circ}K$ (Reed-Hill *et al.*, 1965; copyright AIME,
used by permission).

Effect of Load - Unload Frequency on Hysteresis Loops

It was stated earlier that anelastic strain is time dependent. If the hysteresis loops commonly observed after plastic strain are truly an anelastic effect, their shape should be frequency dependent. Roberts and Hartman (1964) investigated this in a prestrained magnesium crystal over a range of cyclic frequencies from about 0.01 to 0.5 cycles per second. Their results are shown in Fig. 4.4. Notice that, as expected, increasing the frequency raises the apparent elastic modulus, because less time is available for anelastic strain to occur. If the cyclic frequency were to be increased sufficiently, the hysteresis loop should disappear and the modulus would have the same value as that of the unstrained material.

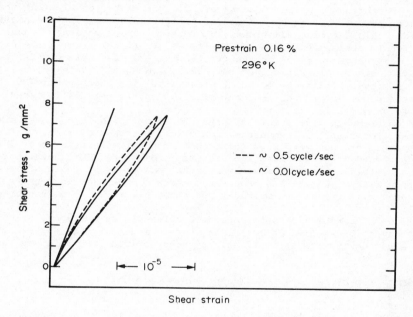

Fig. 4.4.   Effect of cyclic frequency on hysteresis loops (Roberts and Hartman, 1964; copyright AIME, used by permission).

Use of Hysteresis Effects to Study Dislocation Motion

Careful study of hysteresis loops in prestrained metals, both single crystal and polycrystalline, has provided numerous insights into dislocation movement and particularly into the lattice friction forces within a metal that oppose dislocation movement. In fact, it is common in microstrain testing to report a "friction stress" (σ_F) on the basis of measured hysteresis loop area (ΔW) and the amount of anelastic strain (ϵ_A) observed at the top of the load - unload cycle, using the relationship (Roberts and Brown, 1960)

$$\sigma_F = \frac{1}{2} \frac{\Delta W}{\epsilon_A} .$$ (4.1)

Thus, σ_F values can be obtained by measuring the slope of a graph of $\frac{\Delta W}{2}$ versus ϵ_A.

Roberts (1967) points out that this concept of a friction stress is an oversimplification. The correct meaning requires consideration of the precise shape of the entire hysteresis loop.

Friction stress is an important concept in dislocation thery. A crystal lattice, perfect except for a single dislocation, would have an inherent friction (small, but not zero) opposing any movement of that dislocation. This friction stress might be different for an edge dislocation than for a screw dislocation and would depend on both the specific material and crystal structure. This friction stress might be increased by impurity atoms or solute atoms, by other lattice imperfections, by other dislocations, etc. Thus, by making measurements of σ_F through analysis of hysteresis loops, it is possible to deduce something of the nature of the dislocation arrangements in strained materials. Furthermore, by introducing alloying elements, or altering the test temperature, or varying the amount of prestrain, or varying the amplitude of the load - unload cycle, additional information can be obtained concerning this frictional stress. For example, Lawley *et al.* (1966) studied the effect of plastic strains of up to 9% on the friction stress of single crystal beryllium. They reported values ranging from 0.2 to 1.3 MN/m^2 (28 to 185 psi), with no significant effect of amount of prestrain. The major portion of this friction stress was attributed to impurities ($\sim$30 parts per million) and the remainder to inherent lattice friction.

Solomon and McMahon (1971) investigated the effect of solute atoms, prestrain, heat treatment, and temperature on friction stress of iron.* It was observed that σ_F did not vary with the amount of prestrain up to about 7%. Likewise, carbon content and test temperature, from 75 to 300K, had little effect on σ_F. This contrasts with the conventional yield strength of iron and iron - carbon alloys, which is strongly temperature dependent in this range. Silicon, however, increased σ_F markedly and caused it to be temperature dependent as well.

Analysis of hysteresis loops to obtain σ_F values is an extremely valuable tool of fundamental research on materials. As indicated in the next section, there are also some practical implications to be derived from such studies.

Elastic Limit from Hysteresis Loop Measurements

In Chapter 2 and again in this chapter, an elastic limit (σ_E) has been defined as the stress below which closed hysteresis loops are not observed. It has already been noted that σ_E values may depend on the sensitivity of the strain measuring system and that it is probably necessary for some plastic strain to be introduced before closed loops can be observed.

There are at least three methods that can be used to measure σ_E. The first two are straightforward and the third requires an assumption concerning the mechanism responsible for closed loop formation. One method is to record load - unload curves and observe the stress where hysteresis loops can first be detected. The problems with this method are obvious--the value of σ_E will be a function of strain sensitivity.

A second method is to plot the hysteresis loop area as a function of the maximum stress reached during the load - unload cycle and extrapolate to zero loop area. The stress at zero loop area is taken to be σ_E. A third method

---

*Actually, they measured σ_E, the stress necessary to produce closed hysteresis loops. As will be shown later, σ_E is approximately equal to $2\sigma_F$.

is very similar to that described earlier for measuring σ_F, the friction stress. As noted in eq. (4.1),

$$\sigma_F = \frac{1}{2} \frac{\Delta W}{\epsilon_A} \ .$$

If $\sigma_E = 2\sigma_F$ (the rationale for this follows), then

$$\sigma_E = \frac{\Delta W}{\epsilon_A} \qquad (4.2)$$

and σ_E is simply the slope of the curve of loop area versus anelastic strain,* i.e., total strain minus elastic strain.

The rationale for stating that $\sigma_E = 2\sigma_F$ can be summarized briefly. As dislocations move over relatively large distances during plastic prestrain, they pile up against barriers, creating a back stress (σ_B). Because of this back stress, the dislocations will move in the reverse direction when the external load is removed. As they move, the back stress diminishes. Ultimately, a balance will be achieved between the back stress and the friction stress (σ_F). At this point, the dislocations will cease their reverse motion. Now when a stress is reapplied in the same direction as that causing the prestrain, the dislocation will not be able to move until both the back stress and the friction stress have been exceeded. In other words, $\sigma_E = \sigma_B + \sigma_F$. Since σ_B equals σ_F, it follows that $\sigma_E = 2\sigma_F$.

Figure 4.5 illustrates use of the second and third method for obtaining σ_E

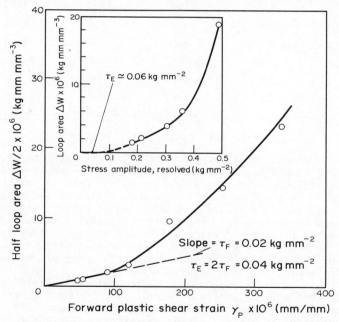

Fig. 4.5. Example of measurement of elastic limit (τ_E) by second and third methods described in text (Lawley *et al.*, 1966; copyright Pergamon, used by permission).

---

*Some references use the term forward plastic strain rather than anelastic strain.

values. Table 4.2 compares results obtained on several low-carbon steels and
an Fe-3Si alloy by all three of the above methods. When testing at relatively
low strain sensitivity, which is often necessary at high stresses because of
the large elastic strain component, the third method in Table 4.2 gave con-
sistently lower, and presumably more accurate, values than the other two
methods.

Table 4.2 Examples of σ_E Determinations by the

Three Methods Described in the Text

(Solomon and McMahon, 1971; copyright Pergamon, used

by permission)

Specimen	Tempera-ture ($^\circ$K)	Method 1 (psi)	Method 2 (psi)	Method 3 (psi)
(F5FC)	300	5200	4800	5300
(low-C steel)	198	6000	5400	5400
	77	6600	6200	6100
F4Q	300	3100	2500	3875
(low-C steel)	198	5450	4800	4375
	77	12 400	11 000	8850
FeSi	300	28 600	24 500	19 575
	198	34 700	31 000	24 675
	77	65 000	60 500	56 375

4.3 OTHER SOURCES OF ANELASTIC BEHAVIOR

It should be noted that there are numerous anelastic effects, in addition
to those arising from dislocations, that can produce closed hysteresis loops
in metals and crystalline ceramics, depending on the rate of cyclic loading
and the sensitivity of the detection techniques. Included are ordering of
solute atoms, grain boundary relaxation, movement of twin-boundaries, movement
of magnetic domain boundaries in ferromagnetic materials, movement of interstitial solute atoms to preferred lattice sites, and adiabatic heating and cooling
(Zener, 1948). Each of these will have its own relaxation time (anelastic
response time) at a particular temperature and, thus, may or may not produce
a hysteresis loop in the stress - strain curve.

In noncrystalline or partially noncrystalline materials, such as glasses,
glass-ceramics, and plastics, anelastic strain is a common occurrence. Such
materials can be viewed as supercooled liquids, i.e., liquids with high viscosity. When subjected to stress, viscous materials exhibit strain at a rate
that is proportional to the stress:

$$\dot{\epsilon} = \frac{1}{\eta} \sigma \qquad\qquad (4.3)$$

where η is the viscosity coefficient, the value of which is a strong function
of temperature. This strain apparently results from the fact that the atoms
and molecules are not closely packed. As discussed in Section 5.1, application

of a stress can cause atoms and molecules to take up new positions (see Fig. 5.16) at a rate that depends on the temperature. When the stress is removed, the atoms and molecules will eventually return to their original locations, i.e., the strain will gradually recover. Examples of such behavior have been reported for low-expansion glasses and glass-ceramics, as described in Chapter 9. Anelastic behavior from this source might also be anticipated in crystalline ceramics containing a small percentage of a glassy phase at the grain boundaries.

Internal friction measurements at very low stresses are commonly employed to study the various anelastic effects over a wide range of frequencies (10^4 to 10^{-12} sec^{-1}). At a particular temperature, say near room temperature, a number of internal friction peaks will be observed, each at a different frequency of cycling and each displaying a different peak amplitude.* Many of these, such as grain boundary relaxation, occur far too slowly at room temperature in most metals to show up at the cyclic rates employed in the load - unload tests normally employed in microstrain testing. On the other hand, certain other anelastic effects, such as motion of magnetic domain walls and movement of interstitial solute atoms, may well occur at these frequencies near room temperature. If they occur extremely rapidly relative to the cyclic loading rate, they will merely lower the apparent elastic modulus. If they occur at such a rate that they lag slightly behind the applied stress, they will contribute to the hysteresis loop. Presumably, in those materials where plastic-strain induced hysteresis loops have been studied, these other anelastic effects have either not been present or were too small to be detected. This is evident from the absence of hysteresis loops prior to plastic strain in most of the microstrain data reported. Nonetheless, they could be important in certain special cases. For example, the anelastic strain associated with magnetic domain wall motion in nickel and certain nickel - iron alloys is not insignificant and far surpasses that in pure iron or low alloy steels (Bozorth, 1951). Thus, domain wall motion might well contribute to hysteresis loops observed for these materials in the type of load - unload test often used in microstrain testing.

4.4 MINIMIZING MECHANICAL HYSTERESIS

Mechanical hysteresis may be of little consequence in some applications, while in others may be extremely troublesome. Where it is troublesome, it would obviously be desirable to minimize it or eliminate it completely, without at the same time compromising other desirable properties of the material.

Major attention has been given in this chapter to "dislocation" hysteresis, i.e., hysteresis believed due to dislocation segments, created by plastic strain, bowing under the action of a cyclic stress. It was noted, however, that similar effects could arise from other sources. These included reversible motion of magnetic domain walls and reversible migration of interstitial impurity atoms (for example, carbon or nitrogen in iron) to new lattice sites as the stress is varied. In noncrystalline materials, viscous flow can lead to mechanical hysteresis. Thus, to minimize hysteresis effects, it is helpful if the source of the hysteresis is known.

If dislocations are the source of the hysteresis, the following guidelines should be helpful in minimizing hysteresis:

---

*Actually, it is more common to study internal friction at a single frequency, near 1 cps, in a torsion pendulum, over a wide range of temperatures. This, too, will provide a spectrum of internal friction peaks that can be traced to a particular anelastic effect.

1. Avoid introducing plastic deformation of any kind into the material (bending, straightening, drawing, rolling, stretching, peening,* forming, machining,* etc.) unless the deleterious effects of this plastic strain can be eliminated or minimized prior to placing it in service. For example, the effects of plastic strain could be annealed out by heating to a sufficiently high temperature for a short period of time. In certain alloys, it may be possible also to effectively immobilize or lock dislocations by a thermal treatment at a temperature below that required for annealing. In alloys in which dislocations can be immobilized in this way, it may be possible to achieve much larger σ_E values by a combination of plastic deformation and thermal treatments than by avoiding plastic deformation entirely.

2. Avoid rapid cooling to room temperature after a thermal treatment, to prevent development of residual stresses. If residual stresses are present, it means that some part of the body experienced plastic deformation during cooling. The dislocation structure associated with this plastic deformation may give rise to hysteresis loops during subsequent load cycling.

3. If high strength is required, use discretion in employing cold working or strain hardening to achieve this. Where possible, try to achieve the necessary degree of strengthening by other means, such as precipitation hardening. Where cold working is employed for strengthening, it should be followed by suitable thermal treatments to immobilize the dislocations created by the working operations.

If reversible motion of magnetic domain walls is the major source of load – unload hysteresis loops, it will be necessary to immobilize the domain walls. Cold working would probably be beneficial but, from previous considerations, would give rise to hysteresis from dislocation motion. Here again, alloying to produce precipitation hardening is probably a good course to follow and a combination of plastic deformation and thermal treatment may give optimum results. An example of this is shown in Fig. 10.15 for Ni-Span-C, a commercially available ferromagnetic alloy (Fe-42Ni-5Cr-2.5Ti-0.5Al) that exhibits an essentially constant elastic modulus over a range of temperatures both above and below room temperature. It has been determined empirically that cold working Ni-Span-C 30 to 50% followed by heat treating at 595° to 650°C (1100° to 1200°F) results in minimum values of mechanical hysteresis, viz., about 0.02%, where percent hysteresis is hysteresis loop width times 100 divided by maximum deflection.

Where interstitial diffusion is the source of hysteresis loops, it is necessary to immobilize the interstitial impurity atoms. This can be done most effectively by getting them to combine with other alloying elements to form precipitates, dispersed throughout the lattice. As discussed in the previous paragraphs, certain combinations of plastic deformation and thermal treatments may produce an arrangement of these precipitates that not only minimize hysteresis but enhance other properties as well.

In glasses and glass ceramics, where hysteresis is a direct consequence of the viscous nature of the materials, the magnitude of the effect will be principally a function of composition.

---

*Peening and machining normally introduce plastic strain only in regions near the surface and, therefore, their effects would be expected to be smaller than that of the others. However, where cross-sections are small, these could prove to be large effects.

4.5 CHAPTER SUMMARY

Anelastic strain can be troublesome in many precision devices. In an altimeter that uses an aneroid capsule to detect changes in barometric pressure, it is clearly desirable that this device show identically the same stress - strain response under increasing pressure as under decreasing pressure. High precision load cells, precision springs, and precision galvanometers fall into the same category.

In this chapter, various sources of anelastic strain are described. Major attention is given to anelastic effects that arise when a crystalline material has been strained plastically. Such effects are readily detected in load - unload tests conducted at moderate rates, employing strain sensitivities in the range of $\sim 10^{-5}$ to 10^{-7}. Other sources of anelastic strain, such as ordering of solute atoms, grain boundary relaxation, etc., may require more sensitive tools for detection.

It is shown that analysis of hysteresis loops in prestrained metals can be used to study the forces that oppose dislocation motion. This provides a valuable tool for studying the effect of alloying additions or changes in test temperature on the lattice friction stress.

Mechanical hysteresis that arises from plastic strain can obviously be prevented by using materials in the annealed condition and avoiding the introduction of plastic strain of any kind, including that resulting from machining. However, there is evidence that extensive plastic strain in combination with suitable thermal treatments to immobilize or lock dislocations can be even more effective than annealing in minimizing hysteresis effects in some materials.

REFERENCES

Bilello, J. C. and Metzger, M. 1969. Microyielding in polycrystalline copper, *Trans. TMS-AIME*, 245, 2279.

Bonfield, W. and Li, C. H. 1965. The friction stress and initial micro-yielding of beryllium, *Acta Met.* 13, 317.

Bozorth, R. M. 1951. *Ferromagnetism*, Van Nostrand, New York.

Brown, N. and Ekvall, R. A. 1962. Temperature dependence of the yield points in iron, *Acta Met.* 10, 1101.

Brown, N. and Lukens, K. F., Jr. 1961. Microstrain in polycrystalline metals, *Acta Met.* 9, 106.

Helgeland, O. 1967. Some comments on the recent discussion of the anelastic strains and hysteresis observed in the microstrain range, *Scr. Met.* 1, 107.

Lawley, A., Breedis, J. F. and Meakin, J. D. 1966. The microstrain behavior of beryllium single crystals, *Acta Met.* 14, 1339.

Lukas, P. and Klesnil, M. 1965. Hysteresis loops in the microstrain region, *Phys. Status Solidi*, 11, 127.

McMahon, C. J., Jr. 1968. Microplastic behavior in iron, *Microplasticity*, C. J. McMahon, Jr., editor, Wiley, New York, 121.

Nowick, A. S. and Berry, B. S. 1972. *Anelastic Relaxation in Crystalline Solids*, Academic Press, New York.

Reed-Hill, R. E., Dahlberg, E. P. and Slippy, W. A., Jr. 1965. Some anelastic effects in zirconium at room temperature resulting from prestrain at 77K, *Trans. TMS-AIME*, 233, 1766.

Roberts, J. M. 1967. The friction stress acting upon moving dislocations as derived from current microstrain studies, *Acta Met.* 15, 411.

Roberts, J. M. and Brown, N. 1960. Microstrain in zinc single crystals, *Trans. TMS-AIME*, 218, 454.

Roberts, J. M. and Brown, N. 1962. Low frequency internal friction in zinc
    single crystals, *Acta Met.* 10, 430.
Roberts, J. M. and Hartman, D. E. 1964. The temperature dependence of the
    microyield points in prestrained magnesium single crystals, *Trans. TMS-AIME*,
    230, 1125.
Rutherford, J. L. and Swain, W. B. 1966. Research on materials for gas-lubri-
    cated gyro bearings, General Precision, Inc., First Technical Summary
    Report to NASA on Contract NAS 12-90.
Solomon, H. D. and McMahon, C. J., Jr. 1971. Solute effects in micro and
    macroyielding of iron at low temperatures, *Acta Met.* 19, 291.
Tinder, R. F. 1965. On the initial plastic behavior of zinc single crystals,
    Acta Met. 13, 136.
Zener, C. 1948. *Elasticity and Anelasticity of Metals,* University of Chicago
    Press, Chicago, Illinois.

Chapter 5.

Applied Stress Effects: Dimensional Instability from Microplastic Strain

As noted in Chapter 2, conventional designs can normally tolerate reasonable amounts of plastic deformation and thus, a conventional yield strength is measured at offset strains of about 0.2% (2×10^{-3}). In this chapter, attention is directed toward much smaller levels of plastic strain--generally near 10^{-6} but covering the range from about 10^{-4} to 10^{-9}. Strain in this range will be referred to as microplastic strain.

The first part of the chapter describes microplastic strain resulting from short duration loading, as in tension, compression, or torsion tests carried out at moderate strain rates. This is referred to as microyield behavior. Then the discussion moves to microplastic strain resulting from prolonged exposure to a constant level of stress. The term microcreep is used to describe such behavior. The chapter concludes with a brief discussion of microplastic strain recovery and the micro-Bauschinger effect.

5.1 MICROYIELD BEHAVIOR

The fact that materials can exhibit plastic deformation at stresses well below the conventional yield strength or even below the "elastic limit" often goes unrecognized. With the advent of dislocation theory (see discussion in Section 2.4), it became evident that plastic deformation is often to be expected at relatively low stresses in crystalline materials. In very pure metals with close-packed crystal structures, according to theory, the stress required to move dislocations through the crystal lattice, i.e., to produce plastic strain, should approach zero. This expectation has been confirmed by various investigators. Tinder and Trzil (1973), for example, observed plastic strain in zinc monocrystals at shear stresses as low as 0.2 kN/m^2 (0.03 psi)!

In engineering materials, as compared to pure metals, dislocation movement may be hindered by various obstacles. Thus, in these materials it might be anticipated that higher stresses are required to produce a given amount of plastic strain. This is well recognized for macroyield behavior. For example, the 0.2% offset yield strength of metals and alloys can be raised appreciably by treatments designed to impede dislocation movement over relatively large distances. For microyield behavior, the situation is less clear. Here, dislocation movement over large distances may not be required. Accordingly, obstacles that prevent long-range dislocation motion, and thereby increase the conventional yield strength, may have a smaller effect on short-range dislocation movement and microyield strength. For this reason, it is usually unwise to attempt prediction of microyield characteristics from measurements of macroyield properties.

Recognition of the importance of microyield behavior in precision design is growing rapidly. Nonetheless, much remains to be learned about the details and mechanisms of microyielding and the effect of important variables. Each of these topics is discussed in subsequent sections of this chapter.

Threshold Stress for Microyielding

In the discussion of hysteresis loops in Chapter 4, evidence was presented that hysteresis loops in annealed metals are observed in load - unload tests only after a finite amount of plastic deformation has occurred. The question was raised whether there is a threshold stress that must be exceeded if plastic strain is to occur in an annealed material.

Data reported by Roberts and Brown (1960) and Marschall and Maringer (1971) suggest that no threshold stress exists, i.e., that some small amount of plastic strain will occur even when the applied stress is very small. Some of their results are shown in Figs. 5.1 and 5.2. Note the linear relationship between

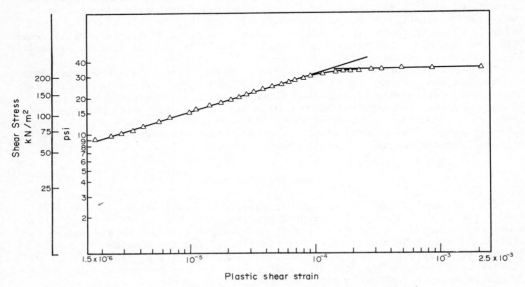

Fig. 5.1.　Logarithmic plot of stress versus plastic strain for an annealed zinc crystal (Roberts and Brown, 1960; copyright AIME, used by permission).

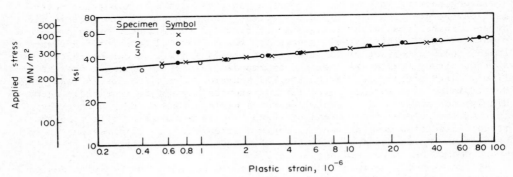

Fig. 5.2.　Logarithmic plot of stress versus plastic strain for 2014-T6 aluminum, stress relieved for 1 hr at 205°C (400°F) (Marschall and Maringer, 1971; copyright ASTM, used by permission).

stress and plastic strain plotted on logarithmic coordinates. Since the linear relationship covers a range of strain from about 10^{-4} to 10^{-7}, it is tempting to conclude that the relationship holds to all smaller strain values as well and that there is no perfectly elastic region. Eul and Woods (1969) reported similar results for beryllium. A logarithmic plot of stress versus plastic strain was linear to strains as small as 10^{-9}.

Other investigators have reported data that support the notion that a threshold stress must be exceeded to produce microyielding. Brown and Lukens (1961) and Brentnall and Rostoker (1965) observed a straight line relationship when stress was plotted versus the square-root of plastic strain. An example of this type of behavior is shown in Fig. 5.3. Extrapolation of such data

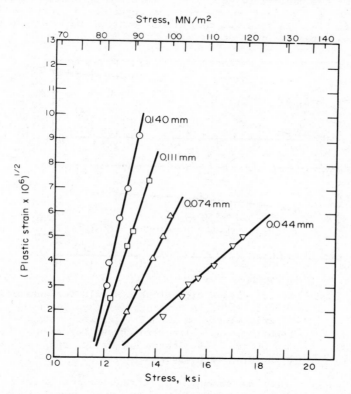

Fig. 5.3. Square root of plastic strain versus stress for various grain sizes in iron (Brown and Lukens, 1961; copyright Pergamon, used by permission).

to zero strain suggests that a threshold stress (σ_0) exists below which no plastic strain will occur. This type of behavior agrees with that predicted in an analysis of microstrain mechanisms (Brown and Lukens, 1961), which indicated that plastic strain (ϵ_p) should be related to stress (σ) by the relationship

$$\epsilon_p = \frac{C\rho d^3 (\sigma - \sigma_0)^2}{G\sigma_0} \tag{5.1}$$

where C is a constant, ρ is the density of dislocation sources, σ_0 is the stress to move the first dislocation, G is the shear modulus, and d is the average grain diameter. Assuming σ_0, ρ, d, and G to be constants in a particular material to be tested, eq. (5.1) can be rearranged to give

$$\sigma \propto \epsilon_p^{1/2}. \tag{5.2}$$

As pointed out by Brentnall and Rostoker (1965), this treatment assumes that plastic flow begins in the interior of each grain at certain appropriate points of origin.

Microyield data have been reported also that indicate a threshold stress (σ_0) but that do not conform to eq. (5.2). Carnahan and White (1964a), for example, observed a sigmoidal stress – strain curve for polycrystalline nickel, as shown in Fig. 5.4. This curve, too, can be extrapolated to zero plastic strain to give a value for σ_0.

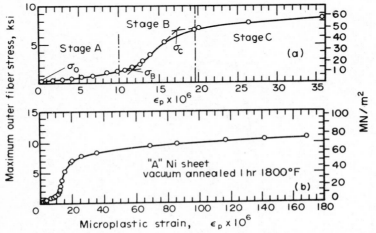

Fig. 5.4.   Stress versus microplastic strain for annealed
             nickel tested in bending: (a) expanded plot
             of initial region illustrating three-stage
             behavior, and (b) entire curve (Carnahan
             and White, 1964a; copyright Taylor and
             Francis, used by permission).

The results suggest that the concept of a threshold stress to initiate microyielding is valid for some materials and not for others. From a practical standpoint, however, even those materials that exhibit linear stress – strain curves on logarithmic coordinates, as in Figs. 5.1 and 5.2, might be said to display an approximate threshold stress. Rather than extrapolating to zero strain to find σ_0, as in Fig. 5.3, such curves could be extrapolated to some small, but finite, strain, say 10^{-10}, to give an approximate σ_0.

Methods Employed to Investigate Microyield Behavior

The most common procedure for investigating microyield behavior utilizes a series of load – unload cycles, with the load increasing incrementally in successive cycles. After each cycle, the specimen gage section is examined for permanent strain at zero load.* In some cases, measurements are made at a

---

*Strain measurement techniques are described in Chapter 8.

small reference load, rather than zero load, to help maintain the alignment of the loading system.

Typical results of such a test are shown in Fig. 5.5 for an aluminum alloy, 2024-T4, at a strain sensitivity of about 10^{-7}. At the strain sensitivity employed, no plastic strain is discernible below about 205 MN/m^2 (30 ksi). Thereafter, plastic strain increases rapidly with increasing stress. Offset yield strength values obtained from Fig. 5.5 include

$$\sigma_y (10^{-6}) = 245 \text{ MN/m}^2 \text{ (35.5 ksi)}$$

$$\sigma_y (10^{-5}) = 310 \text{ MN/m}^2 \text{ (45 ksi)}$$

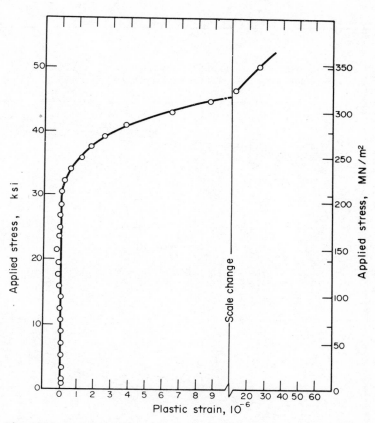

Fig. 5.5.   Microyield data for 2024-T4 aluminum, stress relieved at 205°C (400°F) for 1 hr (Maringer *et al.*, 1968).

Although not shown on the graph, the conventional 0.2% offset yield strength, σ_y (2 x 10^{-3}), in this test was approximately 415 MN/m^2 (60 ksi).

Even though the strain sensitivity in this test was 10^{-7}, it is difficult to obtain a value for σ_y (10^{-7}) from Fig. 5.5. It is possible to obtain an estimate for σ_y (10^{-7}), however, by replotting the data of Fig. 5.5 on logarithmic coordinates, as shown in Fig. 5.6. By extrapolating the straight line

thus obtained, σ_y (10^{-7}) can be estimated at about 195 MN/m^2 (28 ksi). It
has already been pointed out, however, that microplastic strain does not always
show linear behavior when plotted on logarithmic coordinates. The precise
shape of the stress versus microplastic strain curves shown in Figs. 5.1
through 5.6 depends to some extent on the test details. This has received

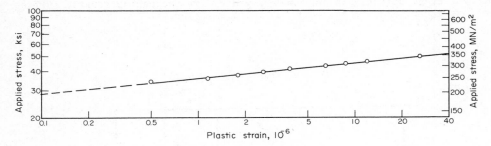

Fig. 5.6.   Logarithmic stress – strain curve for 2024–T4
            aluminum (data from Maringer *et al.*, 1968).

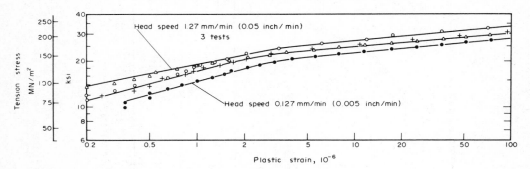

Fig. 5.7.   Effect of loading rate on microyield behavior
            of annealed 304 stainless steel (data from
            Price, 1973).

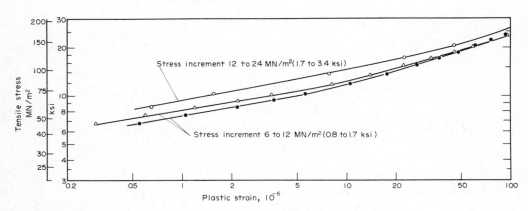

Fig. 5.8.   Effect of stress increment on microyield
            behavior of 304 stainless steel (data from
            Price, 1973).

little attention, but evidence is growing that loading rate, dwell time at
maximum load, unloading rate, time delay between unloading and measurement of
residual strain, and magnitude of the load increment can have some effect on
the results. For example, Price (1973) has shown evidence that increasing
both the loading rate and the load increment tend to raise the stress – microplastic strain curve (see Figs. 5.7 and 5.8). Similar results would be anticipated from decreasing the dwell time at maximum load and increasing the unloading rate. The resulting higher stress values required to produce a given
amount of microplastic strain undoubtedly reflect the decreasing contribution
of time-dependent strain to the measured strain values.

With respect to the delay time between unloading and measurement of residual
strain, two factors appear to be of importance. Certainly if the residual
strain diminishes with time after unloading because of plastic recovery effects,
both the rate and magnitude of the recovery should be noted and reported as
part of the test results. Plastic recovery is discussed in greater detail
in Section 5.3.

A second factor concerning delay time after unloading is perhaps less
obvious. As has long been known, but frequently overlooked, the temperature
of a material changes when it is elastically strained. If a metal is suddenly
extended elastically, it will cool; if compressed, it will heat. Geil and
Feinberg (1969a, b) have demonstrated that this adiabatic heating and cooling
can have significant effects on the results of microstrain testing unless it
is recognized and test procedures adjusted accordingly. Figure 5.9 shows
specimen temperature as a function of time during loading, unloading, and
for some time thereafter, for a prestrained AISI 4340 steel specimen. Note
that the temperature decreases by over $0.6°C$ ($1.1°F$) during loading. Removal
of the load causes the temperature to rise. In fact, it rises above its

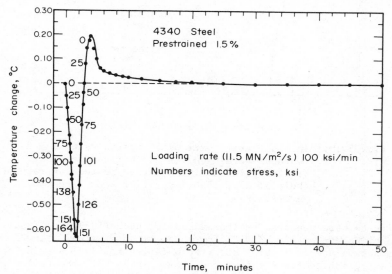

Fig. 5.9.   Effect of loading at a rate of 11.5 $MN/m^2/s$
             (100 ksi/min) to a stress of 1130 MN/m^2 (164
             ksi), with immediate unloading, on the
             temperature change of a prestrained specimen
             of normalized 4340 steel; specimen had been
             prestressed at 1550 MN/m^2 (225 ksi) and
             extended 1.5% (Geil and Feinberg, 1969b).

starting point by 0.2°C. This temperature rise above the starting point is probably due to two factors: (1) during most of the time the specimen was under load it was cooler than its surroundings; hence, heat was flowing into the specimen, and (2) the specimen experienced some plastic strain (45 x 10⁻⁶) at the maximum stress; this generated some heat in the specimen. Approximately 20 min elapse before the specimen returns to its starting temperature. Accordingly, Geil and Feinberg routinely wait about 20 min after unloading before measuring residual strain in microstrain tests.

There is another feature of load - unload microstrain tests that has caused investigators some concern. On certain occasions, and particularly for certain materials, *negative* residual strains are observed, i.e., the specimen appears to be shorter following removal of a tensile load. This was first reported by Muir *et al.* (1955) in tests conducted on hardened carbon-steels. As shown in Fig. 5.10, significant negative residual strains were observed.

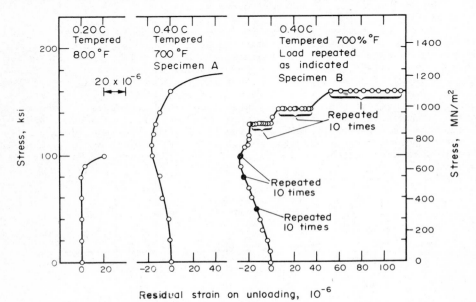

Fig. 5.10. Initial early portions of typical stress - residual strain curves for hardened carbon steels, illustrating negative residual strains (Muir *et al.*, 1955; copyright ASM, used by permission).

Similar effects were reported by Thomas and Averbach (1959) for tests on copper and by Hordon *et al.* (1958) for tests on an aluminum alloy. More recently, Marschall *et al.* (1972) observed negative residual strains for both Ti-6Al-4V and graphite-fiber-reinforced epoxy, while Geil and Feinberg (1969b) found negative strain in prestrained Invar. In all but one of the above examples, strain was measured with bonded electrical resistance strain gages. Accordingly, the negative strain has been attributed variously to creep of the gage-bonding agent, improper bonding of the gages, or exceeding the elastic strain capability of the cement (Thomas and Averbach, 1959; Carnahan and White, 1964b). Though this may provide an explanation in some instances, there are others where these explanations do not appear to suffice.

For example, in the case of hardened steels, Muir *et al.* stated that the negative residual strains reflected a real behavior of the specimens rather than of the strain gages, since precision end-length measurements on several specimens demonstrated that the test bars after tensile loading were, in fact, shorter then their original length. Hordon *et al.* demonstrated that this was the case also for several aluminum alloys, as illustrated in Fig. 5.11.

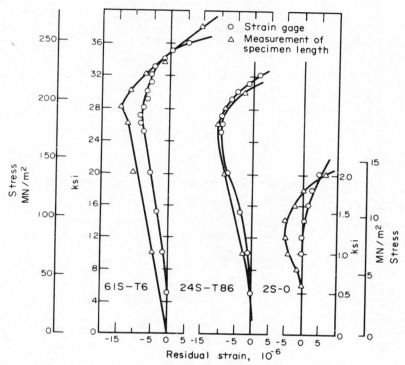

Fig. 5.11.   Applied stress versus residual strain for several aluminum alloys, illustrating negative residual strains (Hordon *et al.*, 1958; copyright Pergamon, used by permission).

Finally, in tests employing capacitance gages with a strain sensitivity of 10^{-7}, Geil and Feinberg observed negative strains in excess of 3×10^{-6} before strain in the expected direction began. This is shown in Fig. 5.12 for a test on prestrained Invar. A similar, though smaller, negative strain effect was noted for 4340 steel, prestrained 1.5%. Geil and Feinberg also reported a "strain reversal" in microyield tests conducted on annealed 4340 steel. As shown in Fig. 5.13 (curves D, E, F, and G), the specimen first experienced some positive residual strain at relatively low stress levels. After exposure to some higher stress level, a portion of this positive residual strain disappeared.

Many questions remain unanswered regarding negative residual strain. Undoubtedly, some reported cases are the result of strain measurement difficulties. In other cases, however, evidence is strong that the negative strains are real.

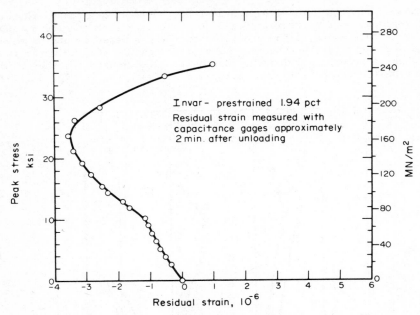

Fig. 5.12. Peak stress versus residual strain observed
            in microplasticity test on a prestrained
            specimen of annealed Invar (after Geil and
            Feinberg, 1969b).

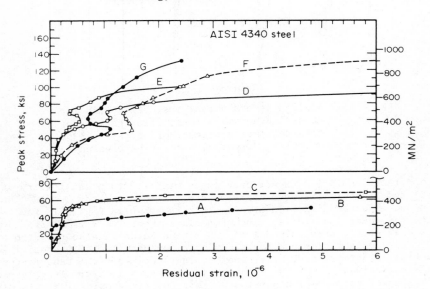

Fig. 5.13. Peak stress – residual strain relationships
            observed in microyield tests on a single
            specimen of normalized 4340 steel (Geil and
            Feinberg, 1969b).

Curve	Maximum stress Previous test ksi	MN/m²	Strain Previous test	Total Prior strain
A	0	0	0	0
B	50	345	4×10^{-6}	4×10^{-6}
C	69	480	24×10^{-6}	28×10^{-6}
D	94	650	181×10^{-6}	209×10^{-6}
E	110	760	19×10^{-6}	460×10^{-6}
F	148	1020	730×10^{-6}	1560×10^{-6}
G	138	950	8×10^{-6}	1568×10^{-6}

No mechanism has been postulated to account for this puzzling behavior, but it might be the result of subtle microstructural changes or subtle redistribution of internal stress in the specimen during exposure to load. From a practical standpoint, negative residual strains are as damaging in a precision component as positive residual strains and, from the limited evidence available, occur at lower levels of applied stress.

Another concern of a number of investigators is that repeated load – unload cycling influences the microyield behavior in some way. Holt (1972) has presented evidence that the microplastic behavior of polycrystalline nickel differs somewhat for repeated loading as compared with continuous loading (Fig. 5.14). He noted also that, even below σ_A, the width and area of hysteresis loops decrease slightly on repeated loading to a given stress level.* Thus, Holt concluded that the repeated loading technique does not give a true re-

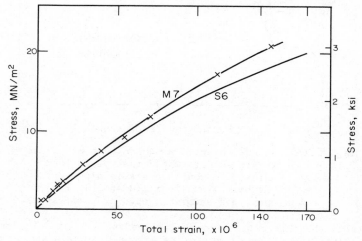

Fig. 5.14. Stress – strain curves for a continuously loaded specimen (S6) and a repeatedly loaded specimen (M7) of polycrystalline nickel (Holt, 1972; copyright Institute of Metals, used by permission).

---

*Many other investigators, however, have reported that hysteresis loops below σ_A are very reproducible under repeated loading.

presentation of microyielding because of a small cyclical hardening effect. Conversely, Bonfield and Li (1964) reported that load – unload testing of QMV beryllium produced microyielding at *lower* stresses than did continuous loading tests (Fig. 5.15). They attributed this to strain recovery that occurred on each unloading cycle, which produced an additional small amount of plastic microstrain on each reloading. Clearly, the data are too sparse to permit any firm conclusions to be drawn.

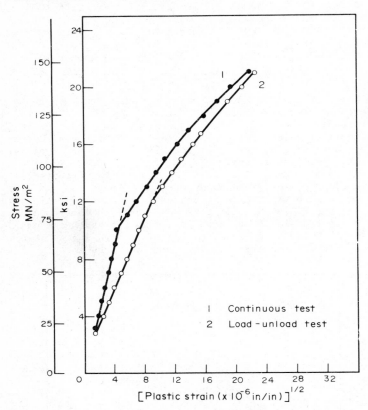

Fig. 5.15. Stress versus the square root of plastic
strain as determined in typical load –
unload and continuous straining tests on
QMV beryllium (Bonfield and Li, 1964;
copyright Pergamon, used by permission).

Unfortunately, the alternatives to cyclic loading also have problems. If continuous loading is employed in microstrain tests, plastic strain can be deduced only by subtracting elastic and anelastic strains from the total strain. Generally speaking, this is a far less accurate method of measuring microplastic strain than that employed in the load – unload technique.

From the discussion in this section, it is clear that measurement of micro-plastic strain behavior has not yet been standardized and that all test results are subject to interpretation. It is clear also that the investigator should be aware of the various factors that can influence test results. To maximize the meaningfulness of the results, the procedures employed in conducting the tests should be reported in detail.

Microyield Mechanisms

As pointed out in Section 2.4, all crystalline materials contain imperfections known as dislocations. The presence of these dislocations permits one plane of atoms to shear over a neighboring plane at a stress several orders of magnitude below that required in a perfect crystal. Such irreversible shearing of one plane of atoms over another constitutes plastic deformation. It is useful in attempting to understand microstrain behavior to consider some of the aspects of dislocation motion that lead to plastic strain. The treatment is oversimplified; dislocation mechanisms are treated in greater depth in many of the references appearing at the end of this chapter.

In extremely pure annealed metals, the only resistance to the initial movement of dislocations is the inherent lattice friction. The stress required to overcome this friction--the Peierls - Nabarro stress--is expected to be quite low, particularly in metals with face-centered-cubic (fcc) and hexagonal close-packed (hcp) crystal structures. As previously noted, this expectation has been confirmed by Tinder and Trzil (1973). Employing strain measurement sensitivities of $\sim 10^{-9}$, values of σ_y (10^{-8}) near 0.2 kN/m^2 (0.03 psi) were reported for high-purity zinc single crystals tested in torsion. In pure body-centered-cubic (bcc) metals, such as iron, lattice resistance is normally somewhat greater than in fcc or cph metals. Crystalline ceramics and intermetallic compounds exhibit even more varied lattice resistance to dislocation motion, depending on crystal structure, type of atom bonding (ionic, covalent, or metallic) and number of possible dislocation glide systems. For example, as dislocations move in a compound to produce shear of one crystalline plane over another, strong covalent bonds between neighboring atoms may have to be broken and nearest-neighbor relationships among atoms may be disturbed. Both effects will create lattice resistance to dislocation motion (Rudnick *et al.*, 1968).

After traveling through a crystal or grain of the pure annealed material, the dislocation approaches a grain boundary. It cannot cross the boundary because the crystal on the other side is not oriented in the same way as the first crystal. Hence, the grain boundary acts as a barrier to further motion of the dislocation, i.e., a so-called back stress will develop. If the applied stress is increased, the dislocation will move slightly closer to the boundary and the back stress will increase accordingly. If the applied stress is removed, the dislocation will move away from the boundary until the back stress has diminished to the value of the friction stress.

The picture will be similar if, instead of speaking of a single dislocation, a dislocation source is considered to be operating on a given plane. This might be a Frank - Read source operating within a grain or a grain boundary source operating at the grain periphery (refer to Section 2.4). Now dislocations, as they move away from the source under the action of applied stress, will "pile-up" at a grain boundary and even larger back stresses will be developed. When this back stress becomes sufficiently large, the dislocation source will essentially cease operating and no more plastic strain will occur until some new source of dislocations is activated. As the applied stress is increased, new sources may be activated on other crystalline planes and the entire process repeated. If similar activities are occurring within each grain, the total strain can multiply rapidly. However, as the strain increases, the number of sources will gradually be used up, i.e., each new increment of strain will require a higher stress than the preceding increment. In other words, the material will be experiencing strain hardening. This differs, however, from conventional strain hardening or macrostrain hardening, where the hardening is normally ascribed to extensive interaction of dislocations rather than to depletion of dislocation sources.

Imgram *et al.* (1968) have shown evidence that little dislocation interaction occurs in the microstrain region. Transmission electron micrographs of a grain boundary region in Ni-Span-C (precipitation hardened Fe-Ni alloy) after a tensile strain of 40×10^{-6} showed little evidence of dislocation interaction. After a plastic strain of 2.5%, however, the interaction of intersecting dislocations was very apparent.

Returning to the mechanism by which microplastic strain can occur in a pure annealed material, as the back stresses created at grain boundaries gradually increase, they may eventually become large enough to activate new sources on the other side of the boundary. Etch pit studies conducted by Vellaikal (1969) on copper have shown that grain boundaries do in fact act as barriers to dislocations in the microstrain region as discussed in Section 8.2. Furthermore, he showed that, as the stress continues to rise, new dislocations could be generated at grain boundaries by the forces created by the piled-up dislocations. In other words, entirely new dislocation sources were created. This suggests that, at higher strain levels, more strain can be produced by an increment of stress than when only existing dislocation sources were being activated. Brown and Lukens (1961) suggested that this changeover from the activation of existing sources to the creation of new sources may constitute the real difference between microstrain and macrostrain behavior. For example, the break in the curve shown in Fig. 5.1 may be indicative of the creation of new dislocation sources.

Wilson and Teghtsoonian (1970), on the other hand, hold a different view. They suggest that in niobium crystals, microflow is associated with movement of edge dislocations while macroflow corresponds to the motion and multiplication of screw dislocations.

Employing this simple model of microstrain, it is possible also to rationalize two other aspects of microstrain behavior--plastic strain recovery and the Bauschinger effect. Both are discussed further in Sections 5.3 and 5.4. If, after removal of stress, all dislocations that had piled up against barriers returned toward their source until the back stress just equalled the lattice friction stress, there would be no further tendency for the dislocations to move in either direction. However, if the dislocations encounter difficulty in returning toward the source, a back stress considerably greater than the normal lattice-friction stress may continue to exist. For example, Vellaikal (1969) suggests that, on loading, various dislocation sources operate at different times, and thus do not interact, even if their paths cross. But on unloading, all the dislocations tend to move back toward their sources at the same time, and many of them cannot reach these sources without interacting with other dislocations. The resulting abnormally great back stress provides a driving force that may cause a gradual further return of dislocations toward their sources, i.e., a gradual recovery of a portion of the plastic strain observed immediately after removal of the stress. Generally speaking, the greater the plastic strain, the greater will be the tendency toward recovery.

Similarly, if a stress is applied in a direction opposite to that originally applied to obtain plastic strain, it is evident that the back stress will aid the applied stress in moving the dislocations back toward their sources and beyond. In other words, the applied stress required to produce plastic strain in a certain direction will be reduced appreciably by prior plastic strain in the opposite direction. This is commonly referred to as the Bauschinger effect. Of course, after the dislocations have moved some distance, they will again encounter barriers, just as in the first case, and similar considerations can be given to achieving additional plastic strain.

What if the metal is not of high purity, as assumed at the beginning of this discussion of microstrain mechanisms? The most important effect is likely to be an increase in stress necessary to cause the dislocations to move. This can be a fairly small effect, arising from a small percentage of dissolved impurity atoms acting to increase the friction stress of the lattice. Or it can be a dramatic effect in which alloying elements, through deliberate treatments, form finely dispersed precipitate particles, often along dislocation lines, that effectively lock existing dislocations and act as barriers to newly generated dislocations. Once a sufficient stress is imposed to move existing dislocations and create new ones, the mechanisms for microstrain may not differ greatly from those in pure metals. However, the stresses required to achieve a given amount of plastic strain may be appreciably greater in the impure and alloyed materials.

Carnahan (1964) has suggested that the microstrain behavior of typical engineering materials can be predicted from a simplified theoretical approach involving dislocation motion. Assuming an applied stress, σ, of 25% of the conventional yield strength and initial dislocation densities, N, of 10^4, 10^6, or 10^8 lines/cm^2, and estimating the dislocation source length, ℓ, from the expression

$$\ell = \frac{Gb}{\sigma}$$

where G is the shear modulus and b is the Burgers vector of the dislocation,* Carnahan calculated the shear strain per unit area, γ, from

$$\gamma = Nb\ell .$$

Table 5.1 gives G, b, σ_y, and σ values for several engineering materials while Table 5.2 shows calculated shear strains for several values of N. Calculated values are compared with experimentally observed values in Table 5.3. Considering the simplifying assumptions involved, Carnahan felt that reasonable agreement between theory and experiment was evident.

Table 5.1 Stress, Shear Modulus, and Burgers Vector Values
for Selected Engineering Materials (Carnahan, 1964)

Materials	σ_y, psi	σ, psi	$G \times 10^{-6}$, psi	$b \times 10^8$, cm
Low carbon steel (SAE 1020)	45 000	11 000	11	2.48
High carbon steel	140 000	35 000	11	2.48
Aluminum	5 000	1 250	4	2.86
Duralumin	18 000	4 500	4	2.86
2024-T36 Al	52 000	13 000	4	2.86
Copper	4 000	1 000	6	2.55
Be-copper	200 000	50 000	7	2.55
"A"-Nickel	10-30 000	2 500-7 500	11	2.49
TD-Nickel	~65 000	15 000	11	2.49

*The Burgers vector (b in Fig. 2.18) specifies the direction and the amount of slip produced when a dislocation moves through a crystal.

Table 5.2

Calculated Shear Strains for Selected Engineering Materials (Carnahan, 1964)

Material	l, cm	γ		
		N, lines/cm^2		
		10^4	10^6	10^8
Low carbon steel	2.5×10^{-5}	6.2×10^{-9}	6.2×10^{-7}	6.2×10^{-5}
High carbon steel	7.8×10^{-6}	1.93×10^{-9}	1.93×10^{-7}	1.93×10^{-5}
Aluminum	9.15×10^{-5}	2.6×10^{-8}	2.6×10^{-6}	2.6×10^{-4}
Duralumin	2.54×10^{-5}	7.26×10^{-9}	7.26×10^{-7}	7.26×10^{-5}
2024-T36 Al	8.8×10^{-5}	2.52×10^{-9}	2.52×10^{-7}	2.52×10^{-5}
Copper	1.53×10^{-4}	3.9×10^{-8}	3.9×10^{-6}	3.9×10^{-4}
Be-copper	3.56×10^{-6}	9.08×10^{-10}	9.08×10^{-8}	9.08×10^{-6}
"A"-nickel	1.1×10^{-4}	2.74×10^{-8}	2.74×10^{-6}	2.74×10^{-4}
	3.6×10^{-5}	8.97×10^{-9}	8.97×10^{-7}	8.97×10^{-5}
TD-nickel	1.8×10^{-5}	4.6×10^{-9}	4.6×10^{-7}	4.6×10^{-5}

Table 5.3

Comparison of Calculated and Experimental

Microstrains (After Carnahan, 1964)

Material	σ ksi	Strain	
		Calculated	Experimental
Copper	1.0	4×10^{-6} (for $N = 10^6$)	1 to 2×10^{-6}
"A"-nickel	2.5 to 7.5	1.5 to 5×10^{-6} (for $N = 2 \times 10^6$)	--
"A"-nickel	2.5	--	1 to 4×10^{-6}
TD-nickel	15	4.6×10^{-5} (for $N = 10^8$)	1.5×10^{-5}
2024-T36 Al	13	2×10^{-6} (for $N = 10^7$)	--
2024-T351	30	--	1×10^{-6}
SAE 1020 steel	11	6.2×10^{-6} (for $N = 10^7$)	--
SAE 1018 steel	11	--	6×10^{-6}
High carbon steel	35	2×10^{-6} (for $N = 10^7$)	--
0.8C tool steel	52	--	1×10^{-6}

In noncrystalline materials such as glasses and plastics, dislocations do not exist. Hence, other mechanisms must be postulated for microyielding of these materials. According to McClintock and Argon (1966), inelastic deform-ation can occur at stresses below the breaking strength because the atoms and molecules are not closely packed. This leaves regions where atom displacement can occur without seriously altering nearest-neighbor relationships, as shown schematically in Fig. 5.16. If the temperature is high enough, thermal fluctuations will cause pairs of atoms to switch back and forth between the indicated positions. Application of a stress creates a preference for the dashed position. When the stress is removed, the atoms will eventually return to their original locations, i.e., any apparent plastic strain will gradually recover with time. Exceptions might be noted if extensive deformation throughout the body deforms the neighborhood of the pair of atoms in Fig. 5.16. In this case the dashed positions may become stable and complete strain recovery unlikely.

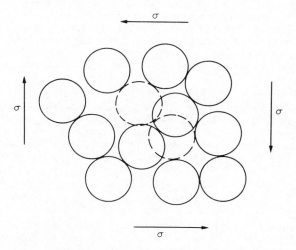

Fig. 5.16.   Mechanism of inelastic deformation in a non-crystalline material (McClintock and Argon, 1966; copyright Addison-Wesley, used by permission).

Nature of Microyield Curves

Some of the examples of microplastic stress – strain curves already pre-sented in Figs. 5.1 through 5.4 illustrate the wide variations of behavior observed. Such results emphasize the importance of presenting stress – strain curves, rather than a single value of yield strength derived from these curves. Presenting curves allows the reader to obtain σ_y values at any offset level of interest and reveals the rate at which strain hardening is occurring in the microstrain region; it can also indicate strain ranges where this rate under-goes marked changes. Furthermore, it reveals any peculiarities in behavior, such as the strain reversals shown in Fig. 5.13.

Effect of Variables on Microyield Behavior

It takes little imagination to see that the microyield behavior of metals can be carried over wide ranges. Introduction of barriers to dislocation movement, however accomplished, should be effective in raising the stress level required to produce a given level of plastic strain. Methods that have been suggested include

- reduction of grain size to increase the number of grain boundary barriers;
- alloying to achieve solid solution strengthening, precipitation hardening, ordering, special grain boundary effects, etc.;
- reinforcement with high modulus fibers or particles;
- cold working or prestraining.

On the other hand, raising the temperature may be effective in *reducing* the stress level required to achieve a given level of plastic strain, either by reducing the effectiveness of dislocation barriers or by thermal activation of additional dislocation sources.

For various reasons, few of the variables that can influence microplastic strain behavior have received systematic investigation. Some limited work has been done in each of the above areas on a few materials, but in many cases the materials studied were high purity metals rather than materials of importance in precision design. Nonetheless, the data are useful in illustrating the general effects of several variables on microplastic strain behavior.

A. *Effects of Grain Size*

One of the first investigations of grain size effects in polycrystalline metals was reported by Brown and Lukens in 1961. They developed a theoretical model which assumed that the only barriers to dislocations in the microstrain region are the grain boundaries. From this model, they developed an equation relating plastic microstrain (ϵ_p) to grain diameter (d) and applied stress (σ):

$$\epsilon_p = \frac{C\rho d^3 (\sigma - \sigma_0)^2}{G\sigma_0} \tag{5.1}$$

where C is a constant, ρ is the density of dislocation sources (assumed to be independent of grain size), σ_0 is the stress required to activate the first dislocation source, and G is the shear modulus.

This equation predicts (1) the stress to first activate a source is independent of grain size, (2) the strain for a given stress varies as d^3; i.e., large grains will show more strain than small grains at any stress above σ_0, and (3) the microplastic strain varies parabolically with the applied stress. In microstrain experiments conducted on relatively pure iron in which every effort was made to minimize any substructure within the grains, Brown and Lukens found that the material behaved almost exactly in accord with these predictions. Furthermore, they examined other data that had been reported for copper (Thomas and Averbach, 1959) and found that they, too, followed the predictions. Figures 5.3 and 5.17 are taken from Brown and Lukens' 1961 paper and demonstrate the excellent agreement between theory and experiment.

In the intervening years, a number of other investigators have examined grain size effects on microplastic properties and have concluded that, frequently, the relationship is not so simple as that proposed by Brown and Lukens. Carnahan and White (1964a), for example, found that curves of stress versus microplastic strain for commercially pure nickel showed three-stage behavior, as illustrated in Fig. 5.4, rather than a simple parabolic relationship. Further, σ_y (50×10^{-6}) was proportional to $d^{-1/2}$, rather than $d^{-3/2}$ as predicted by eq. (5.1). Holt (1973) likewise reported that σ_y (100×10^{-6}) is proportional to $d^{-1/2}$ for 99.98% purity nickel. However, σ_y (20×10^{-6}) for the same specimen was reported to be proportional to $d^{-3/2}$.

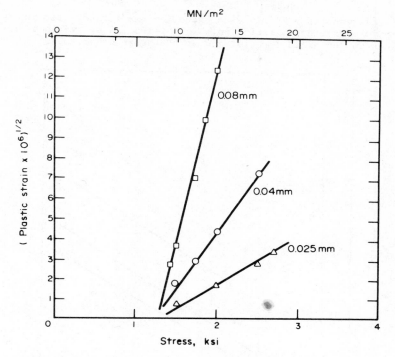

Fig. 5.17.   Square root of plastic strain versus stress
for various grain sizes in copper; data from
Thomas and Averbach (1959) (Brown and Lukens,
1961; copyright Pergamon, used by permission).

Investigations conducted by Brentnall and Rostoker (1965) on Fe-3Si and
commercial purity nickel also showed a $d^{-1/2}$ grain size dependence of microyield strengths, as well as a parabolic relationship between applied stress
and microplastic strain.   For nickel, this $d^{-1/2}$ relationship held for yield
strengths associated with plastic strains of 10^{-6}, 10^{-4}, 10^{-3}, and 10^{-2}, as
shown in Fig. 5.18.   Studies of high purity iron conducted at the same time
failed to show any significant grain size dependence of microplastic properties.
Brentnall and Rostoker attributed this to the presence of numerous inclusions
in the iron, i.e., the type of substructure that Brown and Lukens sought to
avoid in their experiments.

Bilello and Metzger (1969) investigated the microplastic behavior of 99.999%
purity copper at three relatively large grain sizes (0.05 to 0.38 mm).
Although the stress necessary to produce a given amount of microplastic strain
was slightly lower at the largest grain size, there was no indication of a
systematic grain size dependence of the type reported by Brown and Lukens.
They proposed a model for microplastic flow, differing somewhat from that of
Brown and Lukens, to account for the observed behavior.

Some indications of abnormal grain size effects have been reported by
Shemenski and Maringer (1969).   In an investigation of microstrain characteristics of several grades of isostatically hot-pressed beryllium, they observed
that σ_y (10^{-6}) of as-fabricated, low-oxide beryllium (Brush Grade S-100)
decreased as the grain size was reduced.   However, after heating for 30 min at
925° C, σ_y (10^{-6}) was found to increase with decreasing grain size, according

to a $d^{-1/2}$ relationship. This unusual behavior was rationalized in terms of the complex microstructure existing in this material, i.e., various precipitate particles both in the grain interior and at the grain boundaries.

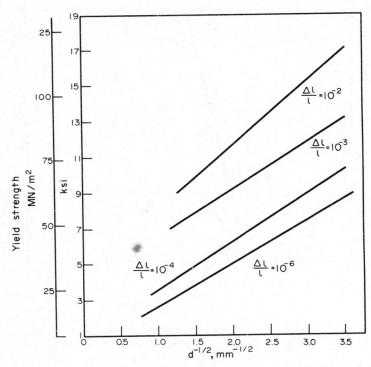

Fig. 5.18. Yield strength of nickel at various plastic
strain levels (Brentnall and Rostoker, 1965;
copyright Pergamon, used by permission).

Perhaps the type of results obtained by Shemenski and Maringer exemplify the complexity that should be anticipated in microstrain behavior of materials that deviate markedly from relatively pure polycrystalline materials that are free from precipitates and other types of substructure. Dislocation models can often help to rationalize such complex behavior but, at the present time, are not adequate to predict this behavior.

Summarizing briefly, there is considerable evidence that decreasing the grain size of polycrystalline materials will increase the stress required to produce a given level of microplastic strain. In some cases, σ_y as a function of d appears to follow a $d^{-3/2}$ relationship and in others a $d^{-1/2}$ relationship. However, as microstructures are made more complex through alloying, heat treatment, and cold working, predictions of grain size effects become increasingly uncertain.

B. Effects of Prestrain and Cold Work

In speaking of conventional mechanical properties, it is common to employ

prestraining or cold working as a method for increasing strength.* It is not
unreasonable to assume that the microyield properties of a material might
also be increased by such methods.

   Some of the earliest work to investigate this was conducted by Roberts,
Carruthers, and Averbach (1952). Specimens of 1020 steel, stress relieved at
650°C (1200°F) were prestrained in tension to plastic strains ranging from
0.8 to 9.4% and the "elastic limit" in tension was then measured as the stress
to produce a permanent strain of 4 x 10⁻⁶. The investigators reported that "the
elastic limit shows a surprising behavior, remaining constant up to nearly
4% prestrain and then decreasing slightly" (see Fig. 5.19). This was, indeed,
a surprising result in view of the fact that the conventional yield strength is
raised by prestrain. However, there is now sufficient evidence to demonstrate
that the results reported by Roberts *et al*. were not anomalous.

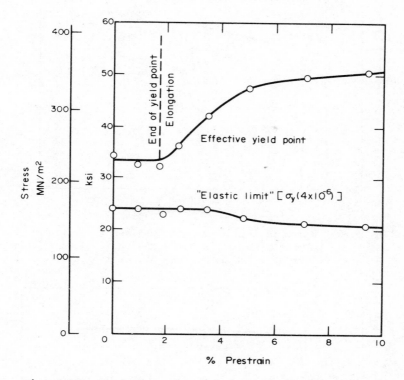

Fig. 5.19. Variation of lower yield and "elastic limit"
            with prestrain for 1020 steel (Roberts *et
            al*., 1952; copyright ASM, used by permission).

---

*Prestraining and cold working have somewhat different meanings, as used in
this chapter. Prestraining refers to strain imparted to a test specimen; for
example, stretching a tensile specimen a known amount prior to measuring
tensile stress – strain properties. Cold working refers to a metalworking
operation, such as rolling, swaging, or drawing, on a quantity of material
from which a test specimen is subsequently removed.

Rosenfield and Averbach (1960) examined the effect of smaller amounts of tensile prestrain on the tensile microyield behavior of high purity copper single crystals and polycrystalline material and of high purity polycrystalline aluminum. Figure 5.20 shows their results for repeated loading of a copper single crystal. Note particularly that each reloading curve begins to show some plastic strain prior to reaching the highest stress attained in the previous loading. Although the resolution of the data in Fig. 5.20 is not adequate to discern the stress at which plastic flow begins on each cycle, Rosenfield and Averbach state that σ_y (2 x 10^{-6}) increases from 0.2 MN/m^2 (30 psi) to about 0.4 MN/m^2 (60 psi) after a prior plastic strain of 160 x 10^{-6}. Further plastic strain has little effect on σ_y (2 x 10^{-6}). Similar results are shown in Fig. 5.21 for polycrystalline copper and aluminum.

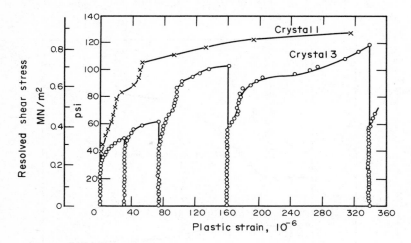

Fig. 5.20.   Repeated load experiments on copper single
             crystals (Rosenfield and Averbach, 1960;
             copyright Pergamon, used by permission).

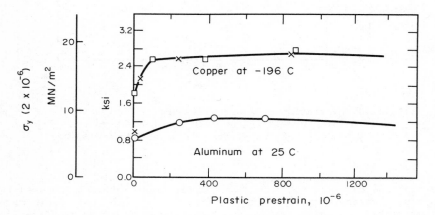

Fig. 5.21.   Effect of plastic prestrain on the 2 x 10^{-6}
             yield of polycrystalline materials (Rosen-
             field and Averbach, 1960; copyright Pergamon,
             used by permission).

On an even finer scale, Tinder and Washburn (1964) have shown the effects of prestrains as small as 4×10^{-8} on subsequent microstrain behavior of copper specimens deformed in torsion. Results of consecutive load - unload cycles are shown in Fig. 5.22. Notice the similarity to the behavior shown in Fig. 5.20.

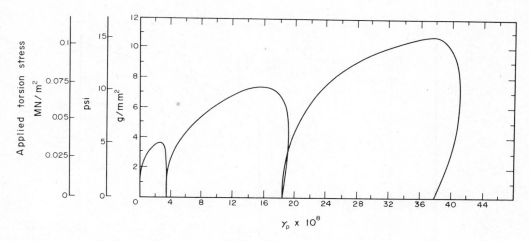

Fig. 5.22. Microyield behavior of OFHC copper (Tinder and Washburn, 1964; copyright Pergamon, used by permission).

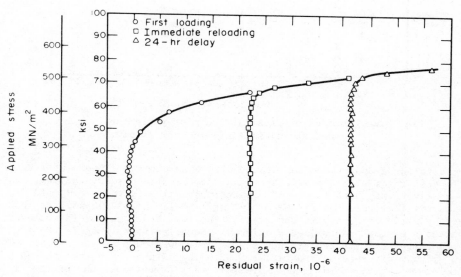

Fig. 5.23. Effect of prestrain on microyield behavior of Ni-Span-C (Imgram *et al.*, 1968).

Imgram *et al.* (1968) conducted similar types of prestrain experiments on two commercial alloys, Ni-Span-C and 440 C stainless steel. Results of repeated loading are shown in Figs. 5.23 and 5.24. It is evident that prestrains of 20 to 50 x 10^{-6} produce a substantial increase in σ_y (10^{-6}). However, the occurrence of some plastic strain on each reloading at a stress below the maximum achieved in the previous loading is also evident.

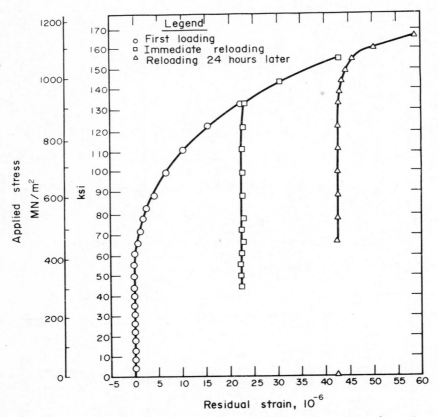

Fig. 5.24.   Effect of prestrain on microyield behavior of
440 C stainless steel (Imgram *et al.*, 1968).

In tests on Ni-Span-C, Imgram *et al.* also examined the effects of larger tensile prestrains—2.5 and 5%. The results are extremely interesting. As shown in Fig. 5.25, both levels of prestrain drastically reduced σ_y (10^{-6}), from its original value of 272 MN/m^2 (40 ksi) to 140 MN/m^2 (20 ksi) or less. In fact, plastic strains were observed at the lowest stress levels employed in the tests on the prestrained specimens. Equally interesting is the fact that the curves for the prestrained specimens show a much steeper slope. Thus, σ_y (10^{-5}) was increased substantially by 2.5 and 5% prestrain even though σ_y (10^{-6}) was greatly diminished. These results demonstrate one of the problems concerning the question of the effect of prestrain on σ_y. The answer depends very strongly on which σ_y is being referred to; i.e., at what level of plastic strain is σ_y measured.

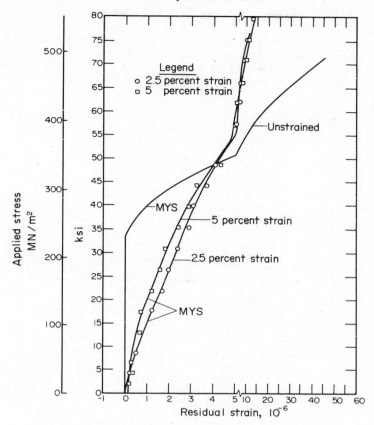

Fig. 5.25. Effect of large prestrains on the microyield
behavior of Ni-Span-C (Imgram *et al.*, 1968).

Another example of the complexity of prestrain effects is demonstrated by
work conducted by Geil and Feinberg (1969b) on normalized 4340 steel. At
the very fine strain resolution of their tests ($\sim 10^{-7}$) they were able to show
some unusual effects of tensile prestrain, as can be seen in Fig. 5.13.
Values of σ_y (10^{-6}) appear to show a substantial increase with small amounts
of prestrain. Thereafter, the values level off or even decrease somewhat.
As noted earlier in connection with the results of Imgram *et al.*, some plastic
strain is observed in Fig. 5.13 at the lowest stress levels employed upon
reloading following prestraining. Thus, σ_y (10^{-7}) is reduced for even very
small prestrains, such as 4×10^{-6}. The strain reversals shown in curves
D, E, F, and G are unexplained; no such effects have been reported by other
investigators. However, Geil and Feinberg also reported that annealed Invar,
prestrained 1.9% in tension, showed a marked strain reversal in a subsequent
microstrain test (see Fig. 5.12).

In contrast to the results of Imgram *et al.*, Kossowsky and Brown (1966)
found no maximum in the curve of σ_y as a function of prestrain for high-
purity iron. Their investigation covered a wide range of tensile prestrains,
from about 10^{-5} to 10^{-1} (0.001 to 10%), introduced at either $-78°$ C or $+25°$ C
($-108°$ or $+77°$ F); no effect of prestrain temperature was detected. Subsequent

microstrain tests were conducted at $-195°$, $-78°$ and $+25°$ C ($-321°$, $-108°$, and $+77°$ F). Results are shown in Fig. 5.26. The plastic strain level at which σ_y was measured was not stated explicitly but is believed to be between 10^{-5} and 10^{-6}, as noted from the strain value at the extreme left side of the figure. Values of σ_E, the stress below which hysteresis loops were not detected, are included for comparison. Prior to prestrain, σ_E and σ_y (10^{-5} to 10^{-6}) were equal.

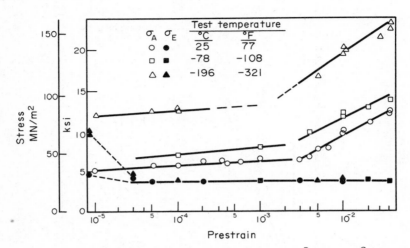

Fig. 5.26.   The effect of prestrain at $-78°$ and $+25°$C on σ_E and σ_A measured at $-196°$, $-78°$, and $+25°$C for high purity iron (Kossowsky and Brown, 1966; copyright Pergamon, used by permission).

As shown in Fig. 5.26, σ_E was reduced slightly by a small amount of prestrain and, thereafter, remained constant with further prestrain. Test temperature likewise had no effect on σ_E. Values of σ_y (10^{-5} to 10^{-6}) increased very slightly with prestrain up to about 0.3% and thereafter increased appreciably more. Similar effects were noted at the three test temperatures.

Parikh and Hay (1971) report results similar to those of Kossowsky and Brown for tantalum single crystals, prestrained and tested in compression. A very slight strengthening effect on σ_y (2×10^{-5}) and σ_y (6×10^{-5}) was observed for prestrains up to about 1%. Thereafter, somewhat greater strengthening was observed.

At least two studies have been made of prestrain effects in commercial beryllium. Rutherford and Swain (1966) examined the effect of very small prestrains (up to 3×10^{-6}) on σ_y (4×10^{-7}) and found a marked strengthening effect (Fig. 5.27). Bonfield and Li (1965) examined a wider range of prestrains (to about 10^{-3}) on σ_y (2×10^{-6}) of QMV beryllium. As shown in Fig. 5.28 for annealed material and in Fig. 5.29 for as-hot-pressed material, very rapid strengthening is observed up to prestrains of $\sim 10^{-4}$ and gradual strengthening thereafter.

Some unusual effects of prestraining have been reported by Hughes and Rutherford (1969) for high purity copper reinforced with continuous tungsten wires oriented in the tensile direction of the test specimens. Figure 5.30 shows σ_y (2×10^{-7}) as a function of tensile prestrain for specimens containing various volume fractions of tungsten fibers. The rate of strengthening

increases markedly with fiber percentage. Also, at a certain level of prestrain, each curve shows a marked increase in rate of strengthening. The
reasons for the transition in the curves is not known but Hughes and Rutherford
speculate that the transition could signify the onset of microplastic flow
in the tungsten fibers.

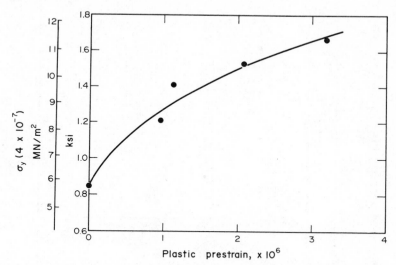

Fig. 5.27. Effect of prestrain on microyield behavior of
            S200 beryllium (Rutherford and Swain, 1966).

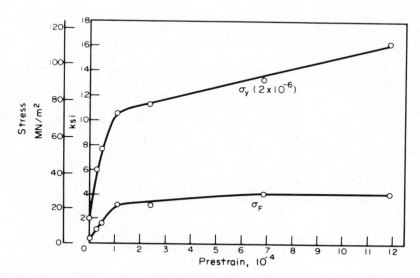

Fig. 5.28. Effect of prestrain on microyield behavior of
            annealed beryllium (Bonfield and Li, 1965;
            copyright Pergamon, used by permission).

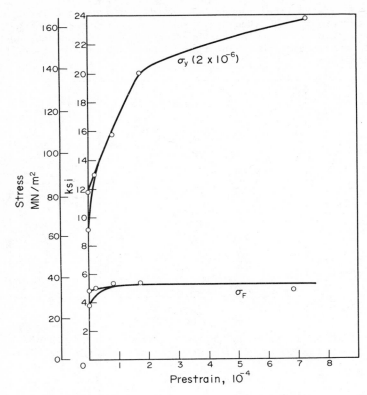

Fig. 5.29. Effect of prestrain on microyield behavior of
            as-hot-pressed beryllium (Bonfield and Li,
            1965; copyright Pergamon, used by permission).

    To this point, the discussion has dealt only with prestrain, as opposed to
cold work. Little has been reported on systematic investigations of the effect
of cold work on microplastic strain behavior. However, Bonfield (1967) has
shown that cold-rolling of solution annealed Cu-1.9Be raises the stress required to produce a given amount of microplastic strain, as indicated in Fig.
5.31. Marschall (1973) has observed that cold drawing of annealed Invar to a
reduction level of ∿35% has a similar effect: the value of $\sigma_y(10^{-6})$ was
increased from about 41 MN/m² (6 ksi) to 310 MN/m² (45 ksi).

    A more complex effect of cold work was reported for silver by Bonfield (1965),
who reversed the processing sequence. He first investigated heavily cold rolled
silver and then removed the cold work by annealing. The results are shown in
Fig. 5.32. Surprisingly, σ_y (2 x 10^{-6}) is greater for the material annealed for
20 hr at 600° C, resulting in substantial grain coarsening, than for the heavily
worked or moderately annealed material.

    In summary, the effects of prestraining and cold working on microplastic
behavior are difficult to generalize. With respect to prestrain, as opposed
to cold work, there appears to be a trend toward the type of behavior shown
schematically in Fig. 5.33. For example, if σ_y is measured at a plastic strain
of, say, 10^{-4} or 10^{-5}, all levels of prestrain will show some strengthening.
If, on the other hand, σ_y is measured at smaller levels of plastic strain

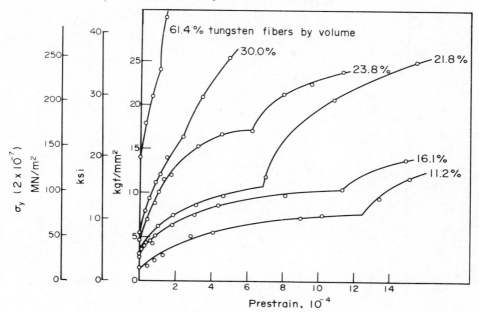

Fig. 5.30.   Effect of prestrain on σ_y (2×10^{-7}) of high
purity copper reinforced with continuous
tungsten wires (Hughes and Rutherford, 1969;
copyright ASTM, used by permission).

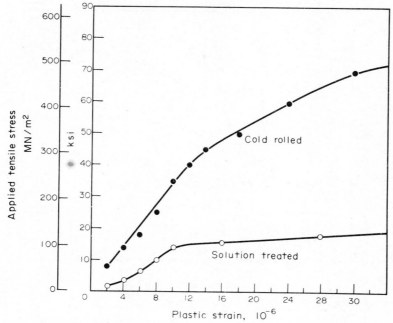

Fig. 5.31.   Effect of cold rolling on microyield behavior
of solution annealed Cu-1.9Be (after Bonfield,
1967; copyright AIME, used by permission).

(10^{-6} or 10^{-7}, for example), distinct maxima will be observed in the strength-ening effects.   There is some physical basis for predicting this general type of behavior, based on dislocation motion and interaction (Imgram *et al.*, 1968).   However, the amount of information available is insufficient to confirm that such a relationship actually exists.

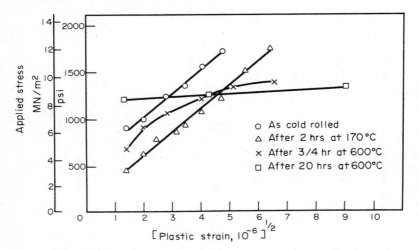

Fig. 5.32.   The effect of annealing after cold rolling on the microyield behavior of silver   (Bonfield, 1965; copyright Pergamon, used by permission).

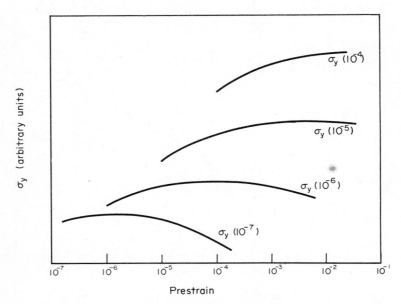

Fig. 5.33.   Schematic illustration of trends observed in studies of effect of prestrain on microyield behavior.

C. *Effects of Impurities, Alloying Additions, and Thermal and Mechanical Treatments*

Without a doubt, the greatest potential for increasing the strength of materials in the microplastic strain region is through alloying in combination with appropriate thermal and mechanical treatments. This area has received little attention in microstrain studies. While it is true that results of microstrain tests have been reported for materials having high microyield strengths achieved through alloying and appropriate treatments, little systematic study has been undertaken to understand the contributions made by the various strengthening mechanisms. The reason for this lack of attention probably can be traced to a statement made earlier. There it was noted that much of the microstrain work conducted to date has had as its objective an improved understanding of the dislocation mechanisms responsible for the early stages of plastic flow. It is far simpler to study these mechanisms in "clean" materials than in materials that are filled with various precipitates, foreign particles, and entangled dislocations. Consequently, investigation of more complex materials and treatments has lagged considerably behind that of relatively pure metals.

Perhaps one of the best examples of what selective alloying and subsequent treatment can accomplish in the way of improving strength in the microstrain region pertains to beryllium.* Bonfield and Li (1963) and Argon and East (1967) have reported $\sigma_y(2 \times 10^{-6})$ values of approximately 17 MN/m^2 (2.5 ksi) for commercially pure beryllium. By selective alloying and special processing, it is possible to increase these values to nearly 700 MN/m^2 (over 100 ksi) as reported by Shemenski and Maringer (1969). The results reported by Shemenski and Maringer were part of a larger study to investigate the effects of chemical composition, powder particle size, fabrication methods, and thermal history on the microstrain behavior of beryllium. Some of their results are presented in Table 5.4, primarily to illustrate compositional effects. Referring to Table 5.4, additions of BeO are seen to impart some strengthening to beryllium, though the 0.7% addition in the S-100 grade produces a $\sigma_y(10^{-6})$ value not much different from that for commercially pure material. In this grade, the principal function of the BeO is believed to be as a grain refiner. With larger quantities of BeO present, as in the P-40 grade, appreciable strengthening is evident. The Be-Ag differs from the S-200 grade only by the 1% silver addition, but the value of $\sigma_y(10^{-6})$ is substantially increased. Little information is available on the mechanism by which silver strengthens beryllium.

The P-50 grade is an unusual material in that it contains very little alloying element but exhibits extremely high microyield properties. The secret to its success lies in the chromium content and in the method of processing. The chromium is deposited on the beryllium powder particles to a depth of a few angstroms. When the particles are consolidated, heat and pressure convert the chromium at the particle interface to an intermetallic compound ($CrBe_{12}$). This compound is very strong and apparently inhibits the activation of dislocation sources. The hot-gas isostatic pressing described in Table 5.4 is also responsible for a portion of the very high reported microyield strength values of the P-50 grade. When conventional hot pressing techniques are employed (temperatures of $\sim 1100°$ C, as compared with $915°$ C for HIP), values of $\sigma_y(10^{-6})$ are reported to be 140 to 165 MN/m^2 (20 to 24 ksi).

---

*Actually, beryllium forms few true alloys, since most other elements can not be dissolved in it. However, the term alloying will be used here in its broadest sense to indicate additions that alter the overall chemistry.

Table 5.4 Microplastic Properties of Selected Beryllium Alloys[a] (Shemenski and Maringer, 1969)

Grade	Vendor	Percentage of Principal Alloying Elements and Impurities					σ_y (10^{-6})	
		BeO	C	Fe	Cr	Ag	MN/m²	lb/in²
S-100	Brush	0.7	0.08	0.09	–	–	13.8	2.0×10^3
S-200	Brush	1.8	0.14	0.14	–	–	22.8	3.3×10^3
P-40	Berylco	7.07	0.48	0.17	–	–	186	27.0×10^3
							98	20.2×10^3
Be-Ag	Honeywell	1.4	0.12	0.14	–	1.0	191	27.7×10^3
P-50	Berylco	<0.5	0.19	0.12	0.27	–	696	100.9×10^3
							714	103.5×10^3

[a]Cylindrical specimen blanks were prepared by vibratory packing and hydropressing powder to 75% density, prior to fitting into thin-walled steel containers. After outgassing the container at elevated temperature and sealing under vacuum, the specimens were further densified (to approximately 100% of theoretical) by hot-gas isostatic pressing (HIP). The HIP process was conducted in an autoclave, employing helium at a pressure of 70 MN/m² (10 ksi) and a temperature of 915°C for 2 hr. Cooling was at a rate of 75°C/hr to 100°C. After machining cylindrical tensile specimens from the blanks, the final diameter of the gage length was achieved by chemically polishing about 0.25 mm (0.010 in) from the radius.

Argon and East (1967) have demonstrated that iron, too, can be effective in raising the microyield strength of beryllium. The effect of adding approximately 0.5 and 1.0% iron to beryllium is shown in Fig. 5.34. In the solution treated condition (1 hr at 1200°C) substantial strengthening accompanies the iron additions. However, even greater strengthening is achieved by a precipitation hardening treatment at 500°C.

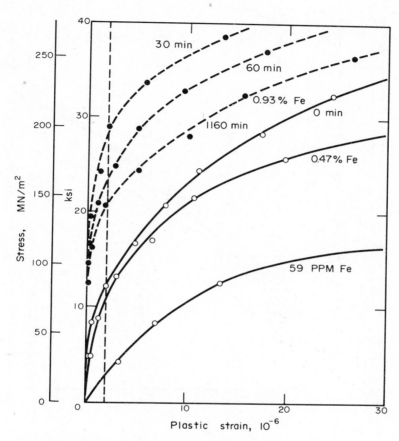

Fig. 5.34. Effect of alloying with iron on the micro-
            yield behavior of beryllium (Argon and East,
            1967; copyright AIME, used by permission).
            Solid lines indicate solution treated con-
            dition; dashed lines indicate specimens aged
            at 500°C for times indicated.

Alloying effects on the microplastic behavior of copper have been the subject of several investigations. Tinder and Washburn (1964) added small amounts of aluminum or iron to high-purity copper. Tubular torsion specimens were annealed at 900°C and cooled very slowly. The resulting grain size was quite large, usually larger than the wall thickness of the tubes. Micro-yield behavior was studied at very low levels of plastic strain, from about 5×10^{-9} to 10^{-6}. The results, shown in Fig. 5.35, indicate that aluminum raises the torsional yield strength at all levels of plastic strains investigated. Iron, on the other hand, reduces slightly the value of σ_y (10^{-8}) and increases slightly the value of $\sigma_y(10^{-7})$.

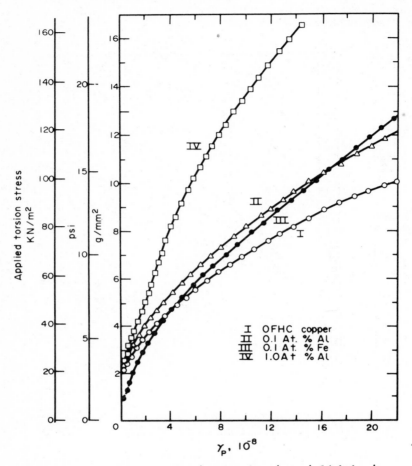

Fig. 5.35. Effect of alloying on the microyield behavior
of copper (Tinder and Washburn, 1964;
copyright Pergamon, used by permission).

Bonfield (1967, 1968) has examined a commercial copper alloy of importance in numerous precision applications. The material investigated was a Cu-1.9 wt.% Be alloy. The beryllium can be retained in solution in the copper by rapid quenching from 800° C. Subsequent reheating in the range 315° to 425° C will cause formation of precipitate particles of a compound of copper and beryllium. To minimize precipitation at grain boundaries and to make it more uniform throughout the grains, Bonfield cold rolled the alloy to 40% reduction prior to subjecting it to the precipitation hardening treatments. Curves of stress versus plastic-strain in the microplastic region are shown in Fig. 5.36.

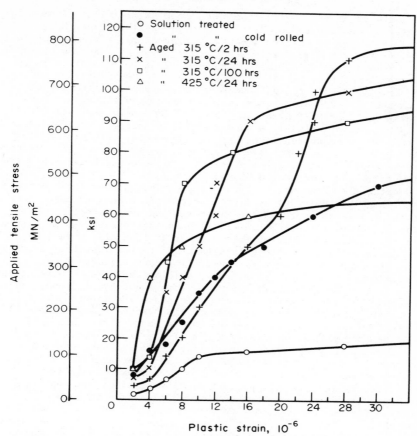

Fig. 5.36.   Effect of cold work and precipitation hardening on the microyield behavior of Cu-1.9Be
(Bonfield, 1967; copyright AIME, used by
permission).

Comparing Bonfield's results with those of Thomas and Averbach for pure
copper (refer to Fig. 5.17), it appears that 1.9% Be in solution does not
significantly increase σ_y (2 x 10^{-6}) of copper. Thomas and Averbach obtained
values in a range of about 8.3 to 12.4 MN/m^2 (1.2 to 1.8 ksi) depending on
grain size. The combination of cold working and precipitation hardening
investigated by Bonfield, however, produced very interesting results.
Strengthening is evident at all levels of plastic strains investigated. But
notice particularly the curves for the specimens receiving the 315°C/2 hr
and the 425°C/24 hr treatments. The first treatment exhibits the lowest
σ_y(2 x 10^{-6}) and the latter treatment the highest σ_y(2 x 10^{-6}) of any of the
precipitation hardened specimens. However, σ_y(30 x 10^{-6}) values show a completely reversed effect. Results such as these serve to reemphasize the importance of providing curves of stress versus plastic strain, rather than
simply a single yield strength value obtained at a given level of plastic
strain.

The relatively low values of σ_y(2 x 10^{-6}) reported by Bonfield for Cu-1.9Be
have not been duplicated by other investigators. Wikle (1970), for example,

found $\sigma_y (2 \times 10^{-6})$ values ranging from 440 to 530 MN/m^2 (64 to 77 ksi) for 1/4-hard material and 680 to 760 $MN/m2$ (99 to 110 ksi) for full-hard material. In both cases, the specimens were aged at 315°C for 2 hr. An even higher value of 945 MN/m^2 (137 ksi) was reported by Hughes and Rutherford (1973) for material that was solution annealed and then aged at 315°C for 3 hr. The reason for the wide disparity in measured values is not known.

Alloying is effective also in increasing the microyield strength of aluminum, though no systematic investigations have been reported. Pure aluminum has a σ_y (2×10^{-6}) of about 5.5 MN/m^2 (800 psi) (Rosenfield and Averbach, 1960). By alloying with such elements as copper, silicon, magnesium, and zinc, and employing appropriate thermal treatments, values of $\sigma_y (10^{-6})$ as large as 275 MN/m^2 (40 ksi) have been reported. An example of this strengthening effect is shown in Fig. 5.2 for precipitation hardened 2014-T6 aluminum which contains 4.4% Cu, 2.8% Si, 0.8% Mn, and 0.4% Mg.

Alloying effects in iron are of great interest because of the commercial importance of steels, Here again, little systematic study has been undertaken concerning effects in the microstrain regime. One exception is an investigation by Solomon and McMahon (1971) on solute effects in iron. The materials investigated and some of the results obtained are summarized in Table 5.5. Table 5.5 also includes results from other investigations of iron base alloys. Solomon and McMahon's results will be considered first.

The materials and treatments selected by Solomon and McMahon were designed to reveal the effect of interstitial carbon atoms and substitutional silicon atoms on σ_E and σ_y.* Water quenching from the annealing temperature was employed when it was desired to keep most of the carbon in interstitial solution whereas furnace cooling was used to remove most of the carbon from solution by permitting it to precipitate out of the iron lattice as an iron carbide. Titanium, because of its great affinity for carbon, was added as an even more effective way to remove carbon from the iron lattice. Prestraining was introduced so that σ_E values could be measured from hysteresis loops, as discussed earlier in Section 4.1. σ_E, it will be recalled, is approximately $2\sigma_F$, where σ_F is the friction stress opposing dislocation motion.

The results of these studies, as shown in Table 5.5, indicate that carbon in solution has very little effect on σ_E and, hence, on σ_F. It does, however, seem to have a slight strengthening effect on σ_y, at least in the plastic strain range from about 10^{-6} to 10^{-5}. Silicon has a substantial strengthening effect on both σ_E and σ_y. Solomon and McMahon discuss their results in terms of effects of solutes on dislocation motion, both short- and long-range.

Table 5.5 also includes selected results of other investigators to provide an indication of the degree of improvement that can be obtained in the microstrain behavior of iron through alloying. It is interesting to compare the results obtained for Invar (nominally Fe-36% Ni) with those for Armco iron (relatively high-purity iron) and Fe-0.002C-0.15Ti. Notice that the large addition of nickel in Invar produces little change in the reported $\sigma_y (10^{-6})$ values of iron. In Ni-Span-C, however, which also contains a large percentage of nickel, other elements have been added to permit precipitation hardening to occur. These produce significant increases in microyield strength.

The data shown for 4340 steel are for the normalized, rather than the fully hardened condition. Had this alloy been quenched and tempered to achieve a more favorable distribution of precipitates, it would undoubtedly

---

*The investigation also included test temperatures as a variable, in an attempt to reveal the reason for the rapid rise in yield strength commonly observed in iron and steel as temperature is reduced below room temperature. Temperature effects are discussed in a subsequent section of this chapter.

Table 5.5 Alloying and Heat Treatment Effects on Microstrain Behavior of Iron

Material	Condition	Approximate value of indicated microstrain parameters, MN/m² (ksi in parentheses)				
		σ_E	$\sigma_y(10^{-6})$	$\sigma_y(3 \times 10^{-6})$	$\sigma_y(6 \times 10^{-6})$	$\sigma_y(10^{-5})$
Solomon and McMahon (1971)						
Fe-0.005% C	Anneal, furnace cool, prestrain[a]	35(5)	–	90(13)	130(19)	–
Fe-0.005% C	Anneal, water quench, prestrain[a]	35(5)	–	105(15)	160(23)	–
Fe-0.035% C	Anneal, furnace cool, prestrain[a]	41(6)	–	95(14)	165(24)	–
Fe-0.035% C	Anneal, water quench, prestrain[a]	35(5)	–	–	290(42)	330(48)
Fe-0.002% C - 0.15% Ti	Anneal, furnace cool, prestrain[a]	28(4)	35(5)	70(10)	–	–
Fe-0.001% C - 3.1% Si	Anneal, furnace cool, prestrain[a]	150(22)	–	–	290(42)	340(49)
Brown and Lukens (1961)						
Armco iron	Anneal	–	75(11)	85(12)	90(13)	95(14)
Geil and Feinberg (1969)						
Fe-0.14% C - 34.3% Ni (Invar)	Anneal, water quench	–	41(6)	62(9)	75(11)	95(14)
4340 steel	Anneal, air cool	–	270(39)	325(47)	365(53)	400(58)
Imgram et al. (1968)						
Ni-Span-C[c]	Anneal, water quench, age 21 hr at 675C	–	260(38)	310(45)	345(50)	380(55)
440 C stainless steel[d]	Anneal, oil quench, temper 1 hr at 260 C	–	470(68)	600(87)	690(100)	770(112)

[a] 3.5 to 7.0% prestrain.

[b] Fe-0.40% C - 1.8% Ni-0.8% Cr - 0.2% Mo.

[c] Fe-42% Ni - 5% Cr - 2.5% Ti - 0.6% Al.

[d] Fe-1.0% C - 17% Cr.

have exhibited higher microyield properties. 440C stainless steel, for
example, after quenching and tempering, shows higher $\sigma_y(10^{-6})$ than any other
material listed in Table 5.5.

The effectiveness of alloying and subsequent treatment on the microyield
properties of iron is further demonstrated by the data presented in Table 5.6.
Recalling from Table 5.5 that pure iron exhibits $\sigma_y(2 \times 10^{-6})$ values near
70 MN/m^2 (10 ksi), Table 5.6 shows that the mere addition of a small amount
of carbon can raise this value by a factor of 10. Note, however, the importance
of heat treatment details. In the as-hardened condition, corresponding to
maximum indentation hardness values, $\sigma_y(2 \times 10^{-6})$ values are relatively low.
Tempering in the range from about 200° to 425° C (400° to 800° F) is required to
develop optimum microyield properties. Even greater microyield properties can be obtained in the highly alloyed maraging steels, as shown in Table 5.6.

From the foregoing discussion, it is clear that alloying, in conjunction
with thermal and mechanical treatments, can produce large improvements in the
microyield strength of metals. Except in a few instances, however, little
imagination has been employed in tailoring alloys and treatments to produce
even greater benefits.

Table 5.6

Alloying and Heat Treatment to Improve

Microyield Properties of Iron

| Material | Condition | $\sigma_y(2 \times 10^{-6})$ | | Reference |
		MN/m^2	ksi	
AISI 1020 steel	As-hardened	166	24	Muir *et al.* (1955)
	Temper 205°C (400°F)	660	96	
	Temper 425°C (800°F)	620	88	
	Temper 650°C (1200°F)	315	46	
AISI 1040 steel	As-hardened	138	20	Muir *et al.* (1955)
	Temper 205°C	660	96	
	Temper 425°C	700	102	
	Temper 650°C	520	75	
AISI 1080 steel	As-hardened	205	30	Muir *et al.* (1955)
	Temper 205°C	505	73	
	Temper 425°C	745	108	
	Temper 650°C	495	72	
300 Maraging steel	Hardened	1840	267	Hughes and Rutherford (1973)
400 Maraging steel	Hardened	2210	321	Hughes and Rutherford (1973)
		to	to	
		2380	346	

D. *Effect of Adding High Modulus Fibers*

Much attention has been given to strengthening of materials by the addition of high modulus fibers. One of the first commercially important examples of this was plastic reinforced with glass fibers. More recently, metal matrix composites have come under investigation, because of the relatively poor elevated temperature characteristics of polymeric matrices. The strengthening fibers in these cases include high modulus metals, graphite, and ceramics.

The simplest type of composite in terms of analysis is one in which the fibers are continuous and oriented in the direction of the applied stress. If the assumption is made that both the fibers and the matrix experience equal strain under the action of a tensile stress applied in the fiber direction and that both behave elastically, then the stress in the fibers can be expressed as

$$\sigma_f = \frac{E_f}{E_m} \sigma_m \qquad (5.3)$$

where σ denotes stress, E denotes modulus of elasticity, and the subscripts f and m denote fiber and matrix, respectively. For example, suppose that continuous fibers with a modulus of 210×10^3 MN/m^2 (30×10^6 psi) are introduced into a matrix with a modulus of 70×10^3 MN/m^2 (10×10^6 psi). At any level of stress applied to the composite material within the "elastic range", the stress in the fibers will be three times that in the matrix. This will be true regardless of the volume fraction of fibers present. The actual stress in the fiber and in the matrix relative to the average stress in the composite will depend on the volume fraction of the fibers (V_f). This can be computed, as shown in the next paragraph.

Assuming that the elastic modulus of the composite obeys the law of mixtures, then

$$E_c = E_m V_m + E_f V_f = E_m(1 - V_f) + E_f V_f \qquad (5.4)$$

where V denotes volume fraction and the subscript c denotes composite. At any stress in the "elastic range", the strain (ϵ) in each phase is equal; i.e.,

$$\epsilon_m = \epsilon_f = \epsilon_c. \qquad (5.5)$$

The stress in the composite is equal to the composite modulus times the composite strain:

$$\sigma_c = \epsilon_c [E_m (1 - V_f) + E_f V_f]. \qquad (5.6)$$

Likewise, the stress in the fiber is

$$\sigma_f = \epsilon_f E_f = \epsilon_c E_f. \qquad (5.7)$$

From eqs. (5.6) and (5.7) the ratio of the stress in the fiber to that in the composite is

$$\frac{\sigma_f}{\sigma_c} = \frac{\epsilon_c E_f}{\epsilon_c [E_m (1 - V_f) + E_f V_f]} = \frac{1}{\frac{E_m}{E_f} (1 - V_f) + V_f}. \qquad (5.8)$$

By similar reasoning, the ratio of the stress in the matrix to that in the composite is:

$$\frac{\sigma_m}{\sigma_c} = \frac{\epsilon_c E_m}{\sigma_c [E_m (1 - V_f) + E_f V_f]} = \frac{1}{V_f \left(\frac{E_f}{E_m} - 1\right) + 1} \qquad (5.9)$$

For the example cited earlier where $E_f/E_m = 3$, eq. (5.8) becomes

$$\frac{\sigma_f}{\sigma_c} = \frac{1}{\frac{1}{3}(1 - V_f) + V_f} = \frac{3}{2V_f + 1} \qquad (5.10)$$

and eq. (5.9) becomes

$$\frac{\sigma_m}{\sigma_c} = \frac{1}{V_f(3 - 1) + 1} = \frac{1}{2V_f + 1} . \qquad (5.11)$$

These relationships are shown graphically in Fig. 5.37. Also shown for comparison are graphs obtained for $E_f/E_m = 5$ and $E_f/E_m = 2$.

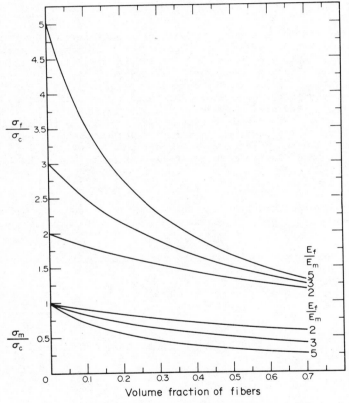

Fig. 5.37. σ_f/σ_c and σ_m/σ_c versus volume fraction of fibers for several ratios of E_f/E_m.

From these curves, it is evident that at a volume fraction of fibers of, say, 0.5, and $E_f/E_m = 3$, the stress in the matrix will be one-half of the average stress in the composite, while the stress in the fibres will be 1.5 times the average. Remember that, up to this point, the assumption has been made that behavior of the composite is entirely elastic.

Consider next the microplastic behavior of a unidirectional composite. Assume that the matrix phase, when tested in the absence of fibers, shows a detectable amount of plastic strain (say, 10^{-6}) at a certain level of stress $[\sigma_{ym} (10^{-6})]$. Assume further that the fibers remain elastic to very high stresses. Then, for the case shown in Fig. 5.37, where $E_f/E_m = 3$, a composite containing 50% fibers should have a microyield strength at least twice that of the matrix phase alone; i.e.,

$$\sigma_{yc} (10^{-6}) = 2\sigma_{ym} (10^{-6}).$$

Figure 5.38 shows the predicted effect of fiber reinforcement on microyield strength, based on the assumptions made earlier. As indicated by the graphs, a linear strengthening effect is predicted as a function of V_f, with the effect being more pronounced with larger modulus ratios (E_f/E_m).

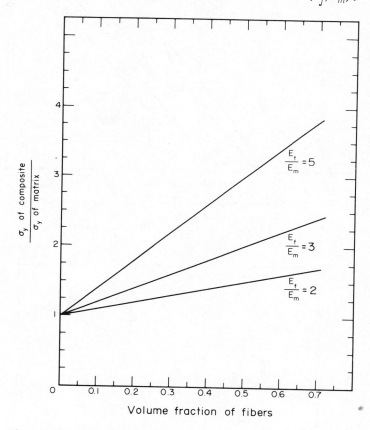

Fig. 5.38. Predicted strengthening effect of continuous
unidirectional fibers.

Several investigations of microstrain behavior have been conducted on metallic unidirectional composites and it is interesting to compare the results obtained with those predicted. Pinnel and Lawley (1969) introduced stainless steel wires (Fe - 15.5% Cr - 4.3% Ni - 2.7% Mo) into a 99.99% purity aluminum matrix. The ratio of E_f to E_m is approximately 3.0. Volume fractions of fibers up to about 0.33 were studied. The investigators measured both σ_E (the stress at which hysteresis loops were first detected) and σ_y (2.5 x 10^{-6}). Typical tensile load – unload curves are shown in Fig. 5.39 to illustrate the shape of the hysteresis loops. Experimentally determined graphs of σ_E versus V_f, and σ_y (2.5 x 10^{-6}) versus V_f are shown in Figs. 5.40 and 5.41, respectively.

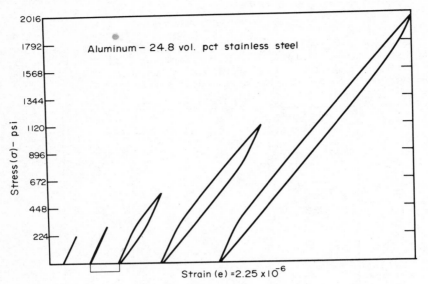

Fig. 5.39.   Load – unload stress – strain curves for an
aluminum – stainless steel composite (24.8%
volume fraction fibers) (Pinnel and Lawley,
1969).

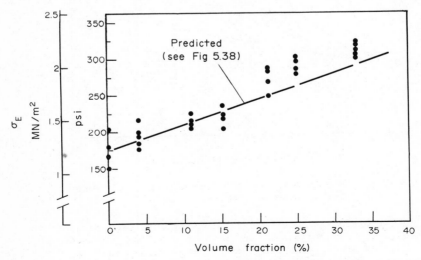

Fig. 5.40. σ_E versus volume fraction reinforcement in
aluminum – stainless steel composites (Pinnel
and Lawley, 1969).

Though some scatter is apparent, the results appear to confirm the linear relationship predicted between microyield strength and volume fraction of fibers. Furthermore, the magnitude of the strengthening effect is very close to that predicted by the equations, particularly for the elastic limit (σ_E), as shown in Fig. 5.40. The slightly greater-than-predicted strengthening effect of fiber reinforcement on σ_y (2.5×10^{-6}), shown in Fig. 5.41, is accounted for by Pinnel and Lawley on the basis of the load - unload curves shown in Fig. 5.39. Above σ_E, the linear elastic behavior assumed in the derivation of eqs. (5.8) through (5.11) no longer applies and the equations must be modified.

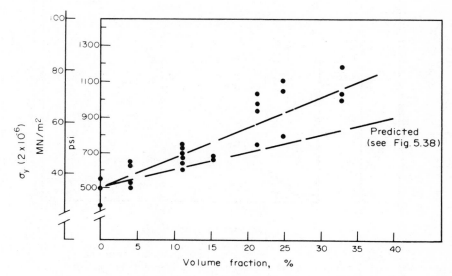

Fig. 5.41. σ_y (2.5×10^{-6}) versus volume fraction rein-
            forcement in aluminum - stainless steel
            composites (Pinnel and Lawley, 1969).

Somewhat different results were reported by Hughes and Rutherford (1969) for unidirectional copper - tungsten composites. The ratio of E_f to E_m in this material is about 3.1. Composite specimens were prepared by infiltrating tungsten wires with molten copper of 99.999% purity. Fiber volume fractions as large as about 0.61 were prepared in this way. Load - unload stress - strain curves were obtained in tension employing a high-sensitivity capacitance extensometer. Measured values included the friction stress σ_F, which is approximately $1/2 \sigma_E$, and σ_y (2×10^{-7}). Hughes and Rutherford observed that σ_F and, presumably, σ_E did not change with increasing volume fractions of tungsten fibers, up to at least $V_f = 0.30$. No σ_f value could be obtained for a specimen containing 61.4 vol.% fibers, since this specimen showed no hysteresis loops until considerable plastic strain ($> 2 \times 10^{-4}$) had been introduced. Up to $V_f = 0.30$, measured values of σ_F were approximately 2.8 MN/m² (400 psi). This differs from Pinnel and Lawley's findings; as noted earlier, they reported that σ_E increased linearly with increasing V_f. Values of σ_y (2×10^{-7}), on the other hand, exhibited a large increase with increasing volume fraction of fibers, as shown in Fig. 5.42. The graph extrapolates to a negative value at $V_f = 0$, indicating that there is a critical volume fraction of fibers (~ 0.10) which must be exceeded if rein-

forcement is to take place. Hughes and Rutherford note that this agrees with
the findings of McDanels *et al.* (1963) and Kelly and Tyson (1964), who reported a similar effect in the macroyield strength versus V_f relationships.

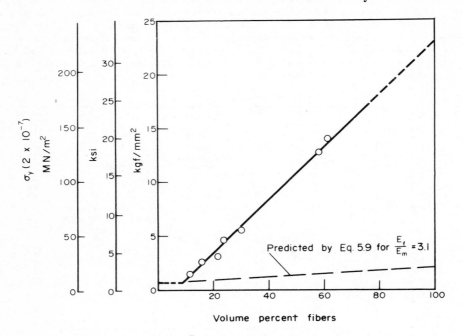

Fig. 5.42. σ_y (2 x 10^{-7}) versus volume percent tungsten
fibers added to copper (Hughes and Rutherford,
1969; copyright ASTM, used by permission).

Of even greater interest is a comparison between the observed results and the
results predicted by eq. (5.9). As seen in Fig. 5.42, the strengthening as a
function of Vf is about ten times that predicted. As already noted, Pinnel and
Lawley, studying a different composite with a similar E_f/E_m ratio, found that
observed and predicted behavior coincided almost exactly. Thus, the assumptions
made in the original analysis appear to be reasonably good in some cases but
poor in others. The reasons for the wide difference in σ_y versus V_f relationship
exhibited by the tungsten - copper composites as compared with the aluminum -
stainless steel composites are not presently known.

Although it does not help to explain the results of Hughes and Rutherford,
it should be noted that the assumption of elastic behavior of the fibers may
not always be applicable. The fibers, if performing properly, do experience
high stresses. Unless the microplastic properties of the fibers are known,
it is not safe to assume that they are behaving elastically.

The entire discussion of fiber strengthening has been based on reducing
the matrix stress below that of the average applied stress by the addition of
high modulus fibers. Fiber additions may also have some other effects on
the matrix. For one thing, they will act as effective barriers to long-range
movement of dislocations in the matrix. They will also give rise to longitudinal residual stresses in the matrix whenever the temperature is changed
(assuming different thermal expansion coefficients for the two phases). Perhaps of greatest significance is the fact that, if the fibers remain elastic

beyond the stress at which the matrix exhibits plastic strain, upon removal of
the stress they will tend to return to their original length. In other words,
they will exert a "back stress" on the matrix, tending to make it, too, return gradually toward its original length.

Based on the limited experimental work conducted to this time, addition of
high modulus fibers to metal matrices produces significant improvements in
microplastic behavior. Attempts to rationalize the degree of strengthening
have not always been successful. In comparing the strengthening obtained
in metals through fiber addition with that obtained by selective alloying and
combined thermal and mechanical processing, the latter appears to be generally
more efficient. Relatively small amounts of alloying element, if properly
dispersed in the form of precipitate particles, can produce greater increases
in microyield strength than can relatively large volume-percentages of fibers.

E. Effect of Temperature

The effect of temperature on conventional mechanical properties has been
the subject of numerous investigations. Yield strength, for example,
because of its importance in design, is frequently reported as a function of
test temperature. No simple description of the yield strength versus temperature relationship will fit all materials. Generally speaking, however,
yield strengths display at least a moderate tendency for lower values with
increasing temperature. This tendency is particularly pronounced in body-
centered-cubic materials, such as iron and many steeels, at temperatures near
room temperature and below.

The effect of temperature on micromechanical behavior is less well known.
In most cases where investigations have been undertaken, the objective was to
obtain basic information on deformation mechanisms. Thus, pure metals of
little interest in precision design were often used. There are exceptions,
however. Weihrauch and Hordon (1964) examined the effect of test temperature
on σ_y (5×10^{-7}) of several commercially important materials, including Invar,
two aluminum alloys, a stainless steel, and a magnesium alloy. The temperature dependence of σ_y (2×10^{-6}) for a Cu-1.9Be alloy was investigated by
Bonfield (1968). These results are discussed later in this section.

In view of the fact that temperature effects are discussed again in connection with microcreep and stress relaxation, it will be worthwhile here to
devote some attention to the mechanisms whereby temperature can influence
deformation. This topic is highly complex in many of its details. It is
discussed in considerable depth by Lawley and Meakin (1968). The oversimplified treatment presented here is primarily to provide the reader unfamiliar
with deformation theories with a physical feel for the role of temperature.

Temperature is indicative of vibrational motion of the atoms in a solid--
the higher the temperature, the greater the motion. This thermal motion can
aid certain reactions in solids, such as diffusion of atoms from one location
to another during heat treatment. In chemical reactions, temperature effects
are equally important. A rule of thumb states that a chemical reaction rate
can be approximately doubled near room temperature if the temperature is increased only about 5° to 10° C (about 10° to 20° F).

With respect to micromechanical behavior, thermal motion can be viewed as
acting in conjunction with an applied stress to aid movement of dislocations,
i.e., to produce plastic strain. For example, an applied stress insufficient
to cause dislocation movement in the absence of thermal motion (absolute zero
of temperature) might be sufficient at some higher temperature. In such an
event, the dislocation motion would be said to be thermally activated.

Detailed theoretical treatments of thermal activation are based on statistical mechanics. Increased temperature and vibrational motion increase the

likelihood that the barrier (activation energy) for a particular process will
be overcome in a given period of time. Longer times further increase the
likelihood of overcoming the barrier. Thus, thermally activated processes
will exhibit rates that are dependent on temperature.

From statistical principles, the rate of a thermally activated process can
be expressed as

$$\text{rate} = Ae^{-u_a/kT} \tag{5.12}$$

where A is a constant that includes the number of activation sites and the
frequency of the process being activated, u_a is the activation energy that
must be overcome if the process is to proceed, k is the Boltzmann constant,
and T is the absolute temperature. Increasing the value of u_a (the energy
required) reduces the rate, while increasing the product kT (the thermal
energy supplied) increases the rate. At room temperature kT has a value of
approximately 1/40 electron volt (eV). Typical activation energies for
processes involving atom movements in solids are of the order of 1 eV.
McClintock and Argon (1966) have indicated that a u_a value of about 1 eV is
the approximate upper limit for thermal activation at room temperature.

Examination of eq. (5.12) indicates several interesting points. If u_a
is relatively large, the rate at a given temperature will be relatively small.
However, the effect of changing the temperature will be very pronounced. For
example, if u_a for a particular process is 1 eV, a temperature increase of only
about 5°C near room temperature will double the rate, an effect similar to
the rule of thumb described earlier. On the other hand, if u_a is relatively
small, the rate at a given temperature will be relatively high but the effect
of a temperature change will be much less pronounced, i.e., the temperature
dependency will be decreased. For example, if u_a is 0.1 eV, the temperature
would have to be increased approximately 60°C from room temperature to double
the rate.

In plastic deformation processes, eq. (5.12) must be modified to take into
account the stresses acting to move dislocations. There is some lack of
agreement on how this should be done but generally it is assumed that u_a in
eq. (5.12) is a function of the effective stress, i.e., the applied stress
minus an internal stress acting in opposition to the applied stress. For
example, one source of internal stress might be the back stress developed
as dislocations pile up at a barrier. Others include the friction stress of
the lattice, stresses due to impurity atoms and precipitate particles, etc.
It is difficult to obtain quantitative values for these internal stresses and,
hence, experimental results are frequently subject to various interpretations.
In any event, the greater the effective stress, the smaller is the value of
u_a that must be overcome by thermal energy.

Adopting this viewpoint, suppose that a small stress is applied to an
annealed metal at room temperature such that the effective stress is well
below that required to move dislocations. Assume further that the thermal
energy available at room temperature will not be sufficient to overcome the
activation energy barrier, at least not in a reasonable time interval. As
the applied stress is increased, the activation energy u_a will decrease and
the likelihood of a thermal contribution to dislocation motion will increase.
This, of course, implies an effect of testing rate. If thermal activation
is occurring, low rates of stressing will tend to result in dislocation
movement (microplastic strain) at lower applied stresses than will high rates
of stressing. Likewise, stresses applied for long time periods may result in
plastic strain at stress levels insufficient to move dislocations in short
times. Microcreep and stress relaxation, discussed in Sections 5.2 and 6.4,
respectively, are examples of such behavior. Finally, at some level of

applied stress, the effective stress will be sufficient to move dislocations without the assistance of thermal energy.

Returning to the question of observed temperature effects on microyielding, it appears from available evidence that the temperature dependence of the early stages of microplastic deformation depends upon the particular microstrain mechanism operative. Generally speaking, for a particular material in a particular condition, there is little basis for predicting in advance the main factors governing initiation of plastic flow. Rosenfield and Averbach (1960) suggest, however, that if the critical step to initiate plastic flow is either (a) overcoming the inherent lattice resistance to dislocation motion, or (b) activating a dislocation generator, such as a Frank – Read source, then the microyield behavior should be essentially independent of temperature, i.e., not thermally activated. Actually, a slight temperature dependence, similar to the temperature dependence of the elastic modulus, would be predicted if these mechanisms were dominant. As discussed in Chapter 10, the elastic modulus of many materials decreases about 1 to 5% with each 100° C increase in temperature.

The results of tests conducted by Rosenfield and Averbach to measure the effect of temperature on microyield behavior of both aluminum and copper are shown in Fig. 5.43. Copper is seen to exhibit a slight temperature dependence

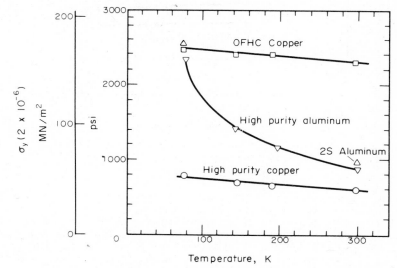

Fig. 5.43. Effect of test temperature on $\sigma_y(2 \times 10^{-6})$ for polycrystalline copper and aluminum (Rosenfield and Averbach, 1960; copyright Pergamon, used by permission).

of $\sigma_y(2 \times 10^{-6})$ while aluminum shows a strong temperature dependence. Arguments were presented by the investigators to suggest that plastic deformation is initiated in both materials by formation of a non-equilibrium amount of stacking fault,* and that this process is thermally activated. Aluminum has

---

*A stacking fault is a faulted layer in close-packed crystal lattices associated with splitting of unit dislocations into component (partial) dislocations.

a much higher stacking fault energy than copper and thus, in line with earlier reasoning, shows a stronger temperature dependence.

Weihrauch and Hordon (1964) studied the effect of test temperature on the microyield behavior of several commercially important alloys. Their results are shown in Fig. 5.44. The temperature dependence of $\sigma_y (5 \times 10^{-7})$ is slightly

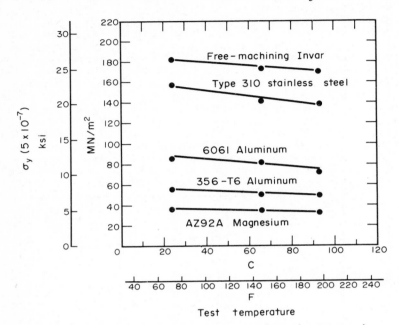

Fig. 5.44.   Effect of temperature on microyield behavior of several commercial alloys (data from Weihrauch and Hordon, 1964).

greater than the expected temperature dependence of the elastic modulus and thus, one can conclude that some thermal activation is occurring.

The complexity of temperature effects on microyield behavior is revealed by investigations conducted on iron (Brown and Ekvall, 1962; Kossowsky and Brown, 1966; Solomon and McMahon, 1971). Each of these investigators studied both σ_E and σ_A at room temperature and below in an attempt to understand the mechanisms controlling plastic flow in iron. It was found that the dependency of microplasticity parameters on temperature depended on:

(a)   the impurity levels in the iron;
(b)   the processing history;
(c)   whether pre-existing dislocations were in the "locked" or "unlocked" states; unlocked dislocations were created by prestraining the specimens;
(d)   the offset strain level at which the yield strength was measured.

For example, Brown and Ekvall reported that σ_E for relatively pure iron, pre-strained to unlock dislocations, was independent of temperature; σ_A, on the other hand, measured with a strain sensitivity of about 4×10^{-6}, was found to depend on temperature, but the dependence lessened with increasing purity. This indicated that an ultra-pure iron should show no temperature dependence of σ_A if measured with sufficiently high strain sensitivity.

Figure 5.45 shows results obtained by Kossowsky and Brown (1966) on prestrained iron of relatively high purity. Note the relatively small temperature

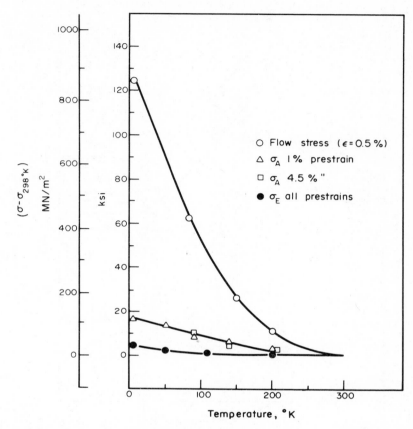

Fig. 5.45. Effect of test temperature on microyield and
           macroyield behavior of iron (Kossowsky and
           Brown, 1966; copyright Pergamon, used by
           permission).

dependence of σ_E, the moderate temperature dependence of σ_A, measured at an
offset strain of about 1×10^{-6}, and the very strong temperature dependence
displayed by the yield strength measured at 0.5% strain. Solomon and McMahon
(1971) verified this type of behavior for iron but showed that iron containing
3% silicon behaved differently. As shown in Fig. 5.46, σ_E, σ_A, and the yield
stress measured at 0.5% strain all showed a very strong temperature dependence,
indicating that the dislocation mechanisms governing microplastic behavior
are different in Fe-3Si than in unalloyed iron.

Microyielding of relatively pure niobium at room temperature and below
has been investigated by Carnahan *et al.* (1967). Their results, some of
which are shown in Fig. 5.47, agree with those reported for iron, in that
the temperature dependence of microyielding depends strongly on the definition
of the microyield strength. The smaller the amount of offset strain at which
microyielding is measured, the smaller the temperature dependence.

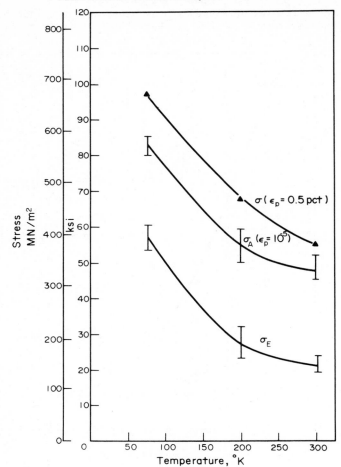

Fig. 5.46. Effect of test temperature on σ_E, σ_A, and
flow stress of Fe-3Si, prestrained several
percent (data from Solomon and McMahon, 1971).

   Stein (1968) developed a theoretical model for microyielding of bcc metals
based on dislocation dynamics. With this model, he computed stress – strain
curves for molybdenum and found reasonable agreement with experimental observations. He then employed his model to predict the effect of temperature on
microyield behavior. Figure 5.48 shows a graph of the calculated microyield
stress at three strain levels for a single crystal of molybdenum.   In
agreement with the experimental findings for the bcc metals and niobium,
Stein's model predicts that the temperature dependence of the microyield
strength is much lower for smaller strains.
   Bonfield (1968) investigated the effect of test temperature on microyielding
in a precipitation hardenable alloy, Cu-2Be. Temperature dependence was
measured for three conditions: (1) solution treated condition, obtained by
water quenching from 800° C (1470° F), (2) precipitation hardened at 315° C
(600° F) for 2 hr; the quenched material was first cold rolled to 40% reduction
to minimize grain boundary precipitation; and (3) precipitation hardened at
425° C (800° F) for 24 hr (overaged).

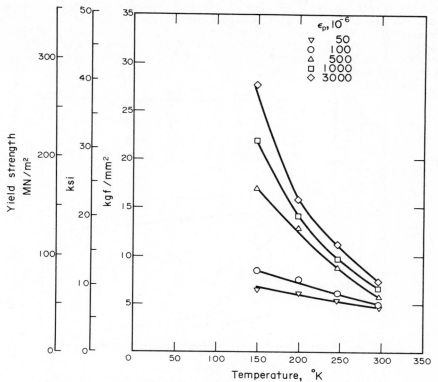

Fig. 5.47. Effect of temperature on yield strength of
niobium measured at various levels of
plastic strain (Carnahan *et al.*, 1967;
copyright AIME, used by permission).

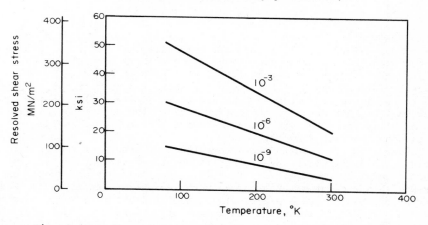

Fig. 5.48. The calculated resolved shear stress necessary
to produce a plastic microstrain of 10^{-9},
10^{-6}, and 10^{-3} as a function of temperature
(Stein, 1968; copyright Wiley, used by
permission).

As shown in Fig. 5.49, $\sigma_y(2 \times 10^{-6})$ for the intermediate precipitation hardened condition showed a strong temperature dependence, particularly below room temperature, while σ_y for the other two conditions exhibited a temperature dependence only slightly greater than the temperature dependence of the elastic modulus. Bonfield discusses the behavior of each condition in terms of microyielding mechanisms.

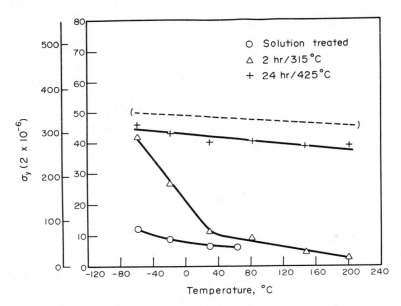

Fig. 5.49.   Temperature dependence of σ_y (2×10^{-6}) for
Cu-2Be in several heat treated conditions;
dashed line indicates temperature dependence
of Young's modulus for copper (Bonfield, 1968;
copyright AIME, used by permission).

Unusual temperature effects on yield behavior have been reported by Thornton *et al.* (1970) and by Mulford and Pope (1973) for Ni_3Al. This is an ordered compound that is often dispersed in nickel-base superalloys to provide strength at elevated temperatures. As shown in Fig. 5.50, the conventional yield strength of this material increases dramatically as the temperature is raised from room temperature to 700° C (1300° F). Microyield behavior, on the other hand, exhibits little or no temperature dependence. Thornton *et al.* discuss this anomalous behavior in terms of changing yield mechanisms with changing temperature.

It is clear from the results shown here that the effect of temperature on microyielding is difficult to predict. It depends on the particular dislocation mechanism by which plastic flow initiates and this apparently can be altered within a given material by changes in heat treatment and processing. From both a fundamental and a practical standpoint, this is a subject of much interest and one that is receiving continuing attention.

For nonmetallic materials (ceramics and glasses), few data are available pertaining to effects of temperature on microyield behavior. For ceramics such as Al_2O_3, which normally fracture at stresses lower than those necessary to cause microyielding (Imgram *et al.*, 1968), it is unlikely that moderate

temperature changes will have much effect on behavior. For ceramics that display ductility, such as lithium fluoride, Kingery (1960) has reported a
strong temperature dependence of the yield strength, similar to that displayed by iron and other bcc metals.

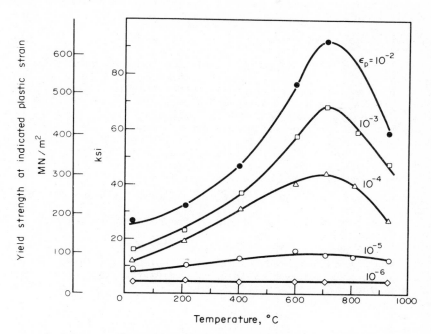

Fig. 5.50.   Microyield and macroyield behavior of Ni3Al
             as a function of test temperature (Thornton
             et al., 1970; copyright ASM and AIME, used
             by permission).

     With respect to glasses and glass-ceramics,* discussion in Chapter 9 in-
dicates that these materials are normally brittle and do not undergo microyielding in the stress ranges in which they would normally be used. They do, however, exhibit marked anelastic behavior and mechanical hysteresis. Although
the authors are unaware of any tests to reveal the effect of temperature on
this anelastic behavior, it is likely that the effect will be relatively pronounced, similar to the effect of temperature on viscosity.
     The question of temperature effects arises also in connection with micro-
creep and stress relaxation. When a particular deformation mechanism is thermally activated, long times will increase the likelihood that the mechanism
will be operative. In fact, because rate effects can be studied relatively
easily in creep and relaxation experiments, such tests are often employed
to study thermal activation parameters.

---

*Glass-ceramics refer to partially crystallized glasses.

5.2 MICROCREEP

Creep is commonly defined as time dependent strain at a constant level of applied stress. Its occurrence in metals operating at elevated temperatures is well documented—it frequently is the limiting factor in design for high temperature applications.

Nature and Definition of Microcreep

Microcreep refers to time dependent strains of small magnitude, occurring under conditions where creep is not normally considered likely, i.e., temperatures near room temperature or below and stresses that are low relative to conventional yield strength values. Compared with other aspects of mechanical behavior, microcreep has received little attention.

Admittedly, it has been known for some time that metals can creep extensively near room temperature and below. Wyatt (1953), for example, investigated the transient creep[†] of pure metals at temperatures ranging from $-196°$ to $+140°$ C ($-320°$ to $+285°$ F) and found extensive creep to occur. He observed several types of transient creep:

(1)   at low temperatures and stresses, creep strain is proportional to the logarithm of time;

(2)   at higher temperatures and stresses, creep strain is proportional to the one-third power of time;

(3)   in an intermediate range, the creep is compound.

However, Wyatt's tests were conducted at stress levels well in excess of those necessary to initiate microyielding--for example, creep strain after 2 sec was always in excess of 100×10^{-6}. This is well outside the range of microcreep as defined here.

Thomas and Averbach (1959) were among the first to report microcreep data that approach the current definition. They observed for high-purity copper that the stress necessary to produce a permanent strain of 1×10^{-6} at room temperature in a load - unload test was about 7 to 10 MN/m^2 (1 to 1.5 ksi), depending on grain size. If the stress was maintained at this level for several minutes, no creep was detected. However, at somewhat higher stresses, approximately twice σ_y (10^{-6}), creep strains of about 1 to 3×10^{-6} were detected within a few minutes. Figure 5.51 shows the microcreep of copper at several stress levels. At the lowest stresses, microcreep strain follows a logarithmic time law, as noted earlier by Wyatt. At the higher stresses, the data clearly deviate from this law.

Moving one step further, Weihrauch and Hordon (1964) investigated the long time microcreep behavior of several commercially important structural materials at stress levels near and below σ_y (10^{-6}). Some of their results are shown in Fig. 5.52. Here it can be seen that significant creep is occurring at stresses well below σ_y (10^{-6}). Though not evident in Fig. 5.52, none of the curves obey a logarithmic time law.

Maringer *et al*. (1968) and Imgram *et al*. (1968) reported microcreep at still smaller fractions of $\sigma_y(10^{-6})$. Shown in Fig. 5.53 is the microcreep behavior of a number of engineering materials, stressed at only 50% of σ_y (10^{-6}). Notice the range of behavior. Several materials exhibit little or no microcreep in 1000 hr at this low stress level, while several others show significant microcreep.

[†]Transient creep refers to a continuously decelerating creep rate; the terms *first stage* and *primary* are sometimes used in place of transient.

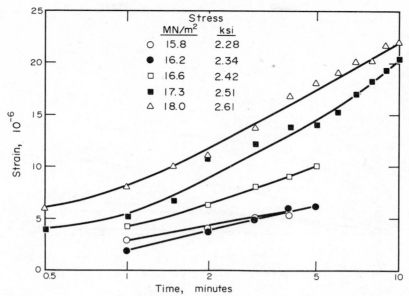

Fig. 5.51. Microcreep of copper at various stress levels
(Thomas and Averbach, 1959; copyright
Pergamon, used by permission).

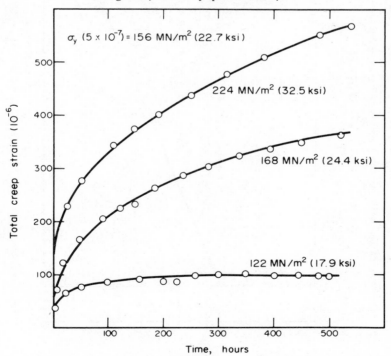

Fig. 5.52. Room temperature creep of 310 stainless steel
at several stress levels (Weihrauch and
Hordon, 1964).

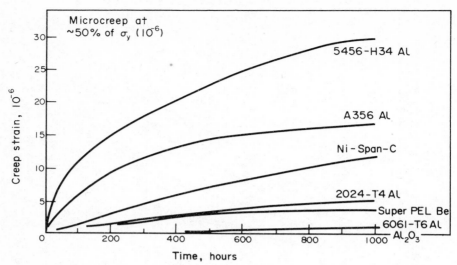

Fig. 5.53. Creep strain versus time at 50% of σ_y(10^{-6}) for several materials (Imgram *et al.*, 1968, and Maringer *et al.*, 1968).

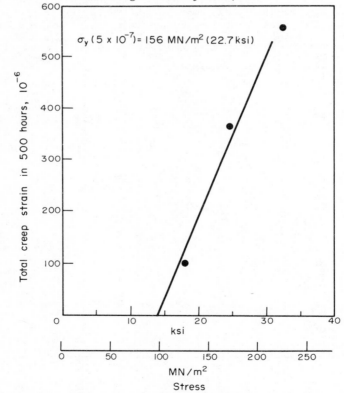

Fig. 5.54. Effect of stress level on creep of 310 stainless steel at room temperature (data from Weihrauch and Hordon, 1964).

In view of these findings, it is reasonable to inquire whether there is a lower limit of stress below which microcreep does not occur. In a practical sense, there undoubtedly is a limit, whose magnitude depends both on the sensitivity with which microcreep can be detected and on the length of time the investigator is willing to wait for detectable creep to occur. An estimate of a lower limiting stress for microcreep can be obtained by replotting data of the type shown in Fig. 5.52 to show creep strain after a certain time as a function of stress level. This has been done in Fig. 5.54. Fitting a straight line to the data and extrapolating the line to zero creep strain indicates that no microcreep should occur in 310 stainless steel in 500 hr at stresses below about 60% of σ_y (10^{-6}). Maringer et al. (1968) have suggested that such data be plotted on logarithmic coordinates. When this is done, the predicted lowest stress for 1×10^{-6} strain in 500 hr in 310 stainless steel is appreciably less than 60% of $\sigma_y(10^{-6})$.

Although the above data clearly indicate that significant microcreep can occur at stresses well below $\sigma_y(10^{-6})$, exceptions have been observed. Experiments conducted in the authors' laboratory indicated that extensive microcreep occurred in low carbon steel only after the applied stress was many times $\sigma_y(10^{-6})$. Load – unload microyield strength tests produced data of the type shown in Fig. 5.55. From this, $\sigma_y(10^{-6})$ is seen to be approximately 40 MN/m^2 (6 ksi). When specimens were exposed to this stress for approximately 100 hr, no significant creep was detected. Incrementally increasing the load on the same specimens revealed little or no creep until a stress of 495 MN/m^2 (72 ksi) was applied. At this point, the specimen underwent almost instantaneous plastic strain of several percent. Subsequent microcreep tests were conducted on other specimens at stress levels of 415, 275, and 140 MN/m^2 (60, 40, and 20 ksi). In no case was the creep strain in excess of 7×10^{-6} after 1000 hr. Clearly this behavior differs substantially from that reported for other materials.

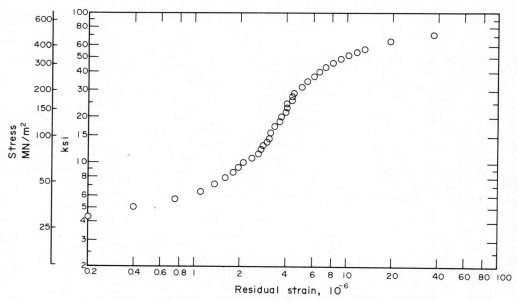

Fig. 5.55. Microyield behavior of low-carbon steel
(Marschall, 1973).

Nonmetallic materials also may exhibit microcreep, though the data available are even fewer than for metals. Crystalline ceramics that exhibit brittleness appear to be highly resistant to microcreep, at least at the relatively low stress levels at which tests have been conducted. As shown in Fig. 5.56 for Al_2O_3, application of a stress equal to 50% of the fracture strength in tension produced no detectable microcreep, within about 2×10^{-6}, in 1000 hr. Rutherford and Swain (1966) also found no microcreep in several ceramic materials stressed to 105 MN/m^2 (15 ksi) for periods of several days.

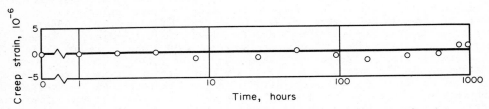

Fig. 5.56.   Microcreep of aluminum oxide at 115 MN/m^2
(16.5 ksi) (Imgram *et al.*, 1968).

Materials investigated included: fine grained, high purity Al_2O_3, hot pressed Al_2O_3, cemented Al_2O_3 (consisting of 90% Al_2O_3, 2% tungsten, and, 8% fluxing agents), and hot pressed BeO.

Rutherford and Swain's results are not surprising, inasmuch as the same materials were loaded in short-duration compression tests to stresses many times greater than those used in the creep tests without evidence of micro-plastic strain.

Materials that are noncrystalline or partially noncrystalline may also exhibit time dependent strain akin to microcreep when stresses are applied for long periods of time. Materials in this category include glasses, glass-ceramics, and graphite-epoxy composites. Microcreep behavior of such materials is discussed in greater detail in Chapter 9. As pointed out there, much or all of the microcreep observed is gradually recovered with time after the load is removed, suggesting that the strain is anelastic rather than plastic.

It is apparent from the foregoing discussion that microcreep data at stresses below $\sigma_y(10^{-6})$ are very limited. This is unfortunate, because it is at these stress levels that many precision devices operate. The reason for the scarcity of data is evident—measurement of extremely small strains over long periods of time is experimentally difficult, tedious and costly. As stress levels decrease and strain sensitivity increases, results can be clouded by other effects occurring concurrently. For example, if a material exhibits a gradual change in dimensions because of a subtle microstructural change, such as a rearrangement of impurity atoms or other effects discussed in Chapter 7, this would be difficult to separate from microcreep. Mikus *et al.* (1960) have described such a problem in studying the dimensional stability of a precision ball bearing steel. Because of the scarcity of microcreep data, it is difficult at this time to generalize concerning microcreep laws and effects of variables on microcreep. Nonetheless, in later sections of this chapter, some of the data currently available are examined to get an idea of the effect of selected variables.

Microcreep Mechanisms

In a discussion of microyield mechanisms in crystalline materials in Section 5.1, dislocation movements were hypothesized. At some stress level, individual dislocations or dislocation generators are activated and the dislocations move until they are stopped by an internal barrier. Before additional plastic deformation can occur, the stress level must be raised sufficiently to activate additional dislocations or to assist those previously activated to overcome the barriers. In Section 5.1 it was pointed out that this microyield behavior is a function of test temperature, thus indicating that microplastic flow can be aided by thermal activation. Accordingly, when a stress in the vicinity of that required to initiate microplastic flow is applied for long periods of time at a temperature above absolute zero, there is a certain probability that thermal fluctuations can assist the applied stress, either in activating dislocations or in overcoming the barriers to the motion of those already activated. The net result is time dependent microplastic strain or microcreep.

The average creep rate will be proportional to the mobile dislocation density ρ_m and the average dislocation velocity $\bar{v}$. From some of the existing microcreep data presented earlier in this section, it appears that microcreep occurs at an ever decreasing rate during the course of a constant stress test. This could be the result of a lowering of either ρ_m, $\bar{v}$, or both, with increasing creep strain. At very low stress levels and very small strains, it is likely that exhaustion of dislocation sources is mainly responsible for the continual decrease of the microcreep rate. At higher levels of stress and plastic strain, however, a reduction in $\bar{v}$ resulting from dislocation interaction may also be a factor in reducing the microcreep rate. Clearly, the actual microcreep behavior displayed by a given material will depend on the particular microplastic flow mechanism operative and the nature of the barriers to dislocation motion. Such details have received little investigation. It should not be surprising, therefore, that laws governing microcreep are not well established and that some materials appear to obey a logarithmic time law and others do not.

A correlation between microyield and microcreep behavior might logically be anticipated. In short duration microyield tests, the slope of the stress-versus-plastic strain curve is indicative of the magnitude of the stress increment needed to activate or reactivate dislocations. A large slope means a large stress increment. In microcreep at a constant stress, thermal fluctuations, rather than increased stress, are responsible for activating or reactivating dislocations. Thus, if the slope of the microyield curve is high, the likelihood of thermal fluctuations activating a given number of dislocations in a given time will be less than if the slope of the microyield curve is low. Accordingly, it might be expected that the microcreep rate of the former would be less than that of the latter. Some evidence in support of this is shown in Fig. 5.57 for several aluminum alloys. Microcreep tests were conducted at a given percentage of $\sigma_y(10^{-6})$ and the results are plotted as a function of a microstrain hardening exponent, n, defined as the slope of a logarithmic plot of microyield data.

In noncrystalline materials, as noted in Section 5.1, inelastic deformation can occur at stresses below the breaking strength because thermal motion of the atoms is continually breaking individual bonds and establishing new bonds in the relatively open structure. Since thermally activated processes are time dependent, it follows that noncrystalline or partially noncrystalline materials may exhibit microcreep or viscous flow under sustained stresses. If the major source of the microcreep is the rearrangement of a finite number of such bonds, it is likely that the material will gradually recover and return to its original dimensions when the stress is removed.

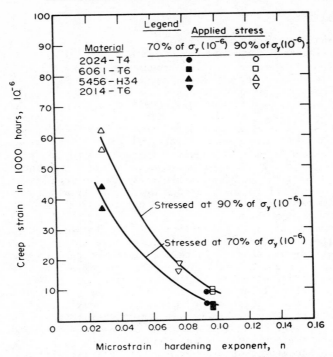

Fig. 5.57. Effect of microstrain hardening exponent on
room temperature creep of aluminum alloys
(Marschall and Maringer, 1971; copyright
ASTM, used by permission).

Effect of Temperature and Stress Level

Both increased temperature and increased stress level are expected to
increase the rate of microcreep for reasons already advanced. Weihrauch and
Hordon (1964) investigated the effect of these variables on five structural
materials: free-machining Invar, 356-T6 aluminum, 6061-T6 aluminum, 310
stainless steel, and AZ92A magnesium. Temperatures investigated were $30°$,
$65°$, and $93°$ C ($85°$, $150°$, and $200°$ F). At each temperature, three stress
levels were used. One was less than $\sigma_y(10^{-6})$, a second was at or slightly
above $\sigma_y(10^{-6})$, and a third was well above $\sigma_y(10^{-6})$. Typical 500 hr creep
curves for stainless steel were shown earlier in Fig. 5.52. The strain shown
on the graph is total plastic strain, i.e., the nearly instantaneous plastic
strain accompanying application of the stress, plus the creep strain. At
stresses below $\sigma_y(10^{-6})$, there appears to be a tendency for creep to reach a
plateau value, whereas at stresses above $\sigma_y(10^{-6})$ creep continues to increase,
although at an ever decreasing rate. The results of tests on the same alloy
at higher temperatures are shown in Figs. 5.58 and 5.59.

It is evident that raising the temperature increases the creep significantly
at each stress level. The data for four of the five materials investigated
are replotted in Fig. 5.60 to show the effect of both stress level and temperature on total plastic strain after 500 hr. A similar graph for free-machining
Invar is shown in Fig. 9.10. It appears that 500 hr creep in these materials
is a linear function of stress above a certain critical stress level. Although

not apparent from these curves, the stress dependence of microcreep for these materials tends to reflect the microstrain hardening behavior observed in microyield tests. In other words, those materials that exhibit extensive microstrain hardening show lower stress dependence of microcreep.

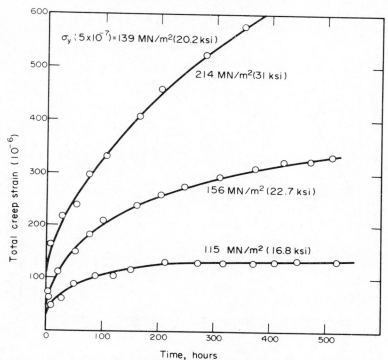

Fig. 5.58.   Effect of stress level on creep of 310 stainless steel at 65°C (150°F) (Weihrauch and Hordon, 1964).

Weihrauch and Hordon analyzed the behavior of these materials in terms of dislocation mechanisms. Equations defining creep as a function of applied stress, temperature, and time were developed for two different stress conditions: (1) applied stresses below, and (2) applied stresses at or above σ_y (10^{-6}). In the first case, creep was found to follow an exponential time dependence, while in the latter case, a $(\text{time})^n$ time dependence was observed, with values of n from 1/3 to 1/2. Others (Maringer et $al.$, 1968; Marschall and Maringer, 1971) have reported $(\text{time})^n$ dependence of microcreep in several aluminum alloys tested at stresses below σy (10^{-6}).

Additional data on the effect of stress on microcreep at stress levels below $\sigma_y(10^{-6})$ have been reported by Maringer et $al.$ (1968). Results are shown in Fig. 5.61. They suggest that creep is related to some power of stress, rather than directly to stress as indicated by Weihrauch and Hordon's data.

The data available at this time are not sufficient to permit the effect of stress level on microcreep to be assessed accurately, particularly at stress levels below σ_y (10^{-6}). In a practical sense, it appears that a limiting stress exists for a particular material below which significant microcreep

will not occur in substantial time periods. An estimate of this can be
obtained by conducting microcreep tests over a range of stresses and extrapolating the results to lower stresses.

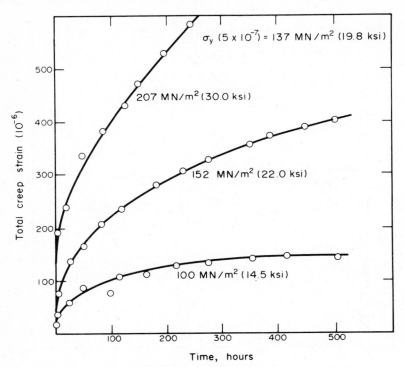

Fig. 5.59.   Effect of stress level on creep of 310
             stainless steel at 93°C (200°F) (Weihrauch
             and Hordon, 1964).

At stresses near and somewhat above $\sigma_y(10^{-6})$, several investigators have
examined the room temperature creep behavior of various titanium alloys.
Hatch *et al.* (1967) were among the first to report that titanium alloys
important in aerospace applications were susceptible to extensive creep at
room temperature at stresses below the 0.2% offset yield strength. Alloys
studied were Ti-5Al-2.5Sn, Ti-6Al-4V, Ti-8Al-1Mo-1V, Ti-6Al-6V-2Sn, and
Ti-4Al-3Mo-1V. Even though strain detection was limited to about 10^{-4},
significant creep was observed in 100 hr tests at stresses as low as 75% of
the yield strength.

Chu (1970) examined Ti-6Al-2Cb-1Ta-0.8Mo at several stress levels ranging
from 80 to 100% of the 0.2% yield strength. Although the value of $\sigma_y(10^{-6})$
is not known for this alloy or for the alloys studied by Hatch *et al.*, it is
probably less than 80% of the yield strength. Chu's results are shown in
Fig. 5.62. Note that creep is extensive, reaching strain levels of about 1000
x 10^{-6} after 500 hr at the lowest stress used. The creep data obey an empirical
expression

$$\varepsilon = At^{0.183}$$

where ε = creep strain, %,

t = time, hours, and

A = a constant that depends on stress $=\left(\dfrac{S}{103.55}\right)3.89.$

In the expression for A, S is the stress in ksi.

Odegard and Maringer (1971) reported similar results for Ti-6Al-4V, tested at 60, 70, 80, and 90% of the 0.2% yield strength for times of 1000 hr. They found that the data obeyed the following expression:

$$\varepsilon =\left(\frac{S}{93.4}\right)^{14.88} t^{0.28}$$

where ε is strain in units of 10^{-6} and the other symbols have the same meaning as before.

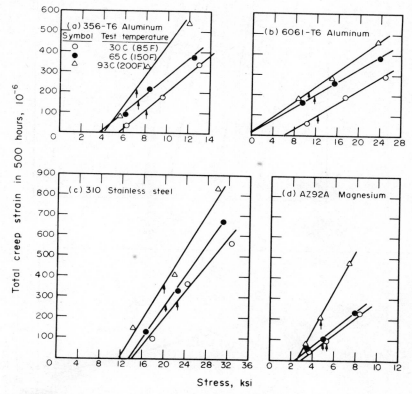

Fig. 5.60. Effect of stress and temperature on microcreep of several engineering materials (data from Weihrauch and Hordon, 1964). Arrows denote σ_y (5×10^{-7}).

Although all the data presented here for titanium alloys were obtained at relatively high stress levels, it is of interest to use the above empirical expressions relating creep and stress level to compute a lower limit for microcreep. For example, assume creep in excess of 1×10^{-6} cannot be tolerated in 1000 hr. What stress will produce this amount of creep? Using Odegard and Maringer's expression for Ti-6Al-4V, the stress is found to be about 53% of the 0.2% yield strength. Employing Chu's expression for Ti-6Al-2Cb-1Ta-0.8Mo, the stress is about 48% of the 0.2% yield strength.

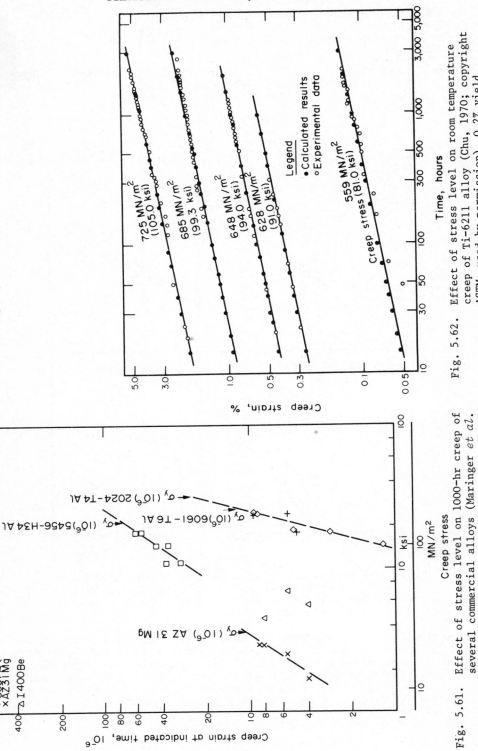

Fig. 5.62. Effect of stress level on room temperature creep of Ti-6211 alloy (Chu, 1970; copyright ASTM, used by permission). 0.2% yield strength = 705 MN/m2 (102 ksi).

Fig. 5.61. Effect of stress level on 1000-hr creep of several commercial alloys (Maringer *et al.* 1968).

With the exception of Weihrauch and Hordon's investigation, few data are
available relating microcreep to temperature. As noted earlier, some effect
should be anticipated. The magnitude of the effect will depend on the
particular microcreep mechanism.operative. In noncrystalline materials, it is
likely that microcreep will be considerably more temperature-sensitive than
in crystalline materials.

Effect of Prestrain and Cold Work

The concept of a back stress developing within a crystalline material as
moving dislocations pile-up against obstacles has been discussed earlier. In
microcreep tests, one way to rationalize the continually decreasing creep
rate is to assume that such back stresses increase with strain and that they
counteract the applied stress. Gasca-Neri *et al.* (1970) have developed a
phenomenological theory for the transient creep portion of conventional creep
tests in which these back stresses play an important role.

If microcreep occurs at an ever decreasing rate due to the development of
internal back stresses, it is logical to assume that prestraining will reduce
the rate of microcreep. Few data are available to verify this expected
effect. Odegard and Thompson (1973) showed that tensile prestraining of
Ti-6Al-4V at a stress of 95% of the 0.2% yield strength lowered the tensile
creep rate in subsequent tests conducted at 70 and 80% of the yield strength.
Their data are shown in Figs. 5.63 and 5.64. For the larger levels of prestrain, negative creep was observed, indicating that the effective internal
back stresses developed by the prestrain exceeded the creep stress in subsequent
tests.* Creep at 90% of the yield strength was diminished by large prestrains
but was apparently accelerated by low and intermediate levels of prestrain.
This latter observation has not been explained.

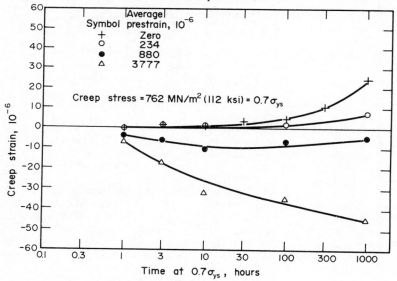

Fig. 5.63. Effect of prestraining Ti-6Aℓ-YV various
amount at 0.95 σ_{ys} on subsequent creep at
0.7 σ_{ys} (Odegard and Thompson, 1973).

*Some uncertainty exists on this point, because creep specimens were unloaded
periodically for strain measurement. Thus, much or all of the negative
creep may have occurred in the absence of an external load and have been
indicative simply of strain recovery.

   Effects of cold working (drawing, rolling, swaging, etc.) on microcreep
have received little attention. Complex effects might be expected—for example,
in a particular instance, cold working may improve microcreep resistance at
room temperature, but reduce it at moderately elevated temperatures. Thermal
treatments following cold working may prove to be particularly beneficial.
Experimental data to verify those expectations for engineering materials are
virtually nonexistent.

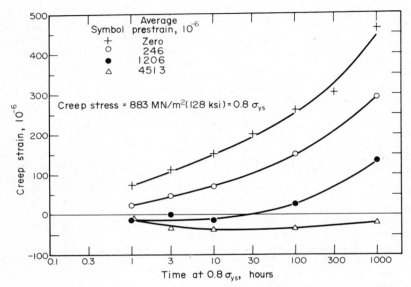

Fig. 5.64. Effect of prestraining Ti-6Aℓ-YV various
            amounts at 0.95 σ_{ys} on subsequent creep at
            0.8 σ_{ys} (Odegard and Thompson, 1973).

5.3 MICROPLASTIC STRAIN RECOVERY

   In the strictest sense, recovery of plastic strain would be more appropriately
discussed in Chapter 6, because it is a manifestation of internal stress
rather than of applied stress. Discussion of the phenomenon is included here
because of its possible implications relative to measurement of microyield
and microcreep properties.
   Some of the earliest studies of microplastic strain behavior of metals
showed conclusively that a portion of the apparent plastic strain gradually
disappeared as a function of time after unloading. This is illustrated in
Fig. 5.65 for zinc single crystals. After prestraining in torsion to the levels
indicated in the graph, the load was removed and strain recovery measured.
The amount of recovery clearly depends on the amount of plastic strain. The
ratio of strain recovered, γ_{pr}, to plastic strain γ_p, is tabulated below:

γ_p	γ_{pr}	γ_{pr}/γ_p
1×10^{-2}	$\sim 10 \times 10^{-5}$	0.01
6×10^{-3}	$\sim 8 \times 10^{-5}$	0.013
1×10^{-3}	$\sim 2 \times 10^{-5}$	0.02
5×10^{-4}	$\sim 1 \times 10^{-5}$	0.02

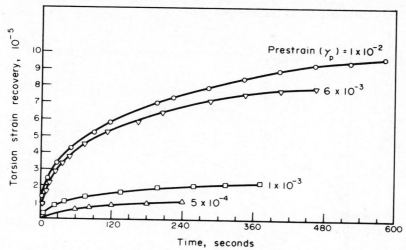

Fig. 5.65. Recovery of torsional microplastic strain in
zinc single crystals (Roberts and Brown,
1960; copyright AIME, used by permission).

For the particular levels of prestrain investigated here, recovery is seen
to represent a rather small fraction (1/50 to 1/100) of the original plastic
strain. This fraction might have been somewhat larger had the observations
been continued for longer times.

Should similar fractional recoveries be anticipated at smaller levels of
plastic strain? This is an important question in connection with investigations
of microyield behavior. For example, if a value of microyield strength is
measured at an offset strain of 1×10^{-6}, recovery of only 1/50 to 1/100 of
this strain with time will have virtually no effect on the reported microyield strength values. No comprehensive investigation of recovery at small
levels of plastic strain has been reported. However, it has been the authors'
observation that in microyield studies at strain sensitivities of 10^{-6} to
10^{-7}, recovery becomes easily observable only after the imposed plastic
strains approach and exceed 10^{-5} to 10^{-4}. This suggests that the recovery
ratios are similar to those reported above by Roberts and Brown (1960). Even
where recovery ratios are larger than these, there seems to be general agreement that *total* strain recovery does not occur in plastically deformed metals.
As noted by Tinder and Washburn (1964), even strains as small as 10^{-8} in
copper were mostly nonrecoverable.

Geil and Feinberg (1969b) have reported somewhat larger recovery effects in
normalized AISI 4340 steel. In a microyield strength test at a strain
sensitivity of 10^{-7}, they followed the strain recovery of a specimen that had
been plastically strained $\sim 5 \times 10^{-6}$ in tension. Over a period of 40 hr after
unloading, approximately 1×10^{-6} (or 1/5) of the initial strain was recovered.

The same investigators prestrained a specimen of the same material
approximately 1.5%, after which it was allowed to rest for 20 days. Subsequent
load – unload microstrain tests revealed that residual strains at zero-load
tended to diminish appreciably with time after unloading. After applying a
stress sufficient to cause a residual strain of 45×10^{-6}, the load was removed and recovery was monitored for nearly a month. As shown by the recovery
curve in Fig. 5.66, the residual strain diminished rapidly for several

minutes after unloading and more gradually thereafter. A small amount of
recovery was still being observed after 650 hr. Note that the ratio of strain
recovery to initial strain is nearly one-half.

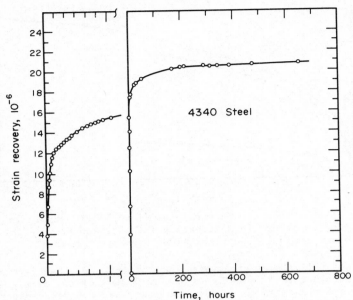

Fig. 5.66. Recovery of microplastic strain observed
after unloading a prestrained specimen of
normalized AISI 4340 steel from the peak
stress of 1130 MN/m^2 (164 ksi); specimen had
been prestressed to 1550 MN/m^2 (225 ksi) and
extended 1.5% (Geil and Feinberg, 1969b).

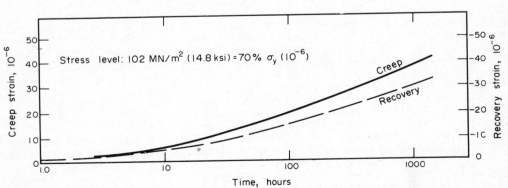

Fig. 5.67. Microcreep and recovery curves for 5456-H34
aluminum, stressed at ~70% of $\sigma_y(10^{-6})$
(Maringer *et al.*, 1968).

TABLE 5.7 Microcreep Data (Maringer *et al*., 1968)

Material	Approximate level of		Percent of σ_y (10^{-6})	Total creep strain (at about 1400 hr). 10^{-6}	Total strain recovery (1400 hr after unloading), 10^{-6}
	Stress, ksi	Strain, 10^{-6}			
2024-Al	18.0	1800	50	5	<1
	24.0	2400	70	11	~1
	24.0	2400	70	6	~1
5456-Al	11.0	1100	50	35	20
	11.0	1100	50	42	22
	14.8	1500	70	40	30
	14.8	1500	70	49	32
	18.0	1800	90	67	46
	17.8	1800	90	62	52
6061-Al	14.8	1500	50	1	1
	18.5	1800	70	6	2
	18.0	1800	70	4	2
	24.6	2400	90	10	4
	24.6	2400	90	11	3
AZ 31-Mg	1.7	270	50	5	<1
	1.7	270	50	2	
	2.5	420	70	7	<1
	3.0	500	90	10	
	3.0	500	90	11	3
TZM-Mo	27.0	630	50	1.8	0
	29.6	700	60	-2.4	0
	29.6	700	60	-0.2	0
	38.0	900	70	1.2	0
	38.0	900	70	-0.2	0
I-400-Be	4.5	100	50	>10	(-6)[a]
	5.6	130	70	5	(-3)
	7.0	160	90	7	(-4)

[a] Recovery strain was in the same direction as creep strain.

Not all materials show such recovery effects. For example, Geil and Feinberg also conducted microstrain tests on annealed Invar (Fe-36Ni) and found no discernible strain recovery effects within about 135 min after unloading. Strain recovery was observed, however, during microstrain tests on Invar that had been prestrained approximately 2%.

Recovery of strain is also observed in microcreep tests conducted in the microstrain region. As before, generalizations are difficult because of the lack of experimental data. In tests conducted on 5456-H34 aluminum, a significant fraction of the microcreep strain observed in 1000 hr was recovered within 1000 hr after unloading, as shown in Fig. 5.67. In other materials tested at the same time, recovery effects were less clearly defined, as shown in Table 5.7. The relatively small magnitudes of both the microcreep strain and the strain recovery make interpretation difficult because of inherent problems in measuring small strains over long periods of time.

From a deformation mechanisms viewpoint, strain recovery is to be anticipated whenever plastic flow occurs in crystalline materials because of the internal stresses developed (Section 5.1). The magnitude and kinetics of the recovery should reveal something of the magnitude of the internal stresses and of the nature of the internal barriers to dislocation motion. The effect of temperature on strain recovery should also provide insights into operative dislocation mechanisms. Some unusual effects of temperature on plastically deformed metals are discussed in Section 6.3.

Viewed practically, strain recovery increases with the amount of plastic strain but appears to be a small fraction (1/100 to 1/5) of the initial strain in most instances. Thus, its effect on microyield strength values measured on test coupons is probably not large. However, in a precision component that is subjected to prestrain or cold working shortly before being placed in service, significant strain recovery may occur over periods of several months, leading to perhaps damaging dimensional changes.

5.4 MICRO-BAUSCHINGER EFFECT

The term Bauschinger effect has long been associated with the observation that, after plastic strain has occurred in one direction, the conventional yield strength for subsequent loading in the reverse direction is less than that for loading in the same direction.

As noted by Abel and Muir (1972), two theories have been advanced to explain the Bauschinger effect. Early theories involved internal stress effects, particularly those resulting from inhomogeneous deformation of the various grains in a polycrystalline specimen. Demonstration of the existence of a Bauschinger effect in single crystals led to other approaches. More recently, dislocation theory has been used to rationalize the Bauschinger effect. It may be the result of back stresses associated with dislocation pile-ups, or may be associated with the ease of moving dislocations in a direction opposite to that associated with the prestraining.

On the basis of a new interpretation of the Bauschinger effect, Abel and Muir showed that the effect is greater, the smaller the prestrain. They proposed to evaluate the effect with the "Bauschinger energy parameter," β_E. As shown in Fig. 5.68, β_E is defined as the ratio of E_s, the energy saved in reverse straining, to E_p, the energy expended in prestrain. The effect of prestrain magnitude on β_E for mild steel is shown in Fig. 5.69. An increase in β_E with decreasing prestrain is clearly evident. According to Abel and Muir, no significant back stresses on dislocations are required to produce this large effect at low prestrains. The creation of mobile dislocations which exhibit directionality in their resistance to further motion is all that is needed. At larger prestrains, significant back stresses are developed and these lead to an increase in the "Bauschinger stress parameter," β_σ (defined in Fig. 5.68), as illustrated in Fig. 5.69.

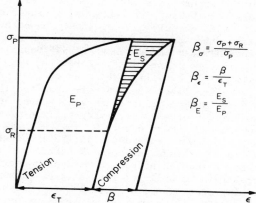

$$\beta_\sigma = \frac{\sigma_P + \sigma_R}{\sigma_P}$$

$$\beta_\epsilon = \frac{\beta}{\epsilon_T}$$

$$\beta_E = \frac{E_S}{E_P}$$

Fig. 5.69. Bauschinger parameters for mild steel as
a function of prestrain (Abel and Muir,
1972; copyright Taylor and Francis, used
by permission).

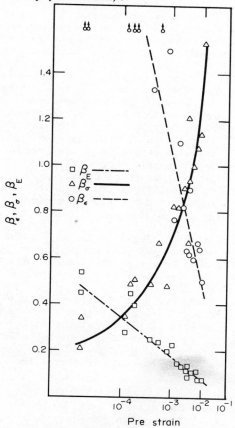

Fig. 5.69. Bauschinger parameters for mild steel as a
function of prestrain (Abel and Muir, 1972;
copyright Taylor and Francis, used by permission).

Although Abel and Muir's experiments were not conducted in the plastic
microstrain region, their results suggest that a Bauschinger effect, or micro-
Bauschinger effect as it is frequently called, should be observed in this
region also. As with many other aspects of microstrain behavior, this has
received little attention. Tinder and Washburn (1964) found evidence of a
micro-Bauschinger effect in copper specimens. Likewise, Kyle *et al.* (1966)
observed that $\sigma_y(10^{-6})$ for I-400 beryllium was reduced from nearly 50 MN/m^2
to less than 30 MN/m^2 (7 ksi to less than 4.5 ksi) when the loading direction
was reversed after the onset of microplastic strain. Carnahan (1964) reported
a substantial micro-Bauschinger effect in polycrystalline nickel, but only after
the microplastic strain exceeded about 4×10^{-6}.

Clearly, the data are too few to make any quantitative statements concerning
the magnitude of the micro-Bauschinger effect. Nonetheless, there is sufficient
evidence of the existence of such an effect to warrant concern in practical
applications. Where prestrain is employed for whatever reason, such as in
straightening or flattening operations, microyield properties in a direction
opposite that of the prestrain will almost surely be lowered.

5.5 CHAPTER SUMMARY

The occurrence of microplastic strain as a result of applied stresses can
be an important source of dimensional instability in precision components.
In this chapter, an attempt is made to summarize current understanding of
microplastic strain--how it occurs, how it is measured, how it is influenced
by important variables, and how it can lead to strain recovery and the micro-
Bauschinger effect.

Microyield behavior (microplastic strain under short duration loading) is
most often measured by a load – unload procedure in which the load is increased
incrementally in succeeding cycles and the test specimen is examined for
residual strain at zero load after each cycle. When residual strain is graphed
as a function of applied stress, various types of microyield curves have been
observed. Many of the data produce a straight line on logarithmic coordinates.
Others produce straight lines when stress is plotted versus the square-root
of residual strain. In some instances, curves of sigmoidal shape are obtained
on standard coordinates. Since standard test procedures for obtaining microyield data have not been developed, it is important that investigators report
test details, such as loading rate, dwell time at maximum load, unloading rate,
time delay between unloading and measurement of residual strain, and magnitude
of the load increment.

Reports by some investigators of negative residual strains in tests to
study microyield behavior have not yet been adequately explained. Although
in some cases they are probably only apparent strains resulting from experimental difficulties, other cases exist in which evidence is strong that
the negative strains are real.

Mechanisms for microyielding are discussed in terms of dislocation movement in crystalline materials. These dislocation concepts make it relatively
easy to understand the occurrence of plastic strain at low stresses, microstrain hardening, strengthening by introduction of obstacles, recovery of
plastic strain, and the micro-Bauschinger effect.

Materials can be altered in various ways to change their microyield
behavior. For example,

(a) The grain size can be changed; generally speaking, fine grained
    materials require somewhat greater stresses than do course grained
    materials to achieve a given level of microplastic strain. However,
    the exact nature of the relationship between microyield behavior and
    grain size depends upon the type of material.

(b) Alloying elements can be added to achieve solid solution strengthening, precipitation hardening, ordering, special grain boundary effects, etc.; this is an extremely effective method for increasing resistance to microplastic deformation. Increases in microyield strengths of 10 to 100 times are not uncommon.

(c) Reinforcing fibers or particles can be added to increase resistance to microplastic deformation; although this method has received little attention, it appears to be less efficient than the alloying method described in (b).

(d) Prestraining or cold working can be introduced; many unanswered questions remain about the effectiveness of such treatments for increasing the resistance to microplastic deformation. Apparently, small amounts of prestrain are effective in this respect, whereas, in some cases at least, larger amounts of prestrain can lead to a reduction of microyield strength. Prestraining and cold working appear to be especially beneficial to microyield strength when they are appropriately combined with alloying and thermal treatments.

Temperature can have an important effect on microyield behavior through thermal activation of dislocation movement. The magnitude of the effect depends on the details of the microyield mechanisms operative in a particular material. For example, the microyield strength of copper exhibits a relatively modest temperature dependence between room temperature and liquid nitrogen temperature, whereas that of aluminum shows a strong temperature dependence over the same range. In body-centered-cubic metals, the temperature dependence of the yield strength is often strongly dependent on the amount of offset strain at which yielding is measured. Conventional yielding, for example (measured at $\epsilon_p = 2 \times 10^{-3}$) is strongly temperature dependent below room temperature, with strength increasing rapidly as temperature is lowered. Microyielding, on the other hand (measured at $\epsilon_p = 10^{-6}$, for example) generally shows very little temperature dependence in this range.

Microcreep (microplastic strain under long duration loading) can occur at stresses well below $\sigma_y(10^{-6})$. This phenomenon has received considerably less attention than microyielding and consequently, the dependence of microcreep on stress level, temperature, and material variables and the relationship between creep strain and time are not well established.

The chapter concludes with brief discussions of microplastic strain recovery and the micro-Bauschinger effect. If a precision component is prestrained or cold worked prior to being placed in service, strain recovery can lead to gradual dimensional changes, while the micro-Bauschinger effect can significantly lower the microyield strength in a direction opposite that of the prestrain.

Despite the acknowledged importance of microplastic strain behavior in precision design, much work remains to be done to (a) provide the designer with adequate design data for existing materials, and (b) develop materials that have improved resistance to microplastic deformation.

REFERENCES

Abel, A. and Muir, H. 1972. The Bauschinger effect and discontinuous yielding, *Phil. Mag.* **26**, 489.

Argon, A. S. and East, G. 1967. The microyield strength of beryllium-iron alloys, *Trans. TMS-AIME*, **239**, 598.

Bilello, J. C. and Metzger, M. 1969. Microyielding in polycrystalline copper, *Trans. TMS-AIME*, **245**, 2279.

Bonfield, W. 1965. Some observations on the microstrain characteristics of silver, *Acta Met.* 13, 551.

Bonfield, W. 1967. Microplasticity in a Cu 1.9 wt pct Be precipitation-hardening alloy, *Trans. TMS-AIME*, 239, 99.

Bonfield, W. 1968. The temperature dependence of microyielding in polycrystalline Cu 1.9 wt. pct. Be, *Trans. TMS-AIME*, 242, 2163.

Bonfield, W. and Li, C. H. 1963. Dislocation configurations and the microstrain of polycrystalline beryllium, *Acta Met.* 11, 585.

Bonfield, W. and Li, C. H. 1964. A transition in the microstrain characteristics of beryllium, *Acta Met.* 12, 577.

Bonfield, W. and Li, C. H. 1965. The friction stress and initial microyielding of beryllium, *Acta Met.* 13, 317.

Brentnall, W. D. and Rostoker, W. 1965. Some observations on microyielding, *Acta Met.* 13, 187.

Brown, N. and Ekvall, R. A. 1962. Temperature dependence of the yield points in iron, *Acta Met.* 10, 1101.

Brown, N. and Lukens, K. F., Jr. 1961. Microstrain in polycrystalline metals, *Acta Met.* 9, 106.

Carnahan, R. D. 1964. Microplasticity, Aerospace Corporation Report No. TDR-269(4240-10)-14, prepared for U.S. Air Force Systems Command.

Carnahan, R. D., Arsenault, R. J. and Stone, G. A. 1967. Effect of purity and temperature on dynamic microstrain of niobium (Cb), *Trans. TMS-AIME*, 239, 1193.

Carnahan, R. D. and White, J. E. 1964a. The microplastic behavior of polycrystalline nickel, *Phil. Mag.* 10, 513.

Carnahan, R. D. and White, J. E. 1964b. Some comments on strain gage techniques for determining microstrain, *Trans. TMS-AIME*, 230, 249.

Chu, H. P. 1970. Room temperature creep and stress relaxation of a titanium alloy, *J. Mater.* 5, 633.

Eul, W. A. and Woods, W. W. 1969. Shear strain properties to 10^{-10} of selected optical materials, Boeing Co., NASA CR-1257.

Gasca-Neri, R., Ahlquist, C. N. and Nix, W. D. 1970. A phenomenological theory of transient creep, *Acta Met.* 18, 655.

Geil, G. W. and Feinberg, I. J. 1969a. Microplasticity. I. Measurement of small microstrains at ambient temperature, U.S. National Bureau of Standards Report 9996.

Geil, G. W. and Feinberg, I. J. 1969b. Microplasticity. II. Microstrain behavior of normalized 4340 steel and annealed Invar, U.S. National Bureau of Standards Report 9997.

Hatch, A. J., Partridge, J. M. and Broadwell, R. G. 1967. Room temperature creep and fatigue properties of titanium alloys, *J. Mater.* 2, 111.

Holt, R. T. 1972. The microdeformation of nickel at room temperature, *J. Inst. Metals*, 100, 376.

Holt, R. T. 1973. The flow stress dependence on grain size during microstrain tests, *Met. Trans.* 4, 875.

Hordon, M. J., Lement, B. S. and Averbach, B. L. 1958. Influence of plastic deformation on expansivity and elastic modulus of aluminum, *Acta Met.* 6, 446.

Hughes, E. J. and Rutherford, J. L. 1969. Microstrain in continuously reinforced tungsten-copper composites, Amer. Soc. Test. Mat. Spec. Tech. Publ. 460.

Hughes, E. J. and Rutherford, J. L. 1973. The Singer Company, Kearfott Division, Little Falls, New Jersey; unpublished data.

Imgram, A. G., Hoskins, M. E., Sovik, J. H., Maringer, R. E. and Holden, F. C. 1968. Study of microplastic properties and dimensional stability of materials, Battelle-Columbus Laboratories, U.S. Air Force Materials Laboratory Report AFML-TR-67-232, Part II.

Kelly, A. and Tyson, W. R. 1964. *Proceedings of the 2nd International Materials Symposium*, Wiley, New York.

Kingery, W. D. 1960. *Introduction to Ceramics*, Wiley, New York.

Kossowsky, R. and Brown, N. 1966. Microyielding in iron at low temperatures, *Acta Met.* 14, 131.

Kyle, P. E., Papirno, R., Tang, C. N. and Becker, H. 1966. Photomechanical investigation of structural behavior of gyroscope components. Task V: Materials and design data, Allied Research Associates, Final report on NASA Contract NAS8-11294.

Lawley, A. and Meakin, J. D. 1968. Thermally activated processes in the microplastic region, *Microplasticity*, C. J. McMahon, Jr., editor, Interscience, New York.

Maringer, R. E., Cho, M. M. and Holden, F. C. 1968. Stability of structural materials for spacecraft application, Battelle-Columbus Laboratories, Final report on NASA Contract NAS5-10267.

Marschall, C. W. 1973. Battelle-Columbus Laboratories; unpublished data.

Marschall, C. W. and Maringer, R. E. 1971. Stress relaxation as a source of dimensional instability, *J. Mater.* 6, 374.

Marschall, C. W., Maringer, R. E. and Cepollina, F. J. 1972: Dimensional stability and micromechanical properties of materials for use in an orbiting astronomical observatory, AIAA Paper 72-325.

McClintock, F. A. and Argon, A. S. 1966. *Mechanical Behavior of Materials*, Addison-Wesley, Reading, Mass.

McDanels, D. L., Jech, R. W. and Weeton, J. W. 1963. Stress-strain behavior of tungsten-fiber-reinforced copper composites, NASA Tech. Note D 1881.

Mikus, E. B., Hughel, T. J., Gerty, J. M. and Knudsen, A. C. 1960. The dimensional stability of a precison ball bearing material, *Trans. Amer. Soc. Metals*, 52, 307.

Muir, H., Averbach, B. L. and Cohen, M. 1955. The elastic limit and yield behavior of hardened steels, *Trans. Amer. Soc. Metals* 47, 380.

Mulford, R. A. and Pope, D. P. 1973. The yield stress of $Ni_3(Al,W)$, *Acta Met.* 21, 1375.

Odegard, B. C. and Maringer, R. E. 1971. The room temperature creep behavior of wrought and as-welded Ti-6Al-4V (STA), Sandia Laboratories Report SCL-DR 710069.

Odegard, B. C. and Thompson, A. W. 1973. Low temperature creep of Ti-6Al-4V, Sandia Laboratories Report SLL-73-5289 (to be published in *Met. Trans.*).

Parikh, P. D. and Hay, D. R. 1971. Effect of plastic prestrain on microstrain behavior of <100> and <110> Ta single crystals, *Scr. Met.* 5, 1039.

Pinnel, M. R. and Lawley, A. 1969. Microstructural characterization of uniaxially strained aluminum-stainless steel composites, presented at AIME Symposium on Metal-Matrix Composites, Defense Metals information Center Memorandum 243.

Price, C. W. 1973. Battelle-Columbus Laboratories; unpublished data.

Roberts, J. M. and Brown, N. 1960. Microstrain in zinc single crystals, *Trans. TMS-AIME*, 218, 454.

Roberts, C. S., Carruthers, R. C. and Averbach, B. L. 1952. The initiation of plastic strain in plain carbon steels, *Trans. Amer. Soc. Metals*, 44, 1150.

Rosenfield, A. R. and Averbach, B. L. 1960. Initial stages of plastic deformation in copper and aluminum, *Acta Met.* 8, 624.

Rudnick, A., Marschall, C. W., Duckworth, W. H. and Emrich, B. R. 1968. The evaluation and interpretation of mechanical properties of brittle materials, Battelle-Columbus Laboratories, U.S. Air Force Materials Laboratory Report AFML-TR-67-361.

Rutherford, J. L. and Swain, W. B. 1966. Research on materials for gas-lubricated gyro bearings, General Precision, Inc., First Technical Summary Report to NASA on Contract NAS 12-90.

Shemenski, R. M. and Maringer, R. E. 1969. Microstrain characteristics of isostatically hot-pressed beryllium, *J. Less-Common Metals*, 17, 25.

Solomon, H. D. and McMahon, C. J., Jr. 1971. Solute effects in micro and macroyielding of iron at low temperatures, *Acta Met.* 19, 291.

Stein, D. F. 1968. A dislocations-dynamics treatment of microstrain, *Microplasticity*, C. J. McMahon, Jr., editor, Wiley, New York, 141.

Thomas, D. A. and Averbach, B. L. 1959. The early stages of plastic deformation in copper, *Acta Met.* 7, 69.

Thornton, P. H., Davies, R. G. and Johnston, T. L. 1970. The temperature dependence of the flow stress of the γ' phase based upon Ni_3Al, *Met. Trans.* 1, 207.

Tinder, R. F. and Trzil, J. P. 1973. Millimicroplastic burst phenomena in zinc monocrystals, *Acta Met.* 12, 975.

Tinder, R. F. and Washburn, J. 1964. The initiation of plastic flow in copper, *Acta Met.* 12, 129.

Vellaikal, G. 1969. Some observations on microyielding in copper polycrystals, *Acta Met.* 17, 1145.

Weihrauch, P. F. and Hordon, M. J. 1964. The dimensional stability of selected alloy systems, Alloyd General Corp., Final Report of U.S. Navy Contract N140(131)75098B.

Wikle, K. G. 1970. Microyield properties of beryllium copper alloy 25, Brush Beryllium Co., Technical Service Report M-84.

Wilson, F. G. and Teghtsoonian, E. 1970. Interpretation of microflow measurements in niobium crystals, *Phil. Mag.* 22, 815.

Wyatt, O. H. 1953. Transient creep in pure metals, *Proc. Phys. Soc., London, Sect. B*, 66, 459.

Chapter 6.

Internal Stress Effects: Dimensional Instability from Alteration of Internal Stress

In Chapter 3 it was pointed out that internal stresses can contribute to dimensional instability if, in service, these stresses are altered in some manner. Two types of internal stresses were described—long-range and short-range. It is important that the distinction between these two types of internal stress be clearly appreciated.

6.1 NATURE OF INTERNAL STRESSES

All internal stress systems have zero resultant force. Short-range internal stresses achieve this balance in distances comparable with the grain size, though in many cases the distances may be much smaller. For example, balanced force fields exist between individual dislocations in a crystal, between dislocations and impurity atoms, between piled-up dislocations and barriers, etc. Long-range internal stresses, on the other hand, achieve a force balance over distances that are large relative to microstructural features—comparable with the gross dimensions of the body. They result primarily from fabrication processes, such as forming, machining, heat treating, and welding.

The literature on internal stress is extensive and terminology is not consistent. Some of the alternate names that have been used for short- and long-range internal stress are:

short-range	*long-range*
microstress	residual stress
microscopic stress	macrostress
tessellated stress	macroscopic stress
stress of the second kind	stress of the first kind
textural stress	type I stress
parasitic stress	first-order stress
Heyn stress	
type II stress	

In this chapter, short-range internal stress will be abbreviated as SR stress and long-range internal stress as LR stress. Because of common usage, the term residual stress will be used interchangeably with LR stress.

A second distinction can be made between SR stress and LR stress. While LR stress can, by control of processing parameters, be reduced essentially to zero, it is doubtful that the same can be said about SR stress. For example, as long as dislocations are present in a crystalline material, their specific locations in a crystal will depend on a force balance with neighboring dislocations. In other words, SR stresses will be present. These forces can be minimized by reducing the number of dislocations; e.g., by annealing.

Yet another distinction between SR and LR stress is in the method of measurement. Techniques are well established for measuring residual stress gradients through a body by sectioning methods or by X-ray diffraction line shift measurements. Short-range stress gradients, on the other hand, cannot be measured directly. In fact, it is possible only to get an idea of an "average"

SR stress, rather than maximum and minimum values. Such information is obtained
from observation of the width of X-ray diffraction lines. When SR stresses
are small, diffraction lines are sharp and narrow, indicating uniform atom
spacing. Broad diffraction lines indicate large SR stresses associated with
abnormally large spacing of the atoms in some regions (internal tension) and
abnormally small spacing in adjacent regions (internal compression).

Although for clarity in discussion, the two types of internal stresses
are treated separately in much of this chapter, it is well to remember that
the effects of the two are additive. In one sense, LR stress can be viewed as
a bias stress that changes magnitude with location rather gradually, superimposed on SR stress, which changes rapidly with location. One may be large
in magnitude and the other small, or vice versa. Shown schematically in Fig.
6.1 are several examples. The pattern shown in Fig. 6.1(a) might be that of a
heavily cold drawn or stretched bar. Here the LR stress is virtually nonexistent,

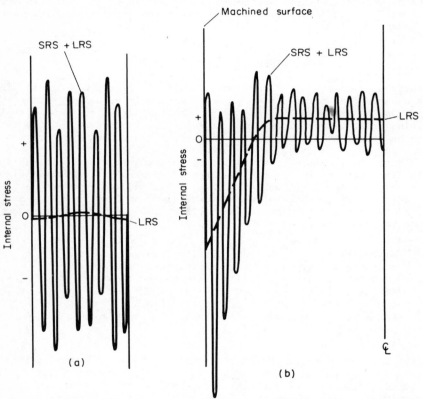

Fig. 6.1.   Schematic representation of short-range
             stresses (SRS) and long-range stresses (LRS):
             (a) plastically stretched bar, and (b)
             annealed bar with machined surface.

but the SR stress is large because of dislocation pile-ups against internal
barriers. A similar pattern would be expected in an annealed-plus-slowly-cooled
polycrystalline material in which neighboring microstructural elements have
different thermal expansion characteristics. The pattern in Fig. 6.1(b) might

be that associated with an annealed specimen with a machined surface. In this
case, both SR stress and LR stress are relatively large near the surface and
small in the interior.

After a brief description of the sources of SR stress in the next section,
much of the remainder of the chapter is devoted to residual stress or LR
stress—its origins and possible ways that it can be altered, either intentionally prior to placing a component in service, or unintentionally during service.
Particularly in the discussions concerning stress alteration by thermal treatments or mechanical vibration, it will be well to remember that SR stresses
can play a very important role in altering LR stress. Indeed, it is only
through microscopic plastic deformation that LR alteration can occur at all
during these treatments. This requires that the short-range force fields
among dislocations and dislocation barriers must gradually achieve a new
balance.

Prior to describing sources of internal stresses, several general points
are worthy of emphasis. First, the presence of internal stress does not
necessarily mean that a part will be dimensionally unstable during service.*
A good example of this is hardened steel gage blocks. Even though they may
contain high residual stress levels as a result of heat treatment, they generally exhibit excellent dimensional stability over periods of many years at
room temperature. Nonetheless, the potential for instability is always
present. Anything that acts to alter the stress pattern will cause dimensional
change, as demonstrated by Eggert (1964). He electropolished about 0.25 mm
(0.010 in) from one side of several commercial gage blocks of AISI 52100 steel,
as shown in Fig. 6.2. This caused the blocks to decrease in length about 200

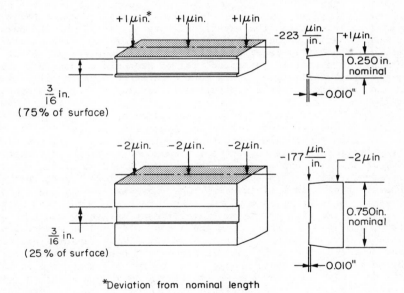

*Deviation from nominal length

Fig. 6.2.   Effects of electropolishing on the length
of experimental gage blocks (Eggert, 1964).

---

*In certain cases, internal stresses may be desirable, to enhance other aspects
of material performance (Van Horn, 1953).

ppm along the etched side, enough to render them useless even for the calibration of hand micrometers. Of course, it is not common practice to etch gage
blocks, but it is not inconceivable that a machinist might etch initials or other
identification symbols into a particular set of blocks. Similar, though
smaller, effects would be anticipated from gradual abrasion and corrosion that
might occur in service.

Secondly, most discussions of internal stress, including the one presented
in this chapter, are considerably oversimplified. For example, it is common
to refer to a part as containing "an internal stress". In fact, it is not
possible to speak in terms of one specific value of internal stress in a part
or specimen. A distribution of stresses is necessary to satisfy the requirements of static equilibrium—the forces in any direction and the moments about
any point must equal zero. Furthermore, it is important to remember the three-
dimensional aspects of internal stress. In real parts, even those with simple
shapes, it is a tedious, experimentally difficult, time consuming, and costly
job to measure LR internal stresses and their distribution. Consequently,
simplifying assumptions are often made to reduce the difficulty of making
measurements.

For example, if it is assumed that the residual hoop stresses in the wall
of a drawn tube vary linearly from a maximum tensile value at one surface to
a maximum compressive value at the other, the maximum values can be obtained
merely by noting the expansion or contraction of the diameter when the tube
is slit longitudinally, and fitting these values into an appropriate formula
(Dieter, 1961). By way of contrast, measurement of the actual three-dimensional stress distribution through the wall requires incremental removal of
material from the tube wall with concomitant precise measurement of longitudinal
and circumferential strain. Such experiments require extremely careful
attention to details. Unless the experiments are automated and the data
collection and reduction computerized, the analysis of the stress distribution
in a single tube can easily require several days for completion. Often, such
measurements simply cannot be justified from an economic standpoint. Consequently, where internal stresses are mentioned in connection with real parts,
more often than not the stress distribution is assumed rather than measured
experimentally.

A vivid example of the complexity and importance of internal stress distribution has been discussed by McClintock and Argon (1966). Coil springs made
from wires are known to contain residual stresses from the coiling operation.
Thermal stress relief treatments frequently cause the springs to change shape.
Intuitively, it would be expected that the springs would partially uncoil as
a result of stress relief. In springs made from stainless steel, this type of
behavior was observed. In music wire springs, however, the springs tended to
coil more tightly when stresses were relieved. This phenomenon was attributed
to a difference in the sign of the maximum residual stress in the two cases.
This, in turn, was shown by McClintock and Argon to depend on material properties
and the ratio of wire diameter to spring diameter. Under appropriate conditions,
either material could be made to open or close as a result of stress relief.
Thus, experiments in which shape change of a specimen is monitored as an indicator of the degree of internal stress relief can be very misleading. In
fact, in certain instances it is possible, though not probable, for appreciable
stress relief to occur with essentially no change in shape or dimensions.

6.2 SOURCES OF SHORT-RANGE INTERNAL STRESS

Two major sources of SR stress are plastic deformation and unequal thermal
expansion of neighboring microstructural elements.

In plastic deformation, SR stress can arise among neighboring crystals or between one phase and another because of unequal amounts of deformation. For example, in a single phase material, certain crystals (grains) are oriented more favorably for plastic flow than are neighboring grains. Thus, after the external deforming forces are removed, a balanced force field or SR internal stress will exist between neighboring grains. A similar effect would be expected in a two-phase material in which one of the phases is more resistant to plastic deformation than the other.

Plastic deformation also can produce SR stress within individual grains. As discussed in Chapter 5, as dislocations move and multiply under the action of an external force, they pile up against barriers, such as grain boundaries, precipitate particles, and other dislocations, and a back stress is developed whose magnitude increases with every dislocation added to the pile-up. When the external force is removed, the back stress acts to return the dislocations toward their source. If plastic deformation has proceeded beyond the micro-strain range, it is likely that barriers, e.g., intersecting dislocations, that act to impede the reverse motion of the piled-up dislocations will also be present. Thus, a SR force balance will be established and the extent to which the dislocations return toward their source will be governed by these forces.

A second source of sizable SR stress is nonhomogeneous thermal expansion. Nonhomogeneous expansion can occur in microstructures of several types:

- matrix phase with dispersed precipitate particles or inclusions;
- two-phase material;
- single phase in which the individual crystals of a polycrystalline microstructure exhibit anisotropic thermal expansivity.

In any of these structures, changing the temperature, either up or down, will cause stresses to be developed among neighboring microstructural elements. In fact, SR internal stresses are probably present even before imposing a temperature change. Thus, changing the temperature will merely superimpose an additional set of stresses. The magnitude of the internal stresses will depend on the temperature change, the differences in expansivity, and the elastic moduli. If the stresses become large enough, microscopic yielding or cracking may occur. Under certain conditions—for example, a modest thermal cycle—it is possible that these stresses are reversible; i.e., upon return to the original temperature, they will be the same as they were originally. Where plastic deformation or cracking occur, however, irreversible changes in SR stress accompany thermal excursions.

Laszlo (1943) was one of the first to attempt calculation of the stresses, sometimes called tessellated stresses, that develop around inclusions with temperature change. Later, Brooksbank and Andrews (1968) measured the thermal expansion coefficients of several types of inclusions found in 1%C-Cr steels. In subsequent papers (Brooksbank and Andrews, 1972; Andrews and Brooksbank, 1972), they reported the use of this expansivity data to compute the stress fields developed around inclusions during heat treatment of steel. They concluded that certain types of inclusions, notably oxides, including calcium aluminates, alumina, and spinel, because of their relatively small expansivity compared with that of steel, could create stresses sufficiently large to cause plastic deformation or microcracking of the matrix. Similar effects would be expected around precipitate particles in precipitation hardened alloys. Generally speaking, however, the relative expansivities of the matrix and the precipitate are not known, thus precluding quantitative estimates of induced SR stresses.

With respect to thermal expansion anisotropy in a single-phase material, such behavior is characteristic of materials that possess noncubic crystal structures. Davidenkov et al. (1960) and Likhachev (1961) examined the problem theoretically to estimate the magnitude of the SR stresses developed as temperature is changed. In one series of calculations in which account was taken of both expansion anisotropy and elastic anisotropy, the stresses developed at the boundary of a bicrystal, oriented to produce maximum thermal expansion mismatch, were computed. The results are shown in Table 6.1 for a variety of noncubic materials, arranged in order of descending stress magnitude. There is little question that in many noncubic materials SR stresses due to expansion anisotropy can reach high levels, even for relatively small temperature excursions.

TABLE 6.1 Maximum Values of the Microstructural Strains
due to Thermal Anisotropy (Likhachev, 1961; Davidenkov *et al*, 1960)

Material	Spatial lattice	psi/deg C	kN/m^2/deg C
Uranium (2 modifications):			
α	Monoclinic (orthorhombic)	360	2490
β	Tetragonal	284	1960
Selenium	Hexagonal	218	1510
Zinc	Hexagonal	177	1220
Aragonite	Rhombohedral	174	1200
Calcite	Rhombohedral (trigonal)	165	1140
Cadmium	Hexagonal	91.4	630
β-Tin	Tetragonal (body-centered)	71.7	496
Tourmaline	Rhombohedral (trigonal)	71.6	494
Beryllium	Hexagonal	63.8	441
Tellurium	Hexagonal	45.7	315
Quartz	Rhombohedral (trigonal)	43.1	298
Zirconium	Hexagonal	38.5	266
Indium	Tetragonal (face-centered)	32.6	225
Antimony	Rhombohedral (body-centered) (trigonal)	21.3	147
Bismuth	Rhombohedral (body-centered) (trigonal)	9.35	65
Magnesium	Hexagonal	2.74	18.9
Graphite	Hexagonal	0.944	6.5

Of particular interest in connection with design of precision devices are the values shown for beryllium in Table 6.1. As discussed in Chapter 5, many grades of beryllium undergo microplastic flow at stresses of the order of 14 MN/m^2 (2 ksi). Intergranular stresses of this magnitude could be reached simply by changing the temperature up or down approximately $30°C$ ($55°F$).

The computed stresses in Table 6.1 are maximum values based on maximum mismatch between adjacent crystals. In an actual polycrystalline material, each grain is surrounded by others having different crystallographic orientations. Hence, on the average, the intergranular stresses will be less than those shown in Table 6.1. Attempts to calculate average or mean stresses developed as a result of thermal expansion anisotropy have been reported by Likhachev (1961). Assuming that an anisotropic grain is surrounded by a matrix which possesses isotropic properties equivalent to those of a polycrystalline body of the same material, Likhachev's estimates of average intergranular stresses for several materials are shown in Table 6.2. Comparison of these values with those in Table 6.1 indicates that the average stresses are only 1/3 to 1/2 the maximum stresses. These average stress values are somewhat greater than those estimated by Davidenkov et al. (1960) for hexagonal polycrystals, where average stresses were computed to be perhaps 1/8 to 1/6 the maximum stresses. Even the average SR internal stresses due to thermal anisotropy, while appreciably less than the maximum values, are seen to be significant in a number of noncubic materials for relatively modest temperature excursions.

TABLE 6.2 Microstructural Strains due to Thermal
Anisotropy in Polycrystals

(Likhachev, 1961; copyright American Institute
of Physics, used by permission)

Material	$\sigma_\perp$		$\sigma_\parallel$	
	psi/deg C	kN/m²/deg C	psi/deg C	kN/m²/deg C
Zinc	69.7	482	−36.8	−255
Calcite	49.6	343	−43.1	−298
Cadmium	44.6	309	−18.0	−124
Tin	21.0	145	−37.6	−260
Tourmaline	17.4	121	−27.9	−193
Quartz	−1.36	−9.3	11.0	76
Antimony	7.48	51.8	−5.20	−36
Bismuth	2.50	17.3	−2.06	−14.2
Magnesium	0.80	5.53	−1.53	−10.6

6.3 SOURCES OF RESIDUAL STRESS

Residual or LR stresses result primarily from fabrication processes—forming, heat treating, welding, machining, plating, etc. They are difficult to avoid during fabrication and can be minimized only by carefully designed processing steps. In this section, various sources of residual stress are described, including thermal treatments, surface treatments, nonuniform cold work, machining, and joints and attachments.

Thermal Treatments

Whenever a body is heated or cooled in such a way that temperature gradients exist, elastic stresses, usually called thermal stresses, will be developed because of nonuniform thermal expansion or contraction. These thermal stresses will persist as long as the temperature gradient exists and their magnitude will depend on the temperature gradient, the thermal expansion coefficient, and the elastic moduli. If these stresses reach sufficient magnitude, the yield strength of the material will be exceeded in certain areas and plastic flow will occur in these regions. As noted by Dieter (1961), residual stresses are produced whenever a body undergoes nonuniform plastic deformation. Thus, in these cases, even after the thermal gradient disappears, residual stresses will be present. On the other hand, if the thermal gradients cause thermal stresses whose magnitudes are insufficient to cause plastic deformation, no residual stresses will result.

Consider, for example, the cooling of a long stress-free metal bar from a high temperature to room temperature. Unless cooling is extremely slow, the outside of the bar will cool faster than the center. Consequently, because of unequal contraction, stresses will be developed in the bar during cooling, tensile at the outside and compressive in the interior. Assume that the center, which is at a higher temperature than the exterior, undergoes plastic flow when the thermal stress becomes great enough, while the exterior remains elastic. Then, as the temperature of the entire bar approaches room temperature, the thermal stresses will diminish. However, because the total contraction of the center in the later stages of cooling will be greater than that of the surface, due to both cooling and plastic deformation, residual stresses will be present at room temperature. These will be compressive at the surface and tensile in the interior. Figure 6.3 shows the type of three-dimensional residual stress pattern resulting from cooling.

The magnitude of the residual stresses developed depends on many factors, including the mechanical and physical properties of the workpiece and several external factors:

Mechanical Properties
Yield strength as a function of temperature
Modulus of elasticity

Physical Properties
Thermal expansion coefficient
Thermal conductivity
Specific heat
Density

External Factors
Section thickness
Temperature range of cooling
Rate of heat removal from surface.

It is not uncommon to find residual stresses with maximum values equal to the yield strength in rapidly cooled parts. On the other hand, if cooling is more gradual and the other controlling factors act to keep the thermal stresses below the yield strength at all times, no residual stresses will be introduced by the cooling operation.

The degree to which residual quenching stresses can be influenced by several external factors is illustrated in Fig. 6.4 for 7075 aluminum plates quenched

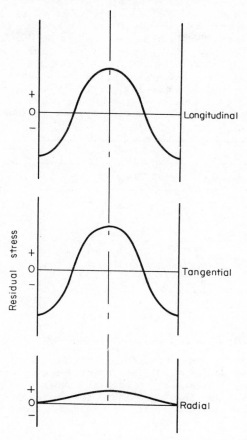

Fig. 6.3.   Schematic illustration of residual stress
patterns in cylindrical bars cooled rapidly
from elevated temperatures.

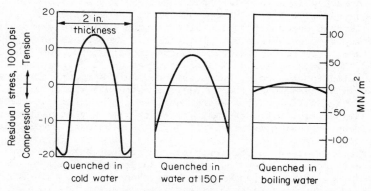

Fig. 6.4a.   Residual stresses in 7075-T6 aluminum plate
specimens quenched in water at different
temperatures (Barker and Sutton, 1967;
copyright ASM, used by permission).

from about 480°C (900°F). At a thickness of 5 cm (2 in), changing the temperature of the water quench from about 20° to 100°C (68° to 212°F) reduced the
maximum longitudinal residual stress by a factor of about 10. Reducing the
section thickness from 5 cm to 0.63 cm (2 to 0.25 in) also reduced the maximum
residual stress, by a factor of about 3.

By reasoning similar to that above, it can be shown that residual stresses,
of opposite sign to those shown in Fig. 6.3, may be introduced by heating.
If, however, heating continues to a temperature where the yield stress is very
low and plastic flow and creep occur readily, the residual stresses will

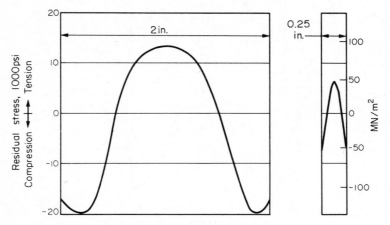

Fig. 6.4b. Residual quenching stresses developed in
            alloy 7075 plate specimens of different
            thickness. Specimens quenched in cold water
            and then aged to T6 temper (Barker and
            Sutton, 1967; copyright ASM, used by
            permission).

quickly vanish. An interesting application of this effect is sometimes employed
in aluminum alloys to reduce the stresses introduced during quenching from the
solution anneal temperature. It is referred to as an "uphill quench". After
the part is quenched in cold water from about 480°C (900°F), it is further
cooled in liquid nitrogen to -196°C (-320°F) in preparation for the uphill
quench. It is then removed from the nitrogen and immediately rapidly reheated
by a high velocity steam blast on all surfaces. The effect of this treatment
on a 7075 plate, 5 cm (2 in) thick, is shown in Fig. 6.5. Note the large
reduction in the range of residual stresses that existed after quenching.

The discussion to this point has considered only symmetrical cooling or
heating. In such cases, even though residual stresses may develop, little
bending or warpage will result. Where cooling or heating is unsymmetrical
(see, for example, Fig. 6.6), thermal gradients will give rise to bending
moments, thereby producing warpage. If any of the warpage involves plastic
strain, permanent distortion will result.

A further complication in the analysis of residual stresses developed by
cooling or heating occurs when the material undergoes a phase transformation
as the temperature changes, accompanied by a change in volume. The best known
example of this occurs in hardenable steel, as discussed in greater detail in
Section 7.4. When a steel workpiece is rapidly cooled from an elevated temperature during the hardening operation, thermal stresses develop because the

exterior contracts at a greater rate than does the interior, just as described
earlier. However, at some temperature in the cooling cycle, the steel begins
to transform from austenite to martensite, with a concomitant increase in
volume that may exceed the contraction due to cooling. The magnitude of
the volume increase depends on steel composition, especially carbon content.

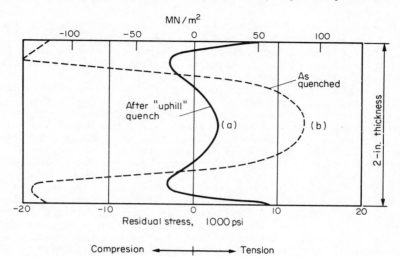

Fig. 6.5.   Response to Uphill-Quench treatment of 7075
             aluminum plate specimen quenched in cold
             water (Barker and Sutton, 1967; copyright
             ASM, used by permission).

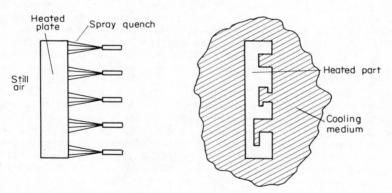

Fig. 6.6.   Examples of nonsymmetrical cooling of heated
             bodies.

As a result of this transformation, the material behaves essentially as one
that expands, rather than contracts on cooling. This produces a residual
stress pattern at room temperature that is the reverse of that described
earlier. The exterior of the workpiece is in tension and the interior is in
compression. When the tensile stresses are sufficiently great, the fracture
strength of the steel may be exceeded and surface cracks appear, particularly
near stress concentrations.

The effects of residual stresses due to hardening, such as cracking, distortion, and potential dimensional instability are a continual problem in steel heat treatment. Accordingly, numerous schemes have been devised to minimize these stresses or to alter their distribution in such a way as to make them less harmful. Some of the methods employed are described below:

(a) *Marquenching*

In this process the quench from an elevated temperature is interrupted at an intermediate temperature, just above that at which the transformation to martensite begins. This permits equalization of interior and exterior temperatures and partial relief of the residual quenching stresses. Once temperatures are equalized, the workpiece is permitted to cool relatively slowly to room temperature during which time it will transform to martensite. Inasmuch as the transformation occurs in the interior and exterior at the same time, the high residual stresses observed in normal hardening, viz., tension at the surface, compression in the interior, will not develop. However, interrupted cooling to alter the residual stress pattern may produce undesirable secondary effects, through a phenomenon known as austenite stabilization. When steels are slowly cooled through, or held isothermally in, a certain temperature range, the austenite phase undergoes stabilization, i.e., it transforms to martensite less readily than would normally be expected. Thus it is commonly observed that rapidly quenched specimens will retain less austenite than specimens cooled more slowly or specimens whose quench is momentarily interrupted. As discussed in Section 7.4, this can be an important consideration from a dimensional stability standpoint.

(b) *Marstressing**

This process involves through-hardening steels whose surface layers are alloyed so as to lower their martensite-start, M_s, and martensite-finish, M_f, temperatures to the extent that the austenite-to-martensite transformation will occur in the interior prior to its occurrence in the exterior. This requires also that the temperature gradients during cooling be reasonably small. When the surface layers finally transform with accompanying expansion, the already transformed interior inhibits the expansion, thus placing the surface layers in compression. This residual stress distribution is the reverse of that obtained in normal hardening.

(c) *Case hardening*

As the name implies, this involves a hard case and a relatively soft core. Two methods are commonly employed to achieve case hardening:

(1) *Surface carburizing*—a low-or medium-carbon steel is surface carburized to render the surface hardenable via transformation of austenite to martensite. Upon cooling from the austenitizing temperature, the low-carbon, nonhardenable core transforms to a ferrite-cementite aggregate at a temperature near 700° C (1300° F) before the case reaches its M_s temperature. When the case finally transforms to martensite, it will be in residual compression, for the reasons described in connection with marstressing.

(2) *Surface heating*—a hardenable steel is heated rapidly by induction or flame so as to austenitize only the surface layers. Upon quenching, the case transforms to martensite but the core, not having been austenitized, undergoes no transformation upon cooling. As before, this gives rise to residual compression stresses in the case.

---

*Developed and patented by General Motors (Patent No. 3,117,041, Jan. 7, 1964).

The above procedures, as well as others, are normally employed to minimize quench-cracking and distortion problems associated with hardening, as well as to improve fatigue behavior. Where these treatments reduce the severity of the residual stress distribution, they should be helpful also in promoting dimensional stability. However, when they merely reverse the sign of the residual stresses, little benefit is to be expected from a dimensional stability standpoint.

Surface Treatments

Several types of surface treatments can produce residual stresses in a workpiece or alter an existing residual stress pattern. Surface treatments in which foreign atoms or ions are diffused into the surface layers of a material will produce residual stresses if the specific volume of the chemically altered zone differs from that of the parent material. This principle is used in strengthening glassware by the Chemcor* process. In this process, some of the surface ions are replaced by other ions which, after heat treatment, produce a new phase of greater volume at the surface. As a result, high compressive residual stresses, of the order of 700 MN/m^2 (100 ksi), are built up in a thin surface layer of the glass.

Similar effects can be produced in steels by nitriding or carburizing. Nitriding is accomplished by heating in an ammonia atmosphere to allow nitride formation at temperatures below those at which transformation to austenite occurs. Compressive surface residual stresses of the order of 1000 MN/m^2 (150 ksi) can be produced in this way, along with appreciable hardening of the surface. Carburizing can be accomplished effectively only at temperatures above those at which transformation to austenite occurs. Consequently, the residual stress pattern after cooling from the carburizing temperature will depend on the details of the processing sequence. Generally, however, the surface residual stresses will be compressive.

Reverse effects may occur if certain elements are removed from a material's surface. For example, decarburization of steel during exposure to elevated temperatures in an oxidizing atmosphere can give rise to tensile residual stresses in a thin surface layer.

Residual stresses can result also when one material is deposited on, rather than diffused into, another material, as in electroplating. According to Kushner (1962), stress levels in the range from 700 to 1400 MN/m^2 (100 to 200 ksi) may build up in the plated layer.

Residual stresses in electrodeposits may be either compressive, e.g., cadmium and zinc, or tensile, e.g., nickel and chromium. The origin of these stresses is not well understood. However, it is known that their magnitude depends on numerous factors, including (a) nature of the substrate surface, (b) bath composition, including type of anion, organic additions, and presence of contaminants, and (c) plating conditions, including bath temperature, current density, and imposition of alternating current on direct current. Table 6.3 shows the relatively large effect of anion type on the magnitude of residual tensile stresses in nickel deposited on copper. The relatively low stress associated with a sulfamate solution can be reduced even further to only a few percent of that shown in Table 6.3 by operating the plating bath at 60° C (140° F) (Kushner, 1962).

---

*Chemcor is a tradename of the Corning Glass Company.

TABLE 6.3 Residual Stress in Nickel deposited from Various Solutions (Kushner, 1962; copyright ASM, used by permission)

Solution anion	Residual tensile stress at indicated plating thickness											
	0.0025 mm (0.0001 in)		0.0051 mm (0.0002 in)		0.0076 mm (0.0003 in)		0.0127 mm (0.0005 in)		0.025 mm (0.001 in)		Limiting Value [a]	
	MN/m^2	ksi	MN/m^2	ksi	MN/m^2	ksi	MN/m^2	ksi	MN/m^2	ksi	MN/m^2	ksi
Sulfamate	110	16.0	83	12.0	73	10.6	66	9.6	59	8.6	59	8.5
Bromide	141	20.4	120	17.5	108	15.7	92	13.4	78	11.3	76	11.0
Fluoborate	179	26.0	141	20.5	128	18.6	123	17.8	118	17.2	117	17.0
Sulfate	193	28.0	187	27.2	172	25.0	165	24.0	159	23.1	157	22.8
Chloride	255	37.0	229	33.2	228	33.1	228	33.0	227	33.0	227	33.0

[a] Extrapolated and calculated values for steady state condition. Temperature 25°C (77°F); current density 325 A/m2 (30 A/ft2). All solutions 1 M in nickel, 0.5 M in boric acid; pH 4; copper substrate.

Nonuniform Cold Working

As noted earlier in this chapter, nonuniform plastic deformation always results in residual stresses, regardless of the manner in which the plastic deformation is introduced. Several examples of mechanically produced nonuniform plastic strains are discussed in this section.

Perhaps the most prevalent procedure, intentionally or unintentionally, for introducing nonuniform plastic strain into a workpiece is plastic bending. Consider a rectangular bar deformed by bending. The convex surface is plastically strained in tension, the concave surface is plastically strained in compression, and the interior regions near the neutral axis experience no plastic strain. When the bending stress is removed, the three regions interact to produce a residual stress of the type shown schematically in Fig. 6.7.

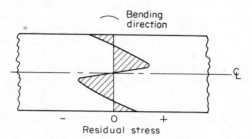

Fig. 6.7.   Residual stress distribution after plastic bending.

Metalworking operations are another frequent source of residual stress (Baldwin, 1949). Rolling, swaging, rod and tube drawing, and deep drawing, for example, can produce residual stresses of large magnitude because of nonuniform plastic strain in the workpiece. The stress at the surface may be either positive or negative and the stress magnitudes high or low, depending on reduction levels, type of tooling, lubrication, etc. In certain instances, compressive surface stresses are intentionally introduced by use of relatively light reductions in rolling or drawing. These tend to concentrate the plastic deformation (elongation) near the surface, thus leaving the surface in a state of residual compression, as illustrated in Fig. 6.8.

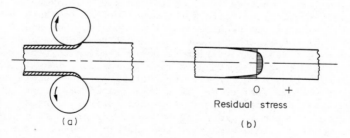

Fig. 6.8.   (a) Nonhomogeneous plastic deformation during rolling, and (b) resulting residual stress pattern.

A method frequently employed for intentionally producing residual compressive stresses at the surface of a metal part is shot peening. The prin-

ciple is the same as for light reductions by cold rolling. Peening plastically elongates the surface layers. The substrate, not being cold worked,
restricts the elongation of the surface layers and places them in residual
compression. Lessells and Broderick (1956) have reported residual compressive stresses of approximately 970 MN/m^2 (140 ksi) near the surface of a
shot peened steel specimen. Koster et $al.$ (1970) reported that the residual
stresses near the surface of hardened AISI 4340 steel could be changed from
700 MN/m^2 (100 ksi) tension after abusive grinding to 1100 MN/m^2 (160 ksi)
compression after shot peening. Additional details of the effect of peening
variables have been discussed by Campbell (1971).

Machining

The term machining is used here to refer to operations in which material is
removed from a workpiece to produce the desired shape and dimensions. It
includes traditional processes, such as milling and grinding, and nontraditional
processes, such as electrochemical machining and laser beam machining.
In line with earlier discussions, whenever a machining operation produces
nonuniform plastic deformation in a workpiece, residual stresses will be present. Such nonuniform deformation is most likely to be present in the traditional processes involving mechanical removal of material. However, it may
be present in certain nonmechanical removal methods as well, particularly
where large thermal gradients occur.
The presence of plastically deformed layers at the surface of metals in
which machining is accomplished mechanically has been amply demonstrated
(Samuels, 1956; Terminasov and Yakhontov, 1959; Graham and Rubenstein, 1966;
Turley, 1968, 1971, 1973). For example, Turley (1971) examined the dislocation
substructure in the surface regions of machined 70/30 brass, employing transmission electron microscopy. He noted a similarity between the structure
associated with machining and that associated with cold rolling. Employing
this correlation, he estimated the plastic strain gradient beneath the machined
surface. Some of his results are shown in Fig. 6.9, and illustrate the effect
of depth of cut. Notice that, regardless of the depth of cut, a fragmented
zone of high plastic deformation was observed just below the machined surface,
equivalent to rolling reductions of over 90%. Below this fragmented
zone, a less heavily deformed zone was observed to extend to a considerable
depth, which depended on the depth of cut.
In view of Turley's observations, the presence of residual stresses after
machining would certainly be predicted. Further, if the assumption is made
that the surface deformation acts to elongate the surface layers relative to
the substrate, then the residual stresses in the surface layers should be
compressive, just as described previously for surface rolling and shot peening.
There is at least one complication, however. Metal removal by cutting or
abrasion can generate sizable quantities of heat, leading to rapid heating of
the surface. This rapid heating can give rise to residual tensile stresses at
the surface. Thus, depending on the various details of the machining operation,
the residual stress at the surface may be either tensile or compressive and
may exhibit a reversal at some distance below the surface. The situation is
complicated further if the heat generated by machining produces a phase transformation with attendant volume change in the surface layers. This can occur,
for example, in steels, as has been demonstrated by Koster et $al.$ (1970) for
grinding of AISI 4340 steel.
Figure 6.10 shows the stress distribution in AISI 4340 steel after grinding
under various conditions. Note the strong influence of grinding conditions
on both the maximum residual stress and the depth to which high stresses exist.
In subsequent metallographic examination of the ground surfaces, it was found

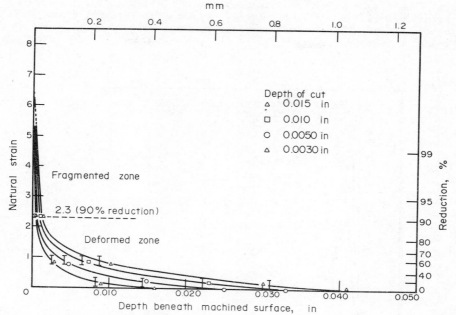

Fig. 6.9. Effect of depth of cut on the strain distrib-
           ution beneath a machined surface in 70/30
           brass. Single cuts made at 0° rake angle
           (Turley, 1971; copyright Inst. of Metals,
           used by permission).

that a layer of fresh martensite, approximately 0.025 mm (0.001 in) thick was
present after abusive grinding (Koster, 1973). This meant that this layer had
been heated to a temperature in excess of about 700° C (1300° F). In conventional
grinding, spotty patches of fresh martensite were detected, while in gentle
grinding, no surface alterations were evident.

     Because of the possible interaction of mechanical deformation, rapid heating,
and transformation, the types of residual stress distribution at a machined
surface are almost limitless. Several examples of the effects of machining
variables on residual stress distribution in commercially important materials
are discussed in the following paragraphs.

     In contrast to the stress distribution shown in Fig. 6.10 for surface
grinding of AISI 4340 steel, Fig. 6.11 shows the stress distribution for face
milling the same material. The effect of varying the degree of tool sharpness
is also illustrated. Notice that, even though the predominant stress is compressive, relatively high tensile stresses are developed in a thin layer near
the surface when the tool is dull. Figures 6.12 and 6.13 show the results of
similar experiments for 250 grade maraging steel and Ti-8Al-1Mo-1V, respectively.

     Decneut and Peters (1973) have examined the effect of feed rate, work
speed, and wheel speed on residual stresses produced in grinding a steel
designated as 100 Cr 6, similar to AISI 52100. Their results are shown in
Figs. 6.14 and 6.15. In all cases, the predominant stresses are tensile.
The severity of the stresses is increased by (1) increasing the feed rate, (2)
decreasing the work speed, and (3) decreasing the wheel speed. A somewhat

different wheel speed effect has been reported by Zlatin *et al.* (1963) for grinding of Inconel 718 and Ti–8A1–1Mo–1V, as shown in Figs. 6.16 and 6.17, respectively. Here, the residual stresses appear to become more severe as wheel speed is increased.

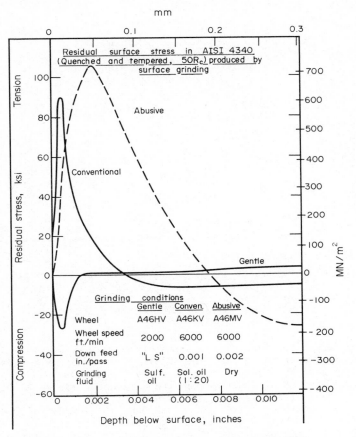

Fig. 6.10.   Residual surface stress in AISI 4340
(quenched and tempered, 50 R_c) produced by
surface grinding (Koster *et at.*, 1970).

   Many more examples could be given of the effect of variables on residual stress distributions associated with mechanical material-removal operations. The type of tool, its sharpness, the type of lubricant or coolant, the metal-lurgical condition of the workpiece, etc., will each have an effect.   The interested reader is referred to a paper by Kahles and Field (1971) in which the general area of machining effects on surface integrity is discussed.   The paper also cites numerous additional references pertaining to residual stresses arising from machining.

   Bellows (1973) has reviewed various nontraditional machining processes, most of which are nonmechanical. He classifies them according to the prin-cipal energy mode involved:

Thermal Processes
   EDM - Electrical Discharge Machining
   LBM - Laser Beam Machining
   PAM - Plasma Arc Machining
   EBM - Electron Beam Machining
Chemical Processes
   CHM - Chemical Machining
   ELP - Electropolishing
Mechanical Processes
   USM - Ultrasonic Machining
   HDM - Hydrodynamic Machining
   LSG - Low Stress Grinding
   AJM - Abrasive Jet Machining
Electrical Processes
   ECM - Electrochemical Machining
  STEM - Shaped Tube Electrolyte Machining
   ECG - Electrochemical Grinding

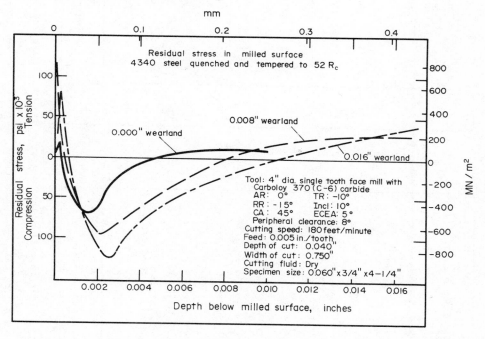

Fig. 6.11.   Residual stress in milled surface; 4340 steel;
             quenched and tempered to 52 R_C (Field and
             Koster, 1968; copyright SME, used by
             permission).

The thermal processes produce heat-affected surface layers and in most
cases surface melting and resolidification. As a result, residual tensile
stresses are frequently formed at the surface, as shown in Fig. 6.18 for
electrical-discharge machined Inconel 718. Chemical processes, on the other
hand, induce no significant residual stresses at the surface. Figure 6.19
shows an electropolished AISI 4340 workpiece to be virtually free of residual
stress.

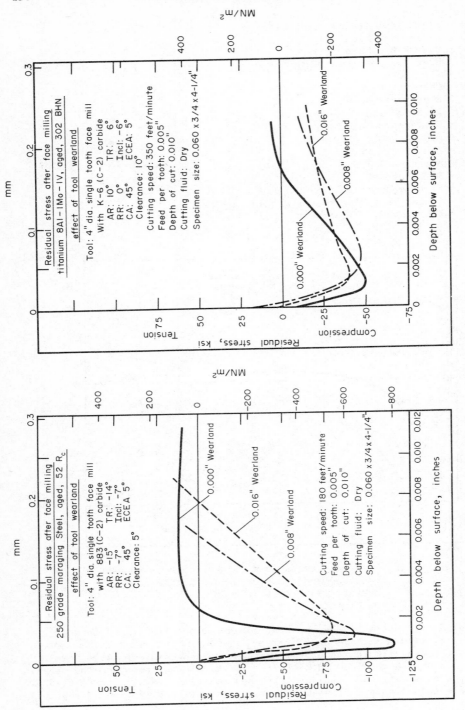

Fig. 6.12. Residual stress after face milling 250 grade maraging steel, aged, 52 R_C; effect of tool wearland (Field and Koster, 1968; copyright SME, used by permission).

Fig. 6.13. Residual stress after face milling titanium 8Al–1Mo–1V, aged, 302 BHN; effect of tool wearland (Field and Koster, 1968; used by permission).

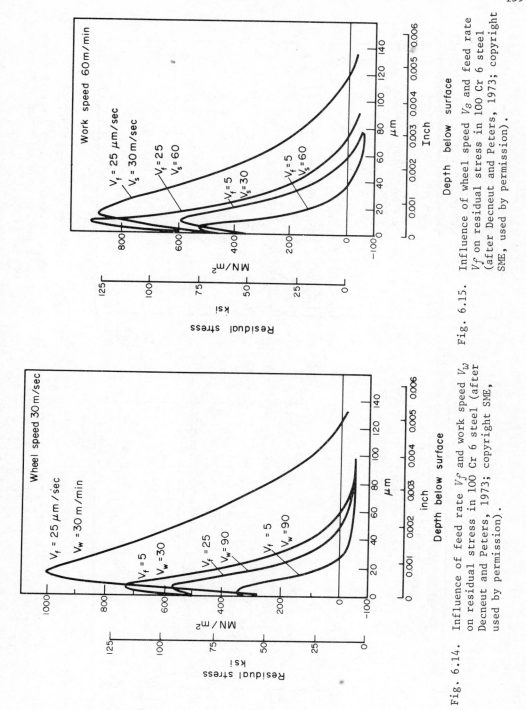

Fig. 6.14. Influence of feed rate V_f and work speed V_w on residual stress in 100 Cr 6 steel (after Decneut and Peters, 1973; copyright SME, used by permission).

Fig. 6.15. Influence of wheel speed V_s and feed rate V_f on residual stress in 100 Cr 6 steel (after Decneut and Peters, 1973; copyright SME, used by permission).

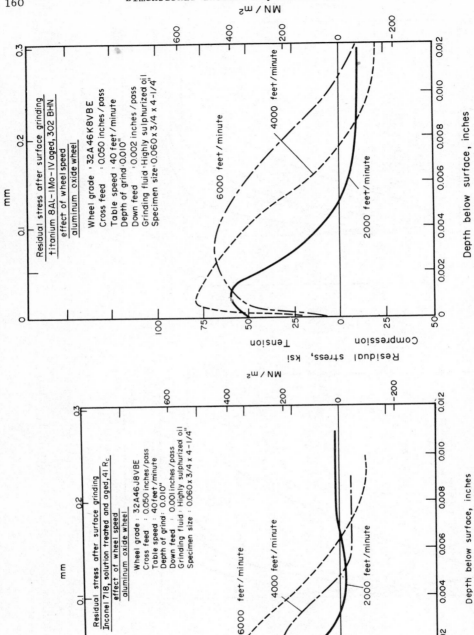

Fig. 6.17. Effect of wheel speed on residual stress after surface grinding titanium 8Al-1Mo-1V (Zlatin et al., 1963).

Fig. 6.16. Effect of wheel speed on residual stress after surface grinding Inconel 718 (Zlatin et al., 1963).

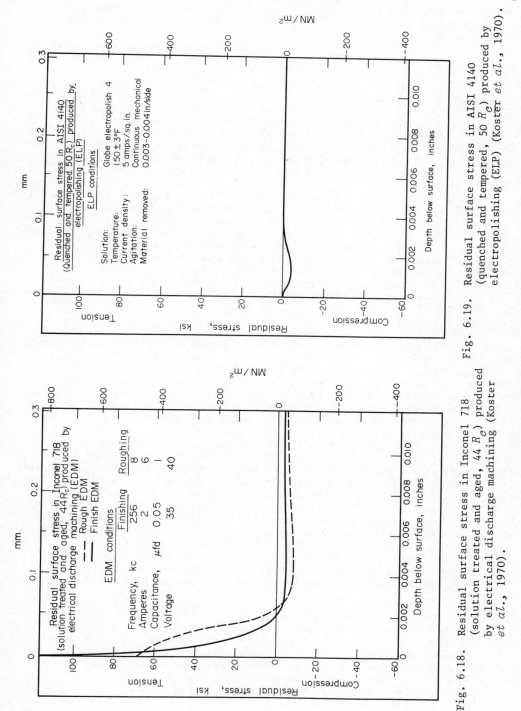

Fig. 6.19. Residual surface stress in AISI 4140 (quenched and tempered, 50 R_c) produced by electropolishing (ELP) (Koster *et al.*, 1970).

Fig. 6.18. Residual surface stress in Inconel 718 (solution treated and aged, 44 R_c) produced by electrical discharge machining (Koster *et al.*, 1970).

Of the nontraditional mechanical processes, only low stress grinding (LSG) has received significant attention from the standpoint of surface effects. LSG produces much less distortion and surface damage than does conventional grinding. Pricipal features of LSG are softer grinding wheels, lower in-feeds, and reduced wheel speeds with oil base cutting fluids. As noted previously in Fig. 6.10, the high tensile residual stresses associated with conventional grinding are often greatly reduced by LSG and, in some cases, the resulting stresses may even be compressive.

Electrical processes such as electrochemical machining, according to Bellows (1973), induce no significant residual surface stresses when properly controlled. For example, a 300 M steel specimen (0.4C–0.8Mn–1.6Si–0.9Cr–0.4Mo–1.9Ni), quenched and tempered to a hardness of 52 Rockwell C, exhibited stresses no greater than about 35 MN/m^2 (5 ksi) to a depth of 0.2 mm (0.008 in) after ECM. Some surface softening of about 2 points, Rockwell C, was observed, however.*

Summarizing briefly, machining processes that remove material by mechanical or thermal means are almost certain to leave residual stresses for some depth below the machined surface. Processes that involve chemical or electrical removal of material, on the other hand, are likely to produce virtually stress-free surfaces. From a dimensional stability viewpoint, the details of the stress distribution will have an important influence on the amount of distortion that occurs in the workpiece immediately upon release from the machining fixture, and on the potential amount of distortion that may occur if these stress patterns undergo alteration under the conditions experienced in service.

Joints and Attachments

Residual stresses are almost certain to be found at joints and attachments. These can arise from several sources, including misalignment, interference fits, tightening of threaded fasteners, and welding.

With the exception of welding, the origins of these residual stresses differ from those described earlier, where the residual stress resulted from non-uniform plastic deformation. No plastic deformation need be involved in misalignment, interference fits, and tightening of fasteners. The stresses develop as a result of elastic interactions between two or more members. For example, when two plates are fastened with a threaded bolt and nut, a tensile stress will be developed in the bolt and a compressive stress in the adjacent areas of the plates as the bolt is tightened. Similar effects will be present also in welded joints if external constraints are present. Although this type of stress may not fall within the generally accepted definition of residual stress, its importance in dimensionally critical applications can be equal to that of residual stresses described earlier. Specifically, alteration of the types of stress described here can lead to shape- and dimensional-changes just as readily as can alteration of other types of residual stress.

In welding, residual stresses develop as the weld metal and the adjacent heat-affected zone attempt to contract as they cool. They are prevented from doing so by the surrounding colder metal, with the result that residual tensile

---

*It is frequently observed that fatigue specimens prepared by stress-free machining (CHM, ELP, ECM, etc.) exhibit poorer fatigue behavior than do specimens prepared conventionally. This is generally attributed to slight surface softening and to the absence of residual compressive stresses at the surface. Such stresses are known to have beneficial effects on fatigue life and are often introduced purposely by shot peening or other methods.

stresses occur in and near the weld, balanced by residual compressive stresses
in surrounding regions. A typical idealized distribution is shown in Fig.
6.20 for a butt weld which shows stresses both parallel and transverse to the
weld. The magnitudes of the stresses developed and the actual stress distribution will depend on welding parameters, including welding current and speed,
rate of heat extraction, number of passes, etc. Digiacomo (1969) has reported
the results of residual stress studies on multipass, constrained butt welds
in 3.8 cm (1½ in) plates of a 0.1% C low-alloy steel. He found compressive
residual stresses of approximately 350 MN/m^2 (50 ksi) at the upper portion of
the welds and the heat-affected zone. The stresses reversed their sign at
about 1 cm (0.4 in) below the weld surface.

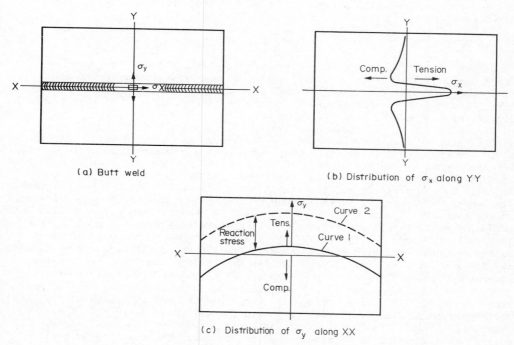

    (a) Butt weld   (b) Distribution of σ_x along YY

                         (c) Distribution of σ_y along XX

Fig. 6.20. Distribution of residual stresses in a butt
weld (Masubuchi, 1965).

6.4   ALTERATION OF INTERNAL STRESS

In this section, attention is given to various ways in which the internal
stresses existing in a component might be altered, with concomitant shape or
dimensional changes. Knowledge of the various ways of altering internal stress
is important from several standpoints in connection with dimensionally critical
applications. First, it will permit comparison of the conditions under which
internal stresses can be altered with conditions to be encountered in service.
This should reveal the likelihood of stress alteration in an actual operating
component that contains internal stresses. Second, it will permit judicious
selection of treatments to reduce the internal stress magnitudes in an operating component to acceptable levels prior to placing it in service.

The discussion is organized as follows. First, alteration of residual stress by a process known as stress relief or stress relaxation is considered. This is basically an effect that occurs internally as the elastic strains associated with residual stresses are replaced by plastic strains. However, external factors, such as temperature or mechanical vibration, may influence the kinetics of stress relief.

Next, alteration of residual stresses by material removal is considered. This is primarily an external effect in that surface layers containing residual stresses are physically removed, leading to a redistribution of the remaining stresses. Following this, mechanical working methods for altering LR internal stresses are discussed. Finally, several "after-effects" are described. These are associated with alteration of SR stresses that arise as a result of directional plastic prestraining.

Stress Relief

Stress relief is often referred to as stress relaxation. The terms are interchangeable. However, in common usage, stress relief usually refers to deliberate attempts to reduce residual stress levels while stress relaxation more often refers to nonintentional reduction of residual stress levels under conditions encountered in service.

Stress relief is a phenomenon closely akin to creep whereby an internal stress existing in a body is reduced in magnitude as the elastic strain responsible for the stress is replaced by plastic strain. Consider a simple example. Suppose a thin flat strip of steel is bent around a circular mandrel, such that the stresses induced in the strip are insufficient to cause microplastic strain. If the strip is released immediately, it will return to its original flat shape. However, if it is allowed to remain bent around the mandrel and at the same time heated to, say, 650° C (1200° F), the elastic strain will gradually be replaced by plastic strain. Upon release from the mandrel, the strip will no longer be flat. If the strip is clamped to the mandrel for a sufficient time at a high enough temperature, the strip will not spring back at all, i.e., the initial elastic strains will have become entirely plastic strains and the initial elastic stresses will have been completely relieved.

As already noted, there is a close relationship between stress relief and creep behavior. In a creep test under constant load, at an appropriate stress and temperature, a material will exhibit a gradual increase in strain with time (Fig. 6.21a). The total strain at any time will be the total of the elastic strain, which is constant, and the creep strain, which is gradually increasing. Stress relaxation, on the other hand, generally refers to a situation in which a stress or strain is imposed and the total strain remains constant (Fig. 6.21b). Thus, if some plastic strain (creep) gradually occurs, the total strain can be held constant only by decreasing the amount of elastic strain. Assuming no change in the modulus of elasticity, this reduction in elastic strain can be accomplished only by a reduction or relaxation of the imposed stress. As the stress is gradually relaxed, the tendency for further creep strain is diminished, leading to stress relaxation rates that decrease with time (Fig. 6.21b).

If stress relaxation is the result of gradual replacement of elastic strain by plastic strain, then it is perfectly logical to assume that stress relaxation should occur under any conditions for which creep can take place. In fact, it is possible to estimate stress relaxation behavior from knowledge of creep behavior. The word *estimate* should be emphasized. Freudenthal (1958), Goldhoff (1971), and Nadai (1963), among others, have pointed out some of the complications in predicting relaxation behavior from creep data. The principles involved in making such an estimate are illustrated schematically in Fig. 6.22.

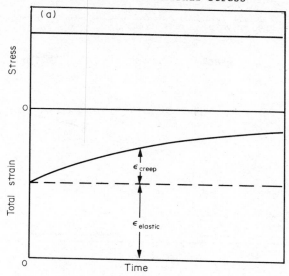

Fig. 6.21a. Creep behavior.

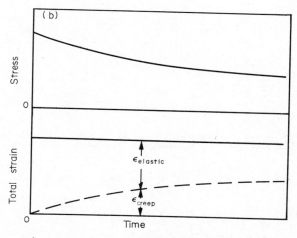

Fig. 6.21b. Stress relaxation behavior.

Assume that the creep behavior of a particular material at a given temperature is known for several levels of applied stress (Fig. 6.22a). If the initial applied stress is σ_1, then creep strain Oa will occur in time t_1. To maintain constant total strain, the elastic strain and, hence, the stress will have to be reduced to σ_2, (Fig. 6.22b), where $\sigma_2 = \sigma_1 - E\varepsilon_{Oa}$. At a stress of σ_2, creep will continue, but at a diminished rate as compared with that at σ_1. The time required to creep an amount bc, equal to Oa, will be t_2. Again, this will force the stress to be diminished to σ_3, where the creep rate is even lower than at σ_2. If this stepwise analysis of the creep curves is continued, an approximate stress relaxation curve can be derived (Fig. 6.22b).

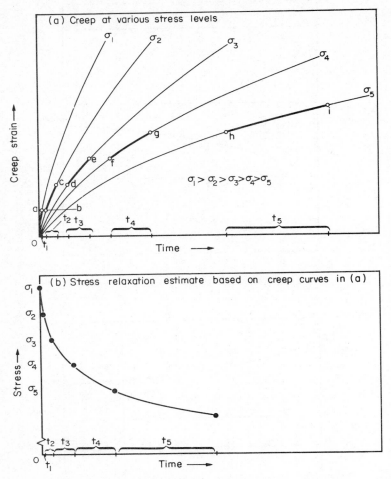

Fig. 6.22. Estimation of stress relaxation behavior from measured creep curves.

The analysis in Fig. 6.22 requires arbitrary selection of the step size. It would be preferable if the analysis could be done mathematically, employing an empirical expression for creep rate as a function of stress. As already noted, in a relaxation test

$$\varepsilon_t = \varepsilon_e + \varepsilon_c = \text{constant} \tag{6.1}$$

where the subscripts t, e, and c refer to total, elastic, and creep strains, respectively. Writing the elastic strains in terms of stress, σ, and elastic modulus, E, gives, after differentiation

$$\frac{d\sigma}{dt} = -E\dot{\varepsilon}_c. \tag{6.2}$$

Combining eq. (6.2) with an empirical expression for $\dot{\varepsilon}_c$ as a function of stress and integrating will then give the relationship between elastic stress and time.

One shortcoming of both the graphical and mathematical analysis is that they do not account for possible creep recovery. As discussed in Chapter 5, it is sometimes observed that a portion of the creep strain in a creep test is recovered with time after removal of the load. In fact, if the load is merely reduced in the course of a creep test, there may be a brief initial period during which the creep rate is abnormally low, or even negative, at the new stress level. This is characteristic of creep recovery. In stress relaxation, the stress is continually diminishing and, if recovery is present, the creep rate, and, hence, the relaxation rate, will tend to be less than that anticipated on the basis of constant stress creep data.

An additional shortcoming is the requirement for accurate constant stress creep data as a function of applied stress, and the reduction of this data into a suitable empirical expression for $\dot{\varepsilon}_c$ as a function of stress. Particularly for a new material, such data are not likely to be available. If such data were to be obtained experimentally in order to predict stress relaxation behavior, it might be a simpler matter to obtain the stress relaxation data directly.

Finnie and Heller (1959) have described various mathematical and graphical methods for predicting stress relaxation behavior from experimentally observed creep behavior, ignoring creep recovery. From comparison of the predicted behavior with experimentally obtained stress relaxation curves, they concluded that stress relaxation behavior can be predicted with reasonable accuracy if suitable empirical expressions for $\dot{\varepsilon}_c$ versus σ are available.

In the following paragraphs, the effects of certain variables on the phenomenon of stress relief are discussed.

A. *Effects of Temperature, Time, and Stress Level*

Stress relief is normally considered a thermally activated process, i.e., it is time dependent and its rate is increased as the temperature is raised. Thus, it is convenient to discuss temperature and time effects together. Similarly, the rate of stress relief is generally raised as the internal stress level is increased. Since temperature and stress act in a like manner, it is useful also to include stress in this discussion.

The role of temperature in deformation of solid materials was discussed in Section 5.1. The point was made with respect to microyielding and microcreep that temperature could have either a large or small effect, depending on the internal mechanisms operating to achieve plastic deformation. These mechanisms are apparently quite sensitive to material condition, i.e., when processed in one way, a given material may display strong temperature dependence of microyield behavior, whereas when processed differently, may exhibit virtual temperature independence. Likewise, the deformation mechanisms may be sensitive to the particular temperature range under consideration. For example, metal deformation is usually strongly temperature dependent above about 0.4 to 0.5 the absolute melting temperature, while at low temperatures, the effect may be appreciably smaller.

In discussing temperature effects on microyield and microcreep behavior of materials for precision devices, major consideration is normally given to temperatures not far removed from ambient. When considering stress relaxation, on the other hand, attention must also be directed toward higher temperatures, because these are the temperatures at which components are subjected to deliberate stress relieving operations. Some information on suitable temperatures and times to achieve substantial relief of residual stresses is available for many materials in various handbooks. In many cases, however, data are difficult to find, particularly for newer materials or for materials whose properties are developed by special processing sequences.

The importance of temperature in stress relieving at elevated temperatures is shown in Fig. 6.23 for a Ti-6Al-4V alloy. At 595°C (1100°F) the residual stress levels can be reduced to half their original values in about 1¼ hr. At 535°C (1000°F) approximately 5 hr are required to accomplish the same amount of stress relief and at 480°C (900°F) approximately 20 hr are required. Such a diagram is extremely valuable in selection of thermal treatments to reduce residual stress levels.

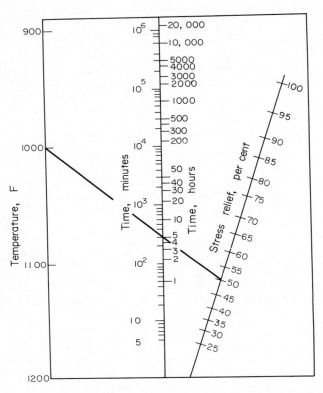

Fig. 6.23. Relationship of time, temperature and percent of stress relief for Ti-6Al-4V alloy (Maykuth, 1968).

Suppose, now, that information was needed on stress relaxation of a dimensionally critical Ti-6Al-4V component near room temperature and none was available. It would be tempting to replot the data contained in Fig. 6.23, or treat it mathematically, in such a way as to permit extrapolation to room temperature and small amounts of stress relief. Such an extrapolation would, of course, assume that the temperature dependence of stress relief remains unchanged between 480°C (900°F) and room temperature. Stress relaxation data can often be fitted to a time – temperature parameter, such as the Larson – Miller parameter,

$$P = T(C + \log t) \tag{6.3}$$

where T is temperature, usually expressed in degrees Rankine, C is an arbitrary constant determined by the data, and t is the time, usually expressed in hours.

In Fig. 6.24 the information from Fig. 6.23 has been replotted in terms of the
Larson - Miller parameter, using a value of 10 for the constant C to provide
reasonable agreement among the curves for different temperatures. The results
can now be examined in an attempt to obtain an estimate of stress relaxation
behavior near room temperature. If the curve in Fig. 6.24 is extrapolated to
the region where stress relaxation is very small (less than 1%), the value of
P from the graph is approximately 12 500. Assuming T is 70° F (530° Rankine),
t is found from eq. (6.3) to be

$$12\ 500 = 530\ (10 + \log t)$$
$$\log t = 23.6 - 10 = 13.6$$
$$t = 4 \times 10^{13}\text{hr}$$

Thus, extrapolation of the elevated temperature results indicates that times
of the order of one billion years would be required at room temperature to
get a small amount of stress relaxation in Ti-6Al-4V. Although little actual
room temperature stress relaxation data is available for this material, such
behavior is highly unlikely based on the observation of significant room temperature creep in this material (Odegard and Maringer, 1971; Wood, 1967; Hatch
et al., 1967; Reimann, 1971; Odegard and Thompson, 1973). Accordingly, it is
generally inadvisable to employ this method to estimate room temperature
behavior from elevated temperature measurements, both because of the extent of
the extrapolation required and the possibility of a different stress relaxation
mechanism operating at the lower temperatures.

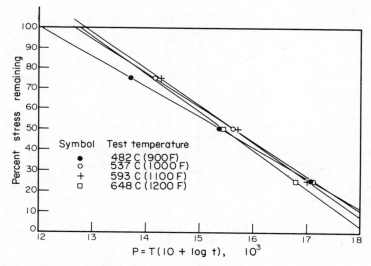

Fig. 6.24. Larson-Miller plot of Ti-6Al-4V relaxation
data from Fig. 6.23.

A safer method for predicting stress relaxation behavior near room temperature is from relaxation or creep data obtained in this temperature range.
Unfortunately, little information of this type is currently available. There
is, however, a growing awareness of the importance of such information in
numerous applications and some progress is being made. In addition to the
work referenced above on room temperature creep in Ti-6Al-4V alloy, Zeyfang
et al. (1971) have studied both creep and stress relaxation of unalloyed titanium near room temperature and the associated thermal activation parameters.

Okazaki and Conrad (1973) have conducted similar studies on Ti-N alloys.
Maringer *et al.* (1968) have reported room temperature microcreep data for
several commercially available alloys, from which stress relaxation can be
estimated.   In a limited investigation on 2014 aluminum, Marschall and Maringer
(1971) showed reasonable agreement between measured stress relaxation behavior
and that predicted from microcreep results.

Some of the results of Fox (1971) on stress relaxation of cold rolled
phosphor bronze strip near room temperature are presented in Fig. 6.25.   These
results leave little doubt of the occurrence of room temperature stress relaxation in this material.   Note particularly in Fig. 6.25 the strong effect of
test temperature—the results agree closely with the rule of thumb cited in
Section 5.1 for the effect of temperature on chemical reactions.   These data
can also be replotted in terms of the Larson – Miller parameter, as shown in
Fig. 6.26 to permit interpolation and extrapolation for various time-temper-
ature combinations.

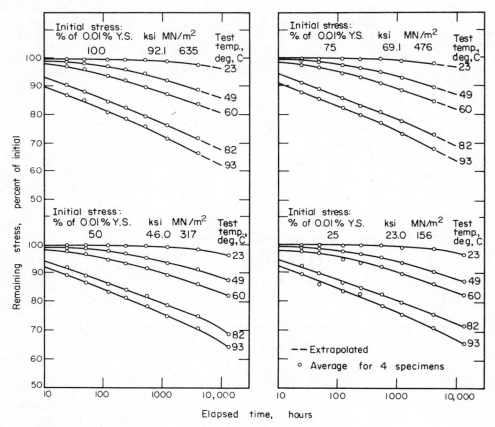

Fig. 6.25.   Stress relaxation characteristics of highly
            cold worked, as-rolled CDA copper alloy 510
            strip.   Reduction in area 97.3%; thickness
            0.254 mm.   Specimen axis transverse to rolling
            direction (Fox, 1971; copyright ASTM, used
            by permission).

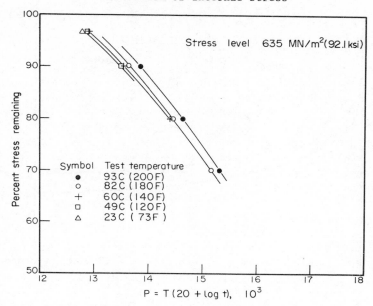

Fig. 6.26.   Larson-Miller plot of CDA copper alloy 510
relaxation data from Fig. 6.25.

In contrast to Fox's results, Nordstrom and Rohde (1972) reported that
stress relaxation of hardened Cu-2Be spring material was virtually unaffected
by temperature over the range from about -50° to + 95° C (-60° to +200° F).
These results lend additional support to observations made earlier relative
to microyield phenomena--with current levels of understanding, it is dangerous
to generalize concerning temperature effects. The effect of temperature on
stress relaxation of a particular material in a particular condition can be
estimated with confidence only if experiments are conducted over the temperature range of interest.

In addition to showing temperature and time relationships, the results of
Fox provide information relative to the effect of initial stress level on stress
relaxation. In an earlier figure (Fig. 6.23), showing the stress relaxation
behavior of Ti-6Al-4V, it was inferred that the initial stress level was unimportant, since it was not specified. The data in Fig. 6.25 give strong
support to the view that initial stress level is relatively unimportant, so
long as stress relaxation is treated on a percentage basis. For example,
from Fig. 6.25, the percentage of stress relaxation at 60°C in 10 hr is nearly
the same for initial stresses that range from 156 to 635 MN/m^2 (23 to 92.1 ksi).
However, the absolute magnitude of the stress reduction is obviously greater
for the higher stress levels.

Additional data illustrating the relatively small effect of initial stress
level on stress relaxation are presented in Figs. 6.27 and 6.28, for an aluminum alloy and a medium carbon steel, respectively.

B.   *Effect of Material Composition and Condition*

Stress relaxation is certain to be affected by both material composition
and condition. For example, alloying to improve elevated temperature creep
resistance will also increase a materials resistance to stress relaxation at
elevated temperatures. Figure 6.29 illustrates the wide difference in stress
relaxation behavior of several aluminum alloys, due, most likely, to compositional differences.

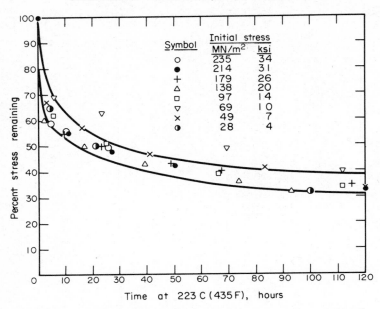

Fig. 6.27. Data for Al-10Cu alloy, aircooled from 500°C,
            illustrating the relatively small effect of
            initial stress level on stress relaxation
            (after Barker and Sutton, 1967).

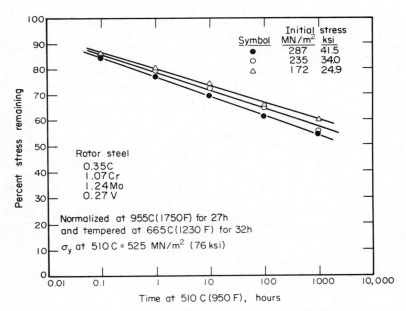

Fig. 6.28. Data for rotor steel, illustrating the re-
            latively small effect of initial stress
            level on stress relaxation (after Manjoine,
            1971).

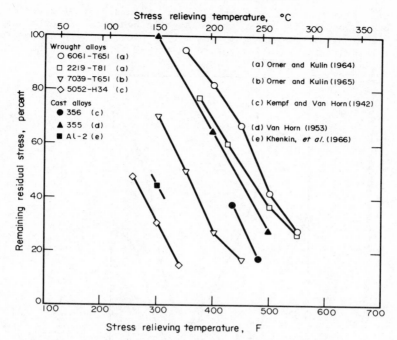

Fig. 6.29. Effect of aluminum—alloy composition on the
response to stress relief annealing (10 hr)
(courtesy of D. N. Williams).

Likewise, mechanical and thermal processing sequences can strongly influence
elevated temperature stress relaxation behavior. For example, in superalloy
springs operating at high temperatures, relaxation can be minimized by avoiding
heavily cold worked microstructures and by overaging to produce strength levels
well below the maximum attainable levels. Freeman and Voorhees (1956) have
compiled a large body of information from various sources pertaining to
relaxation properties of steels and super-strength alloys at elevated temperatures. These compilations reveal effects of both chemical composition and
material condition.

A specific example of the importance of degree of aging in precipitation
hardened alloys has been reported by Doroshek and Tseytlin (1963). They
studied the effect of various tempering treatments on the elevated temperature
stress relaxation behavior of heavily cold worked iron – nickel alloys, similar
in composition to Ni-Span-C (Fe-42Ni-5Cr-2.5Ti-0.5Al), but additionally alloyed
with Cr, Al, and Mo. Prior to working, the materials were solution annealed
at 1050° C (1920° F). The tempering or aging, conducted at temperatures ranging
from 300° to 800° C (570° to 1470° F), produced precipitation hardening, as indicated by hardness and yield strength measurements. Figure 6.30 shows the
results of stress relaxation tests in which the initial stress was stated to
be less than the yield strength, at temperatures from 350° to 500° C (660° to
930° F). At all test temperatures, minimum relaxation is seen to be associated
with tempering in the range from 550° to 600° C. For less highly alloyed
materials of the same general type, Doroshek and Tseytlin reported minimum
relaxation was associated with tempering at about 450° C. Undoubtedly, different levels of cold work or different cold work-plus-temper cycles would have

revealed other optimum tempering temperatures. Figure 6.31 illustrates the
role of chemical composition on stress relaxation behavior of certain spring
steels and alloys after optimum heat treatment. Clearly, both composition and
processing are important variables governing stress relaxation behavior.

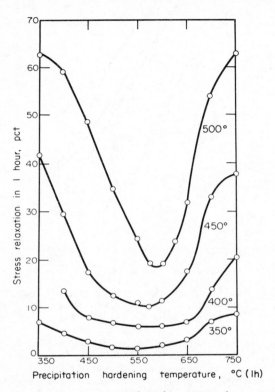

Fig. 6.30. Effect of precipitation hardening temperature
            on stress relaxation at indicated test temperatures
            for Fe-35.5Ni-13.3Cr-5.5Mo-1.7Ti-1.0Al
            (Doroshek and Tseytlin, 1963).

Of perhaps greater importance in most dimensionally critical applications
is the role of material composition and condition in stress relaxation near
room temperature. Here, few guidelines are available. Of course, this dearth
of information is of little consequence if dimensionally critical components
are free of residual stresses when placed in service. Since this will not
always be possible or, in some cases, desirable, information on effects of
composition and material condition could be helpful in material selection and
processing for precision application. For example, in Chapter 5, it was shown
that, in some cases, prestraining or cold working can raise the stress required
to produce a given level of plastic strain, say 10^{-6}. This would be considered a beneficial effect in many precision applications because it would
minimize the likelihood of permanent dimensional change in the event of an
overload in service. However, whether prestraining or cold working increases
resistance to stress relaxation near room temperature is not known with certainty. Odegard and Thompson (1973) have shown that tensile creep of Ti-6Al-
4V at room temperature can be diminished by tensile prestrain. This would

imply that stress relaxation rates in this material can be diminished as well, though no experiments were conducted to confirm this.

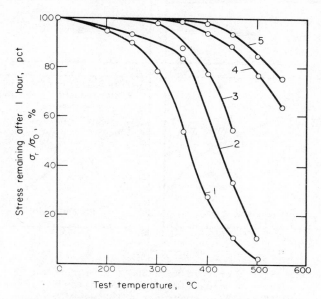

Fig. 6.31.   Effect of alloy composition on stress relaxation of several spring materials after optimum heat treatment (Doroshek and Tseytlin,
1963).

Curve 1: Steels 50KhFA and EI142
Curve 2: Steel EI722
Curve 3: Alloy N41KhT
Curve 4: Alloy N36KhTYuM
Curve 5: Alloy K40TYu

Fox (1971) performed a comprehensive investigation of low temperature stress relaxation behavior of CDA copper alloy 510, a 5% tin phosphor bronze. He compared half-hard strip (20% cold reduction, 0.2% offset yield strength of 382 MN/m^2 [55.5 ksi]) with heavily cold-rolled strip (97.3% cold reduction, 0.2% offset yield strength of 917 MN/m^2 [133 ksi] perpendicular to the rolling direction). Some of his results are shown in Fig. 6.32. They indicate that, at room temperature, the heavily-worked material relaxes less than the half-hard material when oriented transversely to the rolling direction. For the parallel orientation, little difference can be detected. At moderately ele-vated temperatures, however, the heavily worked material exhibits higher stress relaxation rates in both orientations than does the half-hard material. It was further reported that low temperature heat treatments (190° to 274° C for 2 hr) appreciably improved the stress relaxation characteristics of the heavily worked material accompanied by some reduction in the 0.2% offset yield strength.* This is illustrated in Fig. 6.33 in which the as-rolled material is compared with that receiving a 2 hr treatment at 274° C.

---

*Although the yield strength at 0.2% offset was generally reduced by the ther-mal treatments, that at 0.01% offset tended to show a slight increase.

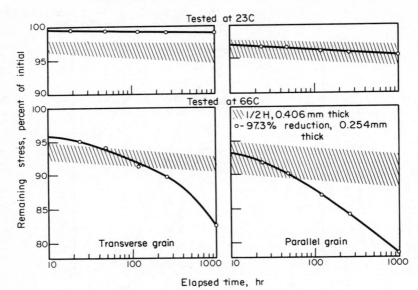

Fig. 6.32. Effect of cold work and temperature on
            stress relaxation of copper alloy 510
            (Fox, 1971; copyright ASTM, used by
            permission).

Initial stress: 0.01% offset yield strength.
Tested in the as-rolled condition.

These results are interesting from several standpoints. They illustrate
that cold working can affect low-temperature stress relaxation in a complex
way. In phosphor bronze at room temperature, cold working is beneficial, while
at slightly elevated temperatures, it is detrimental. Its effectiveness is
greatly increased by a low-temperature heat treatment, whose effect on the
microstructure is unknown.

In attempting to account for the beneficial effect of certain thermal
treatments relative to subsequent room temperature relaxation behavior,
Khenkin (1967) has put forth an interesting idea. He postulates that SR internal stresses in a material act to accelerate relaxation of LR stresses. Thus,
if the SR stresses can be reduced in magnitude, relaxation of the LR stresses
should occur more slowly.* To support his views, Khenkin describes stress
relaxation experiments on an Fe-23Ni alloy in several conditions representing
different degrees of SR internal stress as indicated below:

	Condition	Magnitude of SR internal stresses[a]
(a)	Quench (from 1030° C)	Very high
(b)	(a) plus temper (400° C)	Low
(c)	(b) plus cold work (60%)	High
(d)	(c) plus temper (300° to 400° C)	Low

[a]Based on x-ray line width data of other investigators.

*From a practical standpoint, this discussion pertains primarily to relaxation
of applied stresses in a spring element, and not to relief of LR residual
stresses in a fabricated workpiece. Nonetheless, it is useful for demonstrat-
ing the importance of material condition in stress relaxation phenomena.

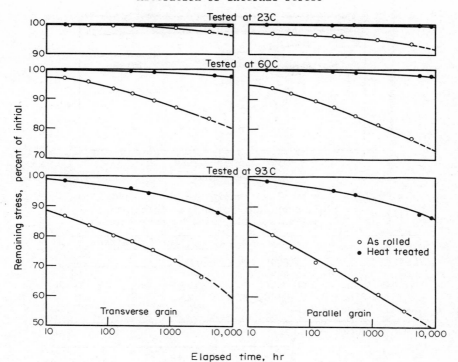

Fig. 6.33. Effect of low temperature heat treatment on
            anisotropic stress relaxation behavior of
            highly cold worked CDA copper alloy 510
            strip (Fox, 1971; copyright ASTM, used by
            permission).
            Reduction in area 97.3%; heat treated for
            2 h at 274°C; initial stress 0.01% offset
            yield strength.

The results, shown in Figs. 6.34 and 6.35, tend to support Khenkin's hypothesis.
Stress relaxation does not occur as rapidly in specimens containing low levels
of SR stress as in those containing high levels. This, plus the observation
that the 0.005% offset yield strength, σ_y (5×10^{-5}), is also raised as SR internal stresses are reduced, indicates, according to Khenkin, the strong
dependence of stress relaxation and microplastic deformation on SR internal
stresses.

Khenkin's data are too limited to be conclusive, but they do offer possible
explanations of the benefits derived from certain treatments. Further, they
indicate a relationship between microyield behavior and stress relaxation which
would be intuitively expected.

Data of the type reported by Fox and by Khenkin are clearly needed if in-
telligent estimates of the role of material variables on low temperature stress
relaxation are to be made. At the present time, little such information is
available.

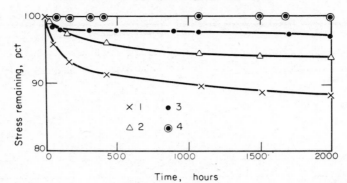

Fig. 6.34. Stress relaxation in Fe-23Ni at 20°C (Khenkin,
           1967; copyright Scientific Information
           Consultants, Ltd., used by permission).

		Initial stress	
Curve	Treatment	MN/m^2	ksi
1	1030°C, oil quench	262	38
2	Same as (1)	169	24.5
3	Same, plus 400°C, 2 hr	725	105
4	Same as (3)	367	53.2

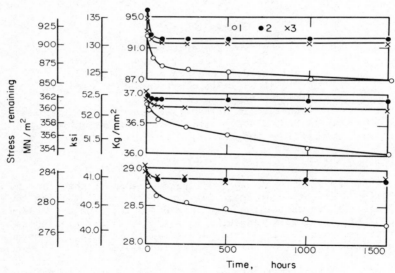

Fig. 6.35. Stress relaxation at 20°C for Fe-23Ni, oil
           quenched from 1040°C, tempered 2 hr at 400°C,
           and cold reduced 60% (Khenkin, 1967;
           copyright Scientific Information Consultants,
           Ltd., used by permission).

Curve	Treatment
1	As cold-reduced
2	Same, plus 300°C, 2 hr
3	Same, plus 400°C, 2 hr

C. Effect of Thermal Cycling

It is not uncommon for manufacturers of precision components to subject them to a thermal cycling treatment prior to placing them in service, frequently without knowing whether any benefits accrue from this operation. In some cases, it may be simply a precautionary measure to expose the component to the temperature extremes it is likely to see in service—a type of proof test or thermal exercising—with the expectation that if small dimensional changes should accompany the cycling, their occurrence will minimize additional changes during subsequent temperature fluctuations in service. On the other hand, the manufacturer may have specific evidence that dimensional stability in service is improved by thermal cycling treatments.

Historically, interest in thermal cycling as a means of improving dimensional stability may have developed from several sources. It has often been reported that castings and other metal products are more stable if they have been allowed to "weather" in the open air for a few months or years. In many parts of the world, this means that they would have been exposed to numerous thermal cycles of appreciable magnitude. Thus, it was only natural to attempt to speed up the weathering process by imposing thermal cycling treatments in place of the natural cycling.

Secondly, much of the early information about dimensional instability on a very small scale came about from studies on hardened steel gage blocks. It was discovered that most of the instability arose from gradual transformation of retained austenite to martensite, manifested by an expansion, or gradual tempering of the martensite, accompanied by a contraction. To minimize these effects, the steel was first cooled below room temperature to force most of the austenite transformation to occur prior to placing the gage blocks in service. This was followed by heating to a moderately elevated temperature to temper the newly-formed martensite. Frequently, the cooling and heating were repeated several times. The effectiveness of this thermal cycling treatment on hardened steels may have led others, unaware of the mechanisms, to believe that it would be effective on all materials.

The question being considered in this section is the effect of thermal cycling on stress relaxation. Is thermal cycling a more effective way of relieving residual stresses than conventional isothermal treatments? Similarly, are components that will be subjected to thermal cycling in service more likely to experience stress relaxation and comcomitant dimensional instability than those operating at a fixed temperature? Thermal cycling might be expected to have an effect on stress relaxation from at least three standpoints (Marschall et al., 1969):

(1) If heating and cooling rates are rapid, temperature gradients can generate thermal stresses. If these stresses, when superimposed on the already existing residual stresses, are of sufficient magnitude to cause plastic strains, the existing residual stress distribution will be altered. An "uphill" quench, as described in Section 6.3 for aluminum alloys, is an example of such a treatment.

(2) If the material exhibits internal differences in thermal expansion coefficients, either with respect to different crystal directions or with respect to different phases in a multiphase alloy or in a composite material, then any temperature change will produce SR stresses. If these, combined with the LR residual stresses, reach a sufficient magnitude, localized plastic strain will occur, with concomitant reduction of the LR residual stresses. To be most effective, the temperature would be cycled both above and below the service temperature to induce SR stresses, first of one sign and then of the other.

(3) If the maximum temperature reached in the thermal cycle is sufficiently
high to promote significant stress relief, and the time at this temperature is sufficiently great, thermal cycling will unquestionably
reduce the magnitude of the residual stresses. However, if this is
the mechanism for stress relief, there is little point in employing
the cycle, particularly the excursion to subambient temperature. It
would be simpler and more effective to employ an isothermal treatment
at the maximum temperature of the cycle.

With respect to item (1), although the principles are well established
whereby residual stresses can be altered by superimposing thermal stresses
from thermal cycling, little experimental work of a quantitative nature has
been done. However, the consequences of altering the residual stress distribution in this manner in an operating component have been described by Henson
and Inouye (1966). They observed gross dimensional changes in forged aluminum
alloy oxidizer valve bodies for rocket engines after thermal cycling between
room temperature and liquid nitrogen temperature. Critical diametral dimensions
of inlet and outlet ports changed to the extent that leakage occurred after
only a few cycles. Both 7075 in the T6 and T73 conditions and 6061-T6 were
susceptible to this instability. Subsequent experiments on valve bodies
repeatedly cooled to 196° C (-320° F) by immersion in liquid nitrogen, followed
by removal from the bath for return to room temperature, revealed diametral
distortions of approximately 0.1% after 24 cycles. The problem was attributed
to residual stresses present from the quenching operation during the solution
treatment cycle and from machining, though no residual stress measurements
were made to confirm this. The presumption was made that these stresses
relaxed or were redistributed by the thermal cycle, though the mechanism
whereby the stresses relaxed was not discussed. Modified processing procedures
aimed at reducing the residual stress magnitudes prior to thermal cycling were
not very successful in reducing subsequent distortion. The investigators
expressed doubt that a processing sequence could be devised that would completely eliminate permanent distortions of the aluminum forgings. This is a
disquieting observation for designers of other aluminum components where thermal cycling is anticipated and where dimensional changes of, say, 10^{-6} are
detrimental to performance.

Although the present discussion pertains to the role of thermal cycling in
stress relaxation, some results of rapid thermal cycling experiments in which
there may have been no initial residual stress are worthy of mention in connection with item (1). As pointed out in item (2) above, thermal cycling of
metals which exhibit crystalline anisotropy of thermal expansion, such as
uranium, zinc, beryllium, and other non-cubic metals, induces SR internal
stresses which can interact with and alter the residual stresses. Bochvar
and coworkers (1958) have found that even metals with cubic crystal structures—
aluminum, copper, iron, and nickel, for example—which exhibit no crystalline
expansion anisotropy, may exhibit large dimensional changes when subjected to
repeated thermal cycling. The major requirements for producing dimensional
changes are that the upper temperatures exceed a certain value, e.g., 200° C
for aluminum, 300° C for copper, 500° C for iron, and that cooling be rapid.
Residual stresses prior to cycling are not required. Cooling rates in their
experiments were generally of the order of 800° C per second. Results of
repeated thermal cycling are shown in Fig. 6.36a for aluminum and in Fig. 6.36b
for nickel. Note particularly the magnitude of the effect. In pure aluminum,
for example, cycling between 15° and 500° C produced longitudinal strains of
approximately 10^{-3} per cycle. When cooling was accomplished very slowly in a
furnace, however, virtually no dimensional change was detected with repeated
cycling.

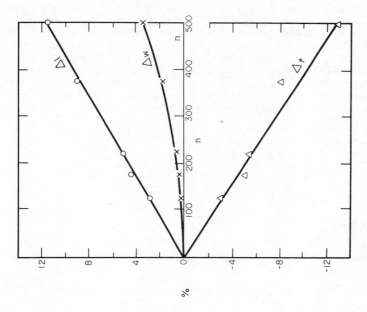

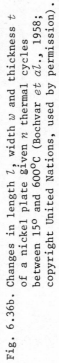

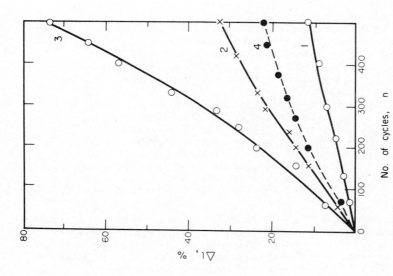

Fig. 6.36a. Length change Δl of plate specimens of pure (1,2,3) and impure (4) aluminum cold rolled to 75%, as a function of the number of cycles n and the thermal cycle temperature range; 1, 15–300°C; 2, 15–400°C; 3, 15–500°C; 4, 15–500°C (Bochvar *et al.*, 1958; copyright United Nations, used by permission).

Fig. 6.36b. Changes in length l, width w and thickness t of a nickel plate given n thermal cycles between 15° and 600°C (Bochvar *et al.*, 1958; copyright United Nations, used by permission).

Davidenkov *et al.*(1961), in further studies on effects of temperature cycling
on cubic metals, discovered complex relationships among the amount of distortion produced by cycling, specimen size and shape, rates of heating and cooling,
temperature range, and material properties, especially the dependence of yield
strength on temperature. In certain instances, significant lengthening was
observed when cylinders were subjected to thermal cycling, while in other
instances, contraction was observed. The effects of several variables are
illustrated qualitatively in Fig. 6.37. When heating or cooling rates are low,
no distortion is observed. When heating is rapid and cooling is slow, contraction is observed. For rapid cooling in combination with slow heating,
either expansion or contraction may occur, depending on the yield strength –
temperature relationship.

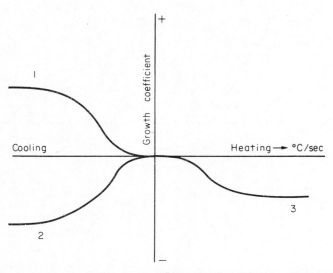

Fig. 6.37.   Influence of rate change of temperature on
             the sign of the growth coefficient:
             1 - rapid cooling in combination with slow
                 heating for materials where there is weak
                 dependence between yield point and
                 temperature;
             2 - rapid cooling in combination with slow
                 heating for materials which have strong
                 temperature dependence of yield point;
             3 - rapid heating in combination with slow
                 cooling for any materials.
             (Davidenkov *et al.*, 1961; copyright Fizika
             Metallov i Metallovedeniye, used by
             permission of Copyright Agency of the USSR).

With respect to item (2), it is well established that SR internal stresses
of high magnitude can develop in a body among neighboring elements whose
expansion coefficients differ, whenever the temperature is changed. In contrast to item (1), no temperature gradients are required. The magnitude of
the stresses developed will depend upon the relative expansion coefficients,
the extent of temperature change, and the elastic moduli. Numerous examples
have been cited of plastic deformation in metals of noncubic crystal structure,

i.e., metals that exhibit anisotropic thermal expansion, when subjected to repeated thermal cycling, even when free of LR residual stresses. This behavior has long been recognized in uranium, where dramatic dimensional changes occur on each cycle and continue to occur for thousands of cycles. Pugh (1957) reported length increases of as much as 0.08% per cycle for 1.25 cm (1/2 in) diameter uranium rods. Figure 6.38 presents photographs of some of the cycled rods. This phenomenon has been studied extensively by others as well, including Bettman *et al.* (1953), Aitchison *et al.* (1962), Boas and Honeycombe (1947), Burke and Turkalo (1958), and Johnson and Honeycombe (1962). Figure 6.39 is a micrograph of zirconium after thermal cycling, showing evidence of plastic deformation.

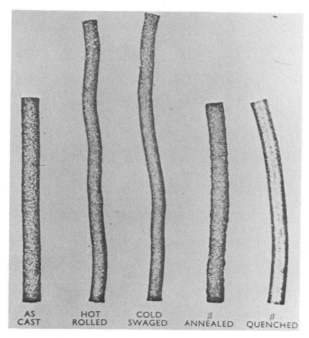

Fig. 6.38. Uranium bars of 1/2 in dia., originally
            smooth and of equal length, after 859 cycles
            between 50° and 600°C (Pugh, 1957; copyright
            Inst. of Metals, used by permission).

If internal stresses sufficient to cause plastic flow can be developed in the manner described above as a result of thermal cycling, then it follows that thermal cycling of such materials should also be capable of reducing the magnitude of any preexisting LR residual stress patterns. Unfortunately, little experimental work has been done to confirm this. Based on a limited investigation, Lokshin (1970) has reported that residual stresses in technically pure beryllium, a noncubic metal, can be reduced in magnitude at a greater rate by thermal cycling than by isothermal exposure to the upper temperature of the cycle. Results of various treatments on the stress relaxation behavior of beryllium bend specimens initially stressed to 78 MN/m^2 (11.2 ksi) are shown in Fig. 6.40. Notice that isothermal treatment at temperatures ranging from 100° to 600°C (210° to 1110°F) produced significant stress relaxation.* However, by thermal cycling at an unspecified rate but, presumably, slowly between 400°C and either -70° or -196°C, the rate of relaxation was greater than that observed for holding at 400°C for the same time periods.

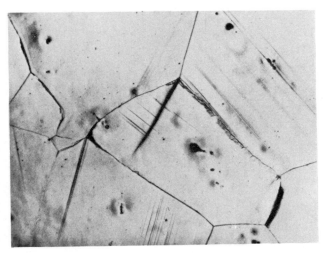

Fig. 6.39.   Microstructure of zirconium after 5 cycles
from 15° to 600°C showing slip, boundary
migration and a typical kink at a grain
boundary triple point (X150) (Johnson and
Honeycombe, 1962; copyright Elsevier
Sequoia, used by permission).

Thus, Lokshin recommends use of thermocyclic stabilizing treatments, TST, to reduce residual stress levels in beryllium if heating to 600°C is technologically undesirable or when strain softening must be avoided.

In addition to Lokshin's stress relaxation results for beryllium, Roberts (1960) showed that creep of noncubic metals can be greatly accelerated by thermal cycling. He subjected alpha uranium helical springs to steady loads, corresponding to less than 10% of the yield strength, while cycling the temperature slowly between 100° and 300°C. As a control, he held other springs under the same loads at a constant temperature of 350°C; no creep was observed in the control specimens. The thermally cycled specimens, however, were unable to support the relatively small applied loads without extending plastically as shown in Fig. 6.41. The plastic deflection after the first few cycles was about 10 times the initial elastic extension. For continued cycling, the rate of extension diminished and finally stopped, provided the amplitude of the thermal cycle was not too large. The cessation of creep was attributed by the author to a gradual redistribution of the intergranular stresses and to strain hardening.

In a more comprehensive study, Likhachev *et al.* (1963) examined the effect of thermal cycling on creep of zinc, a noncubic metal that exhibits crystalline

---

*It is likely that the initial stress was in excess of that required to cause measurable microplastic flow. Hence, the apparent early relaxation at 100°C may simply be an indication of this immediate plastic flow.

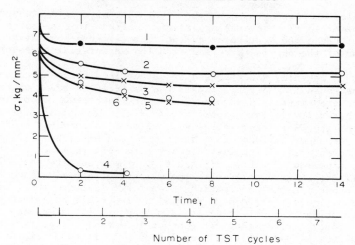

Fig. 6.40. Comparison of thermal cycling with isothermal
exposure relative to their effects on
stress relaxation in pure beryllium (Lokshin,
1970).

Curve	Thermal conditions
1	Isothermal; 100°C
2	Isothermal; 190°C
3	Isothermal; 400°C
4	Isothermal; 600°C
5	Cycled from −70° to +400°C
6	Cycled from −196° to +400°C

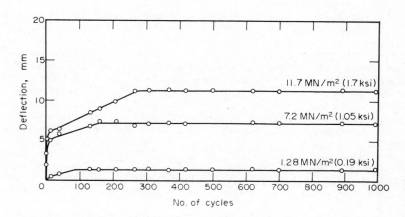

Fig. 6.41. Spring deflection versus number of cycles
for uranium cycled between 100°C and 300°C
and also between 50°C and 300°C (Roberts,
1960; copyright Pergamon, used by permission).

thermal expansion anisotropy. They found that thermal cycling caused a dramatic increase in the creep rate—several orders of magnitude in some cases. The acceleration of creep was attributed to the thermal expansion anisotropy and the interaction of the applied stresses with the SR internal stresses developed during temperature cycling.

The results of Lokshin, Roberts, and Likhachev lend credence to the idea that residual stresses in noncubic metals can be relieved by an appropriate thermal cycling treatment. They do not, however, make clear what constitutes an effective thermal cycle for a specific material. To obtain some idea of the magnitude of the SR internal stresses developed in noncubic metals as temperature is changed, Davidenkov *et al.* (1960) and Likhachev (1961) analyzed the problem theoretically. This was discussed in Section 6.2. Results of their analyses appeared in Tables 6.1 and 6.2. They indicate that, in many noncubic materials, SR stresses due to thermal anisotropy can reach levels sufficient to cause plastic deformation, even for relatively small temperature excursions. Thus, relatively modest thermal cycles would be expected to be effective in reducing residual stresses in many noncubic materials.

It was also noted earlier that differences in expansion coefficients among different phases or microconstituents of a body will introduce interfacial stresses as temperature is changed. The work of Laszlo (1943) and Brooksbank and Andrews (1968, 1972) in calculating the magnitudes of such stresses was described in Section 6.2. The latter investigators concluded that oxide inclusions in steel, because of their relatively small thermal expansivity compared with that of steel, could induce SR stresses sufficiently large to cause plastic deformation or microcracking of the matrix.

Although the findings of Brooksbank and Andrews may be of limited practical use relative to stress relaxation by thermal cycling, they do illustrate that SR internal stresses of appreciable magnitude can develop around second phase particles as a result of temperature changes. Similar effects would be expected around precipitate particles in precipitation hardened alloys. Generally speaking, however, little is known about the expansion coefficients of such particles relative to the matrix. Thus, quantitative estimates of induced SR stress are ruled out. Nonetheless, the possible role of SR stresses around second phase particles in helping to reduce residual stresses during thermal cycling should not be overlooked.

Additional evidence of thermal cycling effects in multiphase materials is provided by Khenkin (1967). In studies on an Al—12Si alloy, in which phases with considerably different expansion coefficients are present, he observed that plastic deformation occurred in the aluminum-rich matrix phase upon cooling to room temperature in the furnace, in air, or in water following a 5-hr anneal at $280°$ C. X-ray diffraction line-width studies indicated the presence of SR internal stresses after cooling. Subsequent thermal cycling from $-190°$ to $+150°$ C affected properties as shown in Table 6.4. Khenkin attributed the improved properties to a relaxation of the SR stresses brought about by thermal cycling. However, the possibility exists that the stresses could have been reduced equally well by simply holding at the upper temperature of the cycle and that the cycling actually had little to do with the observed results. Despite this uncertainty in interpreting the results, the observation that plastic flow occurred during the cooling from $280°$ C is clearly indicative of the magnitude of stresses that can develop around second phase particles during a temperature change.

Fibrous composites in which the thermal expansion coefficient of the fibers differs greatly from that of the matrix are also susceptible to dimensional changes upon thermal cycling. Mueller and Marschall (1973) have shown that a boron – aluminum composite exhibits significant growth when subjected to slow thermal cycling (3 hr per cycle) from $-55°$ to $+55°$ C ($-65°$ to $+130°$ F), as

shown in Fig. 6.42. Similar effects might be anticipated for graphite-epoxy
composites. Forest (1974) has reported that graphite-epoxy mirror substrates
exhibit significant shape changes when cycled repeatedly between $-20°$ and $+65°$ C
($0°$ to $150°$ F).

TABLE 6.4.   Properties of Al-12Si Alloy after Various Treatments
             (Khenkin, 1967)

Treatment	$\sigma_y(5 \times 10^{-5})$ kg/mm²	$\sigma_y(2 \times 10^{-3})$ kg/mm²	Relative width of (511) X-Ray line	Relaxation resistance
Anneal 5 hr at $280°$C, air cool	2.3	8.9	24.3	--
Same, plus three cycles of cooling to $-190°$C (10 min) and heating to $150°$C (1 hr)	3.3	9.5	17.2	Improved

Clearly, the picture that emerges from the evidence available at this time
is that thermal cycling of materials containing neighboring elements of varying thermal expansion will, under appropriate conditions, accelerate stress
relaxation, accelerate creep, and produce permanent dimensional changes.

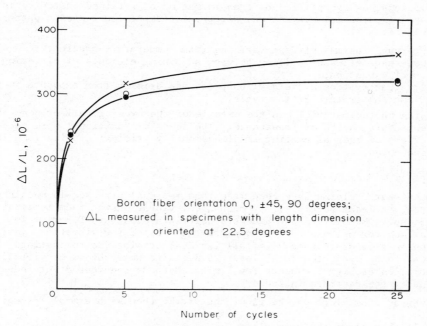

Fig. 6.42.   Length change in composite boron-aluminum
             specimens thermally cycled from $-55°$ to $+55°$C
             ($-65°$ to $+130°$F) (Mueller and Marschall,
             1973).

Finally, turning attention to item (3), thermal cycling may assist relaxation of residual stresses merely by allowing the material to be exposed for some time to the maximum temperature of the cycle. If this is the *only* mechanism by which the cycling helps to relieve stress, it would be more effective simply to heat the material to the maximum temperature of the cycle and hold it for a time sufficient to accomplish the desired stress reduction. There are undoubtedly many cases where thermal cycling is employed, particularly involving cubic metals and low rates of heating and cooling, where this effect predominates.

Marschall and Maringer (1971) have indicated this to be the case for 2014 aluminum, thermal cycled from $-46°$ to $+38°$ C ($-50°$ to $+100°$ F) and $-73°$ to $+93°$ C ($-100°$ to $+200°$ F). Thin strips prepared in such a way as to be virtually free of residual stresses were bent around a mandrel to induce maximum fiber stresses of about 230 MN/m^2 (33 ksi). Previous studies on the same material showed $\sigma_y(10^{-6})$ to be about 260 MN/m^2 (38 ksi). Figure 5.2 shows the micro-yield behavior. While held firmly to the mandrel, some specimens were subjected to temperature cycling and others to a constant elevated temperature. After removal from the mandrel, residual curvature was measured. The degree of curvature was related directly to the relaxation of the original induced stress. Results are presented in Table 6.5. The accuracy of the results was estimated to be ±0.05% stress relaxation. It was concluded that, for the conditions studied, stress relaxation was not accelerated by thermal cycling.

Summarizing, thermal cycling can be of considerable importance in dimensionally critical applications. It may cause unsuspected problems if it occurs in service or it may help to minimize problems if it is employed judiciously prior to placing components in service. Aside from its role in phase transformations, as exemplified by transformation of retained austenite in hardened steels, thermal cycling can alter existing residual stress patterns as a result of:

- long-range thermal stresses arising from temperature gradients;
- short-range stresses arising between adjoining elements of differing thermal expansion;
- thermal activation of relaxation processes arising from exposure to the maximum temperature of the cycle.

Depending on the material and the details of the cycling, all or none of the above factors may be of importance. The need for additional work concerning the effect of thermal cycling in dimensionally critical applications is clearly indicated.

D. *Effect of Stress Cycling and Mechanical Vibration*

For many years, claims have been made that residual stresses in metals can be relieved, at least partially, by mechanical vibrations at room temperature. In a 1943 paper, McGoldrick and Saunders described U.S. Navy Department specifications for certain cast iron parts that called for vibration, jarring, or bumping to accelerate removal of all residual strains due to casting. Each casting was to be subjected to at least 25 definite heavy blows or vibrations, or to vibration at high frequency for 1 min. Methods considered for inducing vibrations in the castings included:

1. Suspend them clear of the floor and strike them with a maul, sledge, or ram.
2. Roll them along the shop floor.
3. Carry them around in a springless wagon over rough pavement.
4. Lift them a short distance and drop them, repeating this operation several times.

TABLE 6.5. Relaxation of Known Stresses in 2014-T6 Aluminum by Thermal Cycling (Marschall and Maringer, 1971)

Specimen identification	Rapid bend and unbend at 68°F	Cycle -50° to +100°F [a]			Hold for 24 hr		Cycle -100° to +200°F [b]		Hold for 72 hr 200°F
		12x	24x	36x	75°F	100°F	12x	24x	
1	–	0.1	0.1	0.1					
2	0	0.15	0.15						
3	0	0.1							
4	0.05				0.25	0.3			
5	0				0.1	0.25			
6	0				0.15	0.3			
7	0						1.55	1.95	
8	0						1.50	1.85	
9	0						1.70	1.95	
10	0								3.35
11	0								3.55
12	0								3.55

a Two hours per cycle.

b Six hours per cycle.

McGoldrick and Saunders noted further that the operation of riveting is believed by some to relieve stresses in partially welded structures, because of the vibrations induced. However, at the time of their paper, there was little factual evidence as to the success of vibrations in achieving significant reductions of residual stress.

In recent years, starting perhaps in the early 1960's, interest in stress relieving by mechanical vibration has reappeared. In fact, several devices for inducing mechanical vibrations for the purpose of stress relief are currently being marketed. The method for inducing vibrations is simply to mount a variable speed eccentric motor to the workpiece.

Proponents of vibrational stress relief cite numerous advantages for this process, in comparison with thermal stress relief treatments:

- Energy requirements of vibrational equipment are far less than those of large thermal stress relieving furnaces operating at high temperatures.
- Initial cost and maintenance costs of vibrational equipment are small compared with those for furnaces.
- Vibration equipment is portable and can be easily taken to the workpiece, even in field installations.
- Turnaround time for stress relieving can be greatly reduced.
- Distortion that may accompany thermal treatments is not a factor in vibratory treatments.
- No scaling or discoloration accompanies vibration.
- Vibration can handle workpieces too large to be thermally treated.
- Vibration is useful in materials where thermal treatments would reduce subsequent strength.

This is, indeed, an impressive list of advantages. In fact, it is so impressive that it is logical to ask why thermal stress relief has not been replaced entirely by vibratory stress relief. The major reason that this has not occurred appears to be serious doubt as to the actual effectiveness of vibrations in relieving stresses. Claims range from substantial stress relief to total stress relief in vibration times of the order of 15 to 30 min at room temperature. However, these claims are seldom substantiated by before-and-after residual stress measurements. Instead, it is stated that vibrated workpieces "performed satisfactorily", or "met tolerances", or "showed less distortion during machining than those not receiving vibration", etc. The final statement certainly implies that residual stresses have been reduced by vibration but the first two say little about the vibrations' effectiveness. There is also some uncertainty among the various manufacturers' claims as to whether vibration is effective on all metallic materials and material conditions. For example, Wick (1971) states that vibration is not effective in "materials that have been cold worked enough to increase their tensile strength and reduce their ductility".

In the scientific community there is a good deal of skepticism concerning vibration and a reluctance to unconditionally recommend the use of vibratory devices for stress relieving, partly because of negative results in a number of controlled experiments designed to test their usefulness and partly because of adverse reaction to pseudo-scientific explanations of how the devices work, contained in promotional literature, patents, and trade magazines. For example, one brochure discusses "equalized" versus "unequalized" residual stresses and states that only unequalized stresses can cause distortion; use of the vibrational device is said to make the stresses the same everywhere (equalized) by realigning the lattice without distortion through the action of vibrational energy.

In spite of the reluctance of scientists to accept many of the claims made by manufacturers of vibrational stress relief equipment, reports of success in employing vibrational techniques to minimize distortion in manufacturing are numerous and use of commercial vibrating equipment is apparently not uncommon. An article in the September 1969 issue of *Heat Treating* estimated that 700 to 1000 units were in use in the United States at that time. The number has undoubtedly increased since then.

There are essentially two explanations offered for reduction of residual stresses by mechanical vibrations. The first one described below represents the view held by most materials scientists while the second is more likely to be voiced by manufacturing personnel.

1. Whenever external forces, as in vibration, are applied to a body containing residual stresses and the sum of the applied and residual stresses is sufficient to cause localized plastic flow, the peak residual stresses remaining after removal of the external force will be smaller than those existing initially. Thus, mechanical vibrations of appropriate magnitude would almost certainly produce stress relief. However, if the combined stresses are not sufficient to cause localized plastic flow, stress relief would not be expected.

2. Introduction of energy, thermal or mechanical, tends to create "harmony" in metals, causing the abnormal atom spacing associated with residual stress to gradually give way to normal spacing and, thereby, to relieve the residual stress. According to this explanation, the metal does not care whether the energy is thermal or vibrational. Thus, from this viewpoint, there is apparently no critical amplitude of vibration required. So long as any residual stress is present, vibrations of any amplitude should produce some stress reduction with time.

The second explanation, while appealing to the layman, has little scientific backing. Low frequency vibrational energy and thermal energy are simply not the same in their effects on the atoms in a metal structure. Thermal energy acts to increase the vibrational activity of individual atoms, thereby increasing the average atom spacing, enhancing diffusion and permitting dislocation motion, i.e., plastic flow and creep, to occur at lower stresses. Mechanical vibrations at frequencies in the range of 10 to 100 cycles per second, on the other hand, produce no such effects. These frequencies are simply orders of magnitude too low to interact with the normal vibrational frequencies of atoms or with the resonant frequencies of pinned lengths of dislocation lines. Thus, the first explanation above, at least at present, is the only sound one from a materials science viewpoint.

Support for explanation 1 was reported by Wozney and Crawmer (1968) who investigated stress relief in SAE 1070 steel strips containing known levels of residual stress introduced by shot peening. Cyclic stressing was accomplished by alternating bending in a fatigue testing machine. From the results of these experiments, the investigators concluded that the residual stresses could be reduced by cyclic stressing, but only when the applied plus residual stresses exceeded the "elastic limit". Thus, based on this viewpoint, if residual stresses are very high—at or near the "elastic limit"— even small amplitude cyclic stresses should produce some stress relief. If, on the other hand, they are far less than the elastic limit, relatively large amplitudes would be required. This coincides with the view held by Lokshin (1965) that useful stress relief effects of vibration treatment are to be expected only when the workpiece undergoes plastic strain with every cycle of vibration.

Certainly the simple condition described above, namely

$$\sigma_{cyclic} + \sigma_{max.\ residual} > \text{"elastic limit"}$$

will produce some stress relief. However, if this is the only requirement for success, then cyclic stressing would not be necessary at all—a single load application would suffice. Two reasons can be advanced for repeated cycling, one involving practical considerations and the other involving little-understood aspects of material behavior.

From a strictly practical point of view, cyclic treatment is attractive because it is often simpler to apply mechanical vibrations to a workpiece, particularly a massive workpiece of complex shape, than it is to apply a known level of steady stress. Large, complicated machines of little versatility would likely be required for the single loading. For vibratory treatment, on the other hand, equipment can be simple, relatively inexpensive, and versatile. Present commercial equipment marketed in the United States generally consists of a low power variable speed d-c motor with an eccentric flywheel (the vibrator or shaker), a variable speed control, a transducer and amplifier, and a display meter or oscilloscope to monitor frequency and amplitude. Frequencies up to about 150 Hz are generally employed.

In considering a vibratory device for dimensional stabilization of castings, Lokshin (1965) noted that low-frequency vibrations at the amplitudes attainable on the table of a vibrating stand cannot possibly set up significant stresses in the material, unless the object is vibrated at its resonant frequency and is free to deform under the vibrations. For this reason, the speed of the eccentric motor in a vibratory unit is usually adjusted during treatment until a resonant condition is reached. Thereafter, in practices followed in the United States, the workpiece is permitted to vibrate at this frequency for from 10 to 40 min, depending on its weight. If more than one resonant frequency is found, the process may be repeated at the other frequencies. In all instances, the workpiece must be supported in such a way that resonance involves deformation of the workpiece itself rather than of the support structures only. Large workpieces are usually placed on rubber load pads or suspended in slings while small workpieces are clamped to spring mounted tables.

In their patent on vibratory stress relieving, Hebel and Hebel (1973) take issue with the practice of vibrating at resonant peaks. They recommend vibrating at a frequency just at the low end of one of the peaks to achieve maximum effectiveness, minimize the possibility of fatigue damage, and minimize noise levels. They claim further that accomplishment of stress relief is indicated by a shift in the peak positions. Once these remain stationary, stress relief is said to be complete.

Thus from a practical viewpoint, mechanical vibration is simply a convenient method for imposing stresses which, when added to the peak residual stresses, will cause localized plastic flow and subsequent stress relief. From a material behavior viewpoint, however, the repeated stress reversals associated with mechanical vibration may produce some stress relieving effects, even at relatively low cyclic stresses, that are at present only poorly understood. Reported successes with commercial vibrating equipment under similar conditions would seem to indicate this. Additionally, certain results reported in the scientific literature pertaining to effects of repeated cyclic stressing provide support for such a notion. Several examples are presented in subsequent paragraphs.

Lubahn and Felgar (1961) have discussed at length a phenomenon referred to as fading of the mean stress, sometimes observed in fatigue testing. When both a mean stress and a cyclic stress are imposed on a specimen and the total is sufficient to cause plastic flow, the first cycle will cause a reduction in the mean stress, as expected. Thereafter, if the earlier discussion holds, no additional reduction of the mean stress should occur, assuming cycling between fixed limits of strain. However, Morrow and Sinclair (1958) observed a gradual decay of mean stress in steel specimens with increasing numbers of cycles, continuing for thousands of cycles. The test conditions were similar to the situation in which a cyclic stress is imposed on a part containing

residual stresses. Similar results were reported by Pattinson and Dugdale
(1962), who found continued reduction of residual stress levels in a steel bar
after 10^7 cycles at a low stress amplitude, and by Boggs and Byrne (1973) in
a Ni-20Co alloy. Lubahn and Felgar rationalized the results of Morrow and
Sinclair's experiments in terms of the Bauschinger effect (see Section 5.4)
and strain hardening. Their analysis indicates that the decay effects are
governed by that portion of the stress – strain curve where plastic strains
are exceedingly small, i.e., the microplastic strain region. Furthermore,
stress reversals appear to be important. Thus, in certain materials at least,
benefits of repeated cyclic stressing as opposed to a single stress cycle to
relieve residual stresses are perhaps to be expected.

Abel and Muir (1973a) also have reported some interesting effects of cyclic
stressing on a low carbon steel. They found that repeated stress cycling of
this material to relatively low stress levels produced significant changes in
its subsequent mechanical behavior in a tension test. The effect was particularly pronounced when the stress cycling included both tension and compression,
rather than tension only. For example, reverse cycling 1250 times to a stress
of only 240 MN/m² (35 ksi) completely eliminated the upper yield point, as
shown in Fig. 6.43, whereas tension cycling 2500 times to a much higher stress
did not. This is all the more remarkable when it is realized that the static
0.0002% offset yield strength of this steel is 300 MN/m² (44 ksi). This seems
to indicate that considerable dislocation activity is taking place, even at
stresses below those normally considered safe from a microyield standpoint, if
stresses are both cyclic and reversed. In another paper, Abel and Muir (1973b)
indicate that microplastic strain under cyclic loading tends to occur in
bursts, rather than uniformly on each cycle.

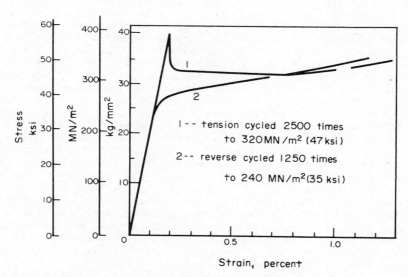

Fig. 6.43. Effect of prior cyclic loading of a
            low-C steel on subsequent yielding
            tension (Abel and Muir, 1973a; copy-
            right Pergamon, used by permission).

Yet another observation that may be pertinent relative to unexpected effects
of cyclic stressing was reported by Lement *et al.* (1951). In experiments on
Invar (Fe-36Ni) bars, relatively large residual stresses were present as a

result of water quenching from high temperatures during heat treatment. It was reported that the magnitude of the residual stresses could be reduced significantly by dropping the bars onto the floor from a height of 1.5 m (5 ft), thereby inducing vibrations of appreciable magnitude in the bars. The residual stress level was reduced additionally each time the bars were dropped, up to 10 drops, where the experiment was terminated.

Though their claims are not well documented, Hebel and Hebel (1973) state that workpieces containing residual stresses exhibit gradually decreasing resonant frequencies as they are mechanically vibrated. After some period of time, of the order of 20 min, corresponding to perhaps 50 000 to 100 000 stress reversals, the resonant frequency reportedly stabilizes, indicating, according to the claims, that stress relief is complete. Such behavior is said to be observed for a variety of workpiece materials, including carbon steel, stainless steel, aluminum alloys, and copper alloys. If, in fact, such a decrease in resonant frequency does accompany mechanical vibration, one possible explanation is a reduction in the effective elastic moduli. These could be reduced only if some plastic strain accompanies the vibration, leading to formation of hysteresis loops in the stress-strain relationships. Such hysteresis loops could conceivably reduce the effective elastic moduli by as much as 5 or 10%. The important point is this—if mechanical vibration at low amplitudes does, in fact, shift resonant peaks, it is quite likely that plastic deformation has occurred in the workpiece. Accordingly, if residual stresses were present initially, it is likely that they will have been reduced by the mechanical vibration.

Several investigators have studied the effect of superimposing cyclic stress on a steady stress in creep tests. Because of the close relationship between creep and stress relaxation, their findings might be helpful in assessing the potential of cyclic stressing in the relief of residual stresses. Some results obtained by Jakowluk (1969a) on Al-1Mg-1Si alloy, solution treated and aged, are shown in Fig. 6.44. Test temperature was $23°$ C ($73°$ F) and the alternating stress frequency was 31 Hz. As seen in Fig. 6.44, for a constant mean stress, σ_m, vibrational creep occurs more rapidly than does steady-load creep, with the effect increasing with greater values of the stress amplitude coefficient $A_\sigma = \sigma_a/\sigma_m$, where σ_a is the alternating stress amplitude. This is to be expected because σ_{max} is progressively greater. For a constant *maximum* stress (Fig. 6.45), vibrational creep occurs *less* rapidly the greater the value of A_σ, except for a very small A_σ value (0.0066). Here the creep is actually accelerated. Jakowluk also reported that, for small A_σ values, the higher the level of stress, the greater is the effect of the vibrations on creep rate. In a subsequent paper (Jakowluk, 1969b), it was reported that vibrational frequency in the range from 0.0033 to 37 Hz is another variable that affects vibrational creep.

Jakowluk interprets the unexpected accelerating effect of small vibrations as a manifestation of increased dislocation mobility, i.e., the vibrations assist dislocations in overcoming obstacles. However, he advises caution in applying the results to other materials tested under other conditions. For example, in certain cases, vibration may accelerate precipitation hardening reactions and thus impede, rather than accelerate, creep.

In studies conducted on lead at $32°$ C ($90°$ F), Kennedy (1956) observed results similar to those reported by Jakowluk. For example, at a mean stress of 7.3 MN/m^2 (1060 psi), the creep rate was increased more by superposing an alternating stress of amplitude ±0.25 MN/m^2 (±35 psi) than by raising the mean stress by 0.5 MN/m^2 (70 psi). In attempting to account for the effects of cyclic stressing, Kennedy (1958, 1963) offers several possible contributing factors, while at the same time cautioning that no really convincing theory exists. He cites evidence, for example, that the yield strength of metals can be reduced by the application of a relatively low amplitude alternating stress (Nevill and Brotzen, 1957). Bhaha and Langenecker (1959) report similar effects and explain them on the basis of activation of anchored dislocations.

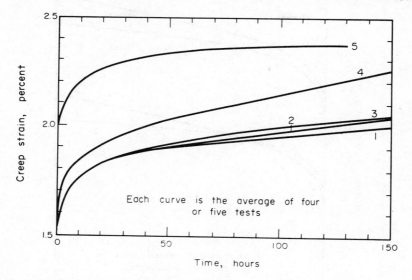

Fig. 6.44. Comparison of the curves of vibrational creep for various A_σ and constant σ_m = 288.6 MN/m^2 (42 ksi) (Jakowluk, 1969a; copyright Polish Society of Theoretical and Applied Mechanics, used by permission).

	Stress parameter, MN/m^2			
Curve	σ_m	σ_α	σ_{max}	A_σ
1	288.6	0	288.6	0.0
2	288.6	3.8	292.4	0.0132
3	288.6	1.9	290.5	0.0066
4	288.6	14.3	302.9	0.05
5	288.6	28.8	317.4	0.10

Kennedy suggests, on the other hand, that additional deformation systems may become active during dynamic stressing. He notes additionally a similarity between irradiation and cyclic stressing effects and suggests that both may be explicable in terms of vacancy generation and aggregation.

In studies unrelated to mechanical behavior, Damask *et al.* (1969, 1973) have noted a significant decrease in electrical resistivity of alpha brass (30% Zn) when subjected to cyclic stressing within the "elastic limit" at 95° C (203° F) at frequencies from 2 to 20 Hz. The resistivity decrease is associated with an increase in SR order, which, in turn, is attributed to diffusion enhancement from vacancies generated by the cyclic stressing.

Summarizing, the picture that emerges is this. Reduction of residual stresses can be accomplished by applying an external load, which, when added to the peak residual stresses, produces local plastic flow and concomitant

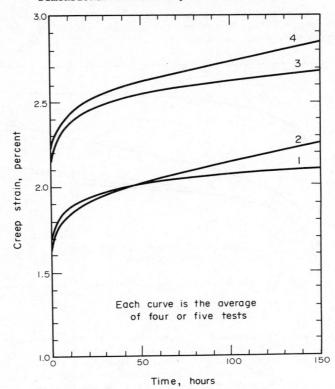

Fig. 6.45. Comparison of the vibrational creep curves
for various A_σ and constant $\sigma_{max} = 302.9$
MN/m² (47.7 ksi) (Jakowluk, 1969a; copy-
right Polish Society of Theoretical and
Applied Mechanics, used by permission).

	Stress parameter, MN/m²			
Curve	σ_m	σ_a	σ_{max}	A_σ
1	275.4	27.5	302.9	0.10
2	288.6	14.3	302.9	0.05
3	302.9	0.0	302.9	0.0
4	300.9	2.0	302.9	0.0066

stress relief. On this, there is general agreement. In large complex parts,
it is simpler and more economical to apply the external load in a cyclic manner, employing vibratory devices operating at or near resonance of the workpiece, than it is to impose static loads. Intuitively, the degree of stress
relief is expected to be a function of the magnitude of the cyclic stress component and not to change significantly after the first few cycles. However,
there is evidence in the scientific literature that repeated cycling may have
a continuing effect. Furthermore, under cyclic stress conditions, the stress

magnitude necessary to produce microplastic deformation may be considerably less than supposed heretofore. Meanwhile, arguments continue as to whether mechanical vibrations of gentle amplitude can actually accomplish 80 to 90% stress relief in short times at room temperature as claimed. Proponents answer affirmatively, citing the alleged similarity between mechanical energy and thermal energy and naming numerous manufacturers who are currently using vibrational treatments. If not successful, say the proponents, why would the manufacturers continue to use them? Opponents answer negatively, basing their answer on (a) the lack of experimental evidence that extensive stress relief accompanies vibration, and (b) a belief that the stresses developed in a work-piece by gentle vibrations are too small to have the effects claimed. Further-more, say the opponents, if the vibration amplitude were to be increased to develop stresses of appreciable magnitude, failure by fatigue might become a problem. Clearly, any subject of this importance in everyday metal fabrica-tion operations and for which so much confusion exists is in need of further careful investigation.

E. *Effect of Neutron Irradiation*

Neutron irradiation can have significant and complex effects on material behavior, primarily as a result of the introduction of vacancies and inter-stitial atoms into the crystalline lattice. In addition, in fissile materials, gaseous fission products can cause large internal pressures as they concentrate in voids and pores. The introduction of defects can drastically alter mechanical behavior and enhance diffusivity such that certain metallurgical reactions can proceed more readily at a given temperature. Eggert (1964) has shown that neutron irradiation of hardened SAE 52100 steel gage blocks can cause reduction of residual stresses, transformation of some of the retained austenite, and relatively large dimensional changes.

In view of these marked effects of neutron irradiation, it should be ex-pected that stress relaxation phenomena will also be affected. Dubrovin *et al.* (1962) have examined this question. They investigated the relaxation of both LR and SR stresses in a number of materials, both fissile and nonfissile. Two methods were used. To investigate relaxation of LR stresses, the residual curvature of an elastically stressed flat strip as a function of irradiation time was measured; to investigate relaxation of SR stresses, the shape of X-ray lines of cold worked specimens was measured and analyzed, before and after irradiation. Neutron irradiation was accomplished in a reactor at 80-200°C to doses of 10^{17} to 4×10^{19} fast neutrons per square centimeter. In the case of the bent strips, unirradiated control specimens were maintained under stress at slightly higher temperatures than those being irradiated.

Under the influence of irradiation, LR stress relaxation was observed in all the materials investigated. These included steel, nickel, nichrome, zirconium, and U–Mo alloys. The greatest effect was seen in the fissile materials. Figure 6.46 shows the results obtained for a U–0.91% Mo alloy, using bent strip specimens. The effect is, indeed, large. Somewhat smaller effects are shown in Fig. 6.47 for nonfissile materials. Although control specimen results are not shown, Dubrovin *et al.* indicated that no stress re-laxation was observed in the absence of irradiation.

It was reported also that SR stresses, introduced by cold working, were likewise reduced by irradiation. Materials studied included U, U–9Mo, Ni, Mo, Pt, Zr, W and Cu–14Sn. Figure 6.48 shows the shape of an X-ray diffraction line peak, before and after irradiation. The reduced width and the resolution of a second peak are indicative of reduced SR internal stresses.

Dubrovin *et al.* also reported an interesting reversion phenomenon. Follow-ing irradiation, the strip specimens that had undergone permanent strain in the relaxation tests showed a tendency to gradually revert to their original

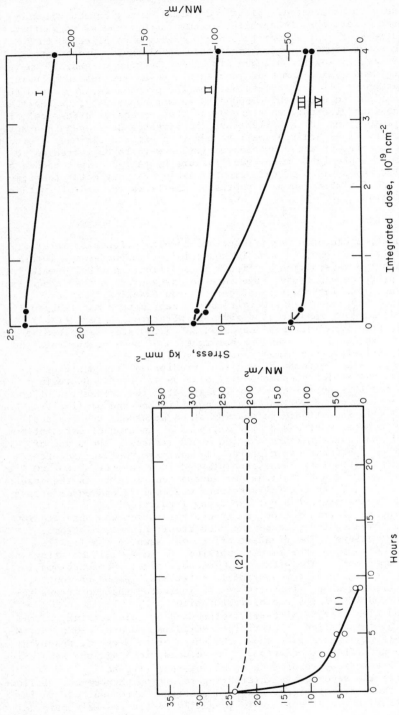

Fig. 6.47.   Stress relaxation under the influence of
irradiation: (1) steel; (II) Nichrome;
(III) nickel; (IV) zirconium (Dubrovin
et al., 1962; copyright Butterworths, used
by permission).

Fig. 6.46.   Stress relaxation in U-0.91% Mo alloy:
(1) irradiated at 80° to 200°; (2) held
at 200°C without irradiation (Dubrovin et
al., 1962; copyright Butterworths, used by
permission).

shape. This was also observed as a gradual broadening of X-ray lines. The explanation advanced by the investigators for this unusual behavior involves the creation and movement of vacancies and interstitials. When these point defects are mobile under the irradiation conditions imposed, they can diffuse in a manner determined by the elastic stress gradients, i.e., interstitials will be attracted to regions of tensile stress and vacancies to compressively stressed regions, thereby minimizing the stress gradients. If a new stress distribution is imposed, such as would occur in the bent strip specimens when the external forces are removed, the defects would tend to redistribute themselves, leading to reversion. It is not clear, however, how this explanation can account for reversion of the SR stresses.

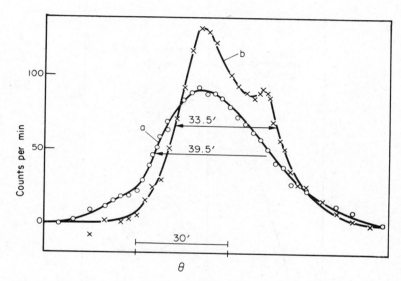

Fig. 6.48. Effect of irradiation on line breadth for alpha-uranium: (a) cold-worked; (b) cold-worked and irradiated to 0.7×10^{17} n/cm^2 (Dubrovin et al., 1962; copyright Butterworths, used by permission).

Although the mechanisms by which neutron irradiation influences stress relaxation remains somewhat in doubt, the implications of the findings of Dubrovin are far reaching with respect to dimensionally critical applications. Irradiation can induce stress relaxation, with concomitant dimensional changes. Following irradiation, gradual reversion can occur, again accompanied by dimensional changes. Where precise dimensional control is crucial, each of these effects must be considered.

F. Effect of an Alternating Magnetic Field

Cullity and Allen (1965) reported that the rate of stress relaxation in nickel at room temperature can be increased by the application of a magnetic field alternating at 60 Hz. An annealed ring was wound with a magnetizing coil and loaded in compression, as indicated in Fig. 6.49. At a load sufficient to produce some plastic strain, the crosshead was stopped. After normal stress relaxation had proceeded for a few seconds, an alternating current was applied to the windings. As shown in Fig. 6.49, the rate of stress relaxation immediately increased. The observed change in slope is about 8 to 1. Cullity and Allen ascribe the effect to magnetostrictive strain (see Chapter 10) within moving domain walls. This magnetostrictive strain can interact with the strain fields of weakly pinned dislocations, causing the dislocations to move, thereby making a contribution to stress relaxation. Effects of this nature would be anticipated only in ferromagnetic materials that exhibit appreciable magnetostrictive strain with magnetization.

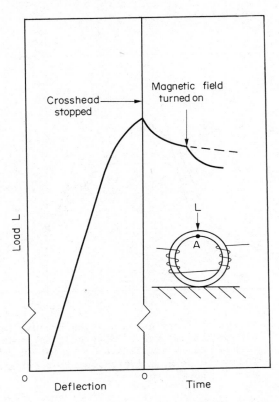

Fig. 6.49. Effect of an alternating magnetic field on stress relaxation in a compressed nickel ring (Cullity and Allen, 1965; copyright Pergamon, used by permission).

Stress Alteration by Surface Removal

As noted earlier, residual stresses within a body must be balanced so that the net force and the net moment are zero. Thus, if tensile residual stresses are present in one region, they must be balanced by compressive residual stresses in an adjacent region. This means that residual stress gradients will exist. If, now, material is in some way removed from the surface, the residual stresses will have to be redistributed to satisfy the conditions of static equilibrium. This will be accompanied by dimensional or shape changes of the body.

A greatly oversimplified example will help to illustrate this. Referring
to Fig. 6.50, assume that three plates, two of equal length and one slightly
shorter (Fig. 6.50a), are arranged side by side and brought to equal lengths
by imposing an elastic compressive load on plates 1 and 3 (Fig. 6.50b). With
the load still applied, the plates are bonded together, as shown, employing
a cement whose shear properties are the same as those of the blocks. Upon
removal of the load, the assembly will lengthen and residual stresses will
develop (Fig. 6.50c), tensile in the center and compressive at the sides.

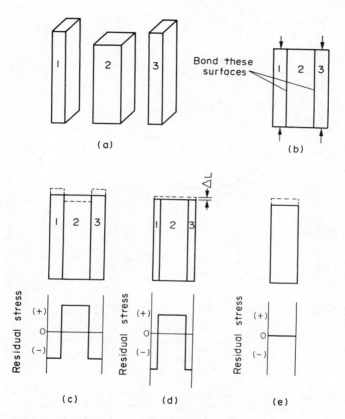

Fig. 6.50. Schematic illustration of stress relaxation by
            surface removal.

Next, suppose that material is removed from the sides of the block in some
manner; for example, by etching. Regions 1 and 3, which are acting to elongate
region 2, are now less effective in this regard because of their reduced cross-
section. Consequently, the block will shorten by an amount ΔL (Fig. 6.50d)
and the residual stress in region 2 will decrease. Note, however, that the
residual stress in regions 1 and 3 will increase as a result of the shortening.
As material removal continues, the block will continue to shorten.
Finally, when regions 1 and 3 have been removed completely, the block will
reach the original length of block 2 and the residual stresses will have disappeared (Fig. 6.50e). The magnitude of the effect will depend upon the magnitude and distribution of the residual stresses and the elastic modulus of the

material. High peak stresses, deep penetration, and low modulus will promote
large dimensional changes as material is removed.

A stress distribution similar to that shown in Fig. 6.50c is typical of
certain types of machining operations. Thus, machined specimens would be
expected to change length as material is etched from the machined surfaces.
Figure 6.51 shows this for Ni-Span-C material that had been lathe-turned prior
to etching. Note particularly the magnitude of the length changes. Once the
damaged layer is removed, no further length change occurs.

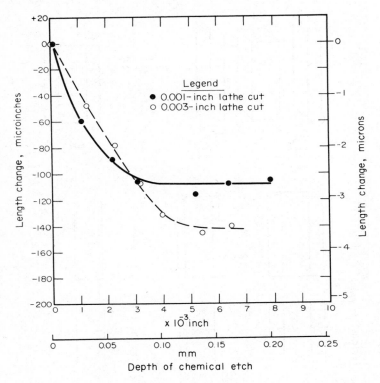

Fig. 6.51.   Dimensional changes in Ni-Span-C on chemical
removal of residual stresses due to machining
(Marschall and Maringer, 1971; copyright ASTM,
used by permission).

Specimen length 75 mm (3 in)

The situation depicted in Figs. 6.50 and 6.51 involves material removal in
such a manner as to maintain a symmetrical distribution of residual stresses.
Such would not have been the case if material had been removed preferentially
from one side. In that instance, warping would have accompanied the length
change, because of unbalanced moments created by material removal.

Alteration of residual stress patterns by material removal has a number of
implications relative to precision devices.

1. If, in service, a dimensionally critical part containing residual
stresses experiences gradual surface removal, whether as a result of

   corrosion, erosion, abrasion, or wear, dimension and shape changes will
   undoubtedly occur.
2. In manufacturing operations in which flatness or straightness is critical, it is important that residual stresses, if present, be kept symmetrical. This means that material removal operations should be accomplished as symmetrically as possible.
3. Certain material removal operations can be employed to reduce the magnitude of residual stresses introduced by previous operations, prior
   to placing a component in service. For example, rough machining or
   grinding may be followed by etching, electropolishing, or low-stress
   grinding to a depth sufficient to remove the previously deformed layer.
   In some cases, this might be an easier way to reduce stresses in a part
   than would subjecting it to other stress relief treatments.

Stress Alteration by Mechanical Working

   In Section 6.3, various sources of residual stress were described. Included
was mechanical working. Thus, if residual stresses from other sources are
present, they can be altered by mechanical working.
   This discussion pertains only to mechanical working procedures that can be
employed to reduce residual stresses. These include stretching, compression,
and reverse bending. Generally speaking, these methods are applicable only
to products of simple shape and uniform cross-sectional area.
   Stretching is a method of stress relief applicable primarily to rolled,
drawn, and extruded products of essentially uniform cross-section. The operation is very simple—the product is merely stretched a preset amount, usually
of the order of a few percent. The mechanism by which residual stress is reduced is also very simple. Referring again to the block depicted in Fig. 6.50c,
region 2 of this block, being originally shorter than regions 1 and 3, is in
a state of residual tension. If now a tensile force is applied to the block
parallel to the long dimension until plastic deformation occurs, region 2 will
deform plastically prior to regions 1 and 3 because of the higher tensile
stress in that region (residual plus applied). Once region 2 has elongated
plastically and the applied load released, the residual stresses will be less
than before because of the reduced length-mismatch between the inner and outer
portions of the block.
   The amount of stretching required to produce virtual elimination of residual
stress ranges from approximately 1 to 2%, as indicated in Fig. 6.52. Even a
fraction of a percent stretching can cause substantial reduction, however.
Stretching in excess of a few percent has no additional benefit. The effectiveness of the treatment can be seen in Fig. 6.53 for 7075 aluminum plate that
was stretched 2% after a cold-water quench. Prior to stretching, the residual
stress range was approximately 200 MN/m^2 (30 ksi).
   Stress relieving by compression is identical in principle to stress relieving
by stretching. Generally, it too is used on products with uniform cross-section, such as cylinders and rectangular blocks. However, it can be adapted to
parts of unsymmetrical shape and variable cross section, such as die forgings
(Barker and Sutton, 1967). For large products where the cross-sectional area
would require extremely large forces, compression is sometimes accomplished in
a series of "bites".
   A somewhat less effective method of reducing residual stresses is by reverse bending, such as is often employed in flattening or straightening operations. Forrest (1947) analyzed this problem for the case of a non-strain-
hardening material, with the results shown in Fig. 6.54. After an initial
plastic bend in one direction, the residual stress distribution is as shown in
Fig. 6.54. As shown in Fig. 6.54b, a small amount of reverse bending
(straightening) reduces the stresses in the outer fibers. However, stress
reductions in the interior are relatively slight.

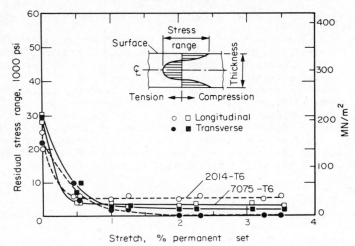

Fig. 6.52.   Effect of stretching on 4.45 cm (1.75 in) 2014-
             T6 and 7075-T6 aluminum plate (Barker and
             Sutton, 1967; copyright ASM, used by
             permission).

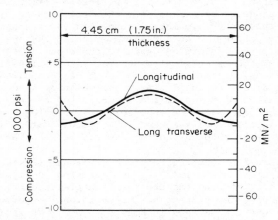

Fig. 6.53.   Residual stresses measured in a 7075
             aluminum plate stretched 2% after cold-water
             quench to T651 condition (Barker and
             Sutton, 1967; copyright ASM, used by
             permission).

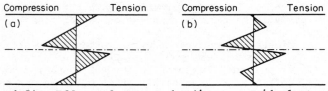

Fig. 6.54.   Effect of reverse bending on residual stress
             distribution: (a) stress distribution after
             plastic bending in one direction, and (b)
             stress distribution after slight restraighten-
             ing (Forrest, 1947; copyright Inst. of Metals,
             used by permission).

After-effects Following Plastic Deformation

It was pointed out in Section 5.3 that strain recovery is often observed
for some period of time following plastic deformation in crystalline materials.
The mechanism for this recovery appears to involve the SR internal stresses
that develop as a result of the plastic strain. Moving dislocations pile up
against obstacles, creating a back stress. Upon removal of the applied stress,
the back stress tends to return the dislocations toward the source, but intersecting dislocations effectively prevent their return. Thus, a balanced stress
system exists to keep the piled-up dislocations from moving in either direction.
The fact that some strain recovery is often observed to occur gradually with
time after plastic deformation suggests that the barriers opposing return of
the dislocations toward their source are more easily overcome by thermal activation than are the obstacles that halted the dislocations initially. Thus,
gradual recovery, i.e., strain in a direction opposite to the prestrain, is
considered to be a "normal" after-effect.

Conditions also exist under which "abnormal" after-effects have been observed, i.e., after removal of the applied stress, strain is observed to continue in the same direction as the prestrain. Barrett (1953) and Edelson and
Robertson (1954) observed this in twisted wires. Following torsional prestrain,
the wires gradually untwisted (recovered) in a normal manner. However, when
the surface layers of the wire were removed by etching, the specimen began to
twist in the direction of the prestrain, as shown in Fig. 6.55. Shortly thereafter, untwisting resumed but at a lesser rate than before etching. This behavior was explained in terms of a coherent oxide film on the wire surface.

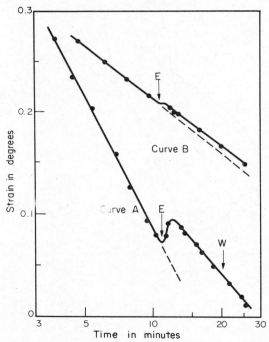

Fig. 6.55. After-effect in polycrystalline zinc, pre-
            twisted to 40 degrees: Curve A; zinc of 99.99%
            purity, containing nitrogen -- etchant applied
            at E, replaced by water at W. Curve B; zinc of
            99.999% purity, melted without contact with
            nitrogen -- etchant applied at E (Barrett, 1953;
            copyright Pergamon, used by permission).

During prestraining, dislocations piled up against the oxide film, as well as against other internal barriers. When the applied stress was removed, strain recovery occurred for reasons advanced earlier. However, when the oxide layer was removed, the dislocation pile-ups previously blocked by the film were able to escape to the surface. This movement of dislocations in the same direction as they had been moving during the prestraining produced additional strain in the same direction. Following this strain burst, normal recovery followed at a reduced rate because the average back-stress on the dislocations was lowered by removal of the surface film.

Other examples of abnormal after-effect have been observed when plastically strained specimens are heated. Figure 6.56 shows results obtained by Vasil'yev and Pal'mova (1966) on copper wires twisted to a surface strain of about 9% and then heated at the rates indicated in the figure. Initially, normal strain recovery is observed. However, as the temperature approaches $200°C$ ($390°F$), the specimen begins to twist in the same direction as that of the prestrain and continues to do so as the temperature is raised additionally. Koiwa and Hasiguti (1965) have reported similar findings for copper specimens twisted at $-195°C$ ($-320°F$) or $-72°C$ ($-100°F$) and then warmed to room temperature. The mechanism for this abnormal after-effect is again traced to the internal stresses arising from the prestrain. Apparently, as the temperature is raised,

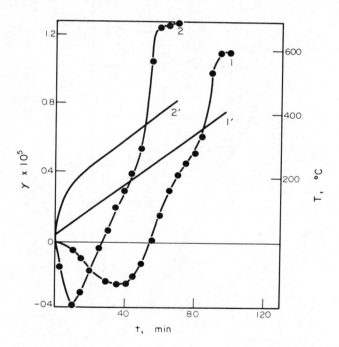

Fig. 6.56.   After-effect in copper specimens after
              twisting to $\gamma = 0.09$: 1' and 2' indicate the
              temperature conditions of the experiment; 1
              and 2 indicate the corresponding strain after-
              effects (Vasil'yev and Pal'mova, 1966;
              copyright Fizika Metallov i Metallovedeniye,
              used by permission of Copyright Agency of
              the USSR).

the barriers to forward dislocation motion are more readily overcome by thermal activation than are the barriers to the return of dislocations toward their source. The result is additional forward strain. Vasil'yev and Pal'mova (1965) assert that such an abnormal after-effect will always be observed in pure metals and alloys where there are no dispersed precipitate particles as obstacles to moving dislocations. However, in precipitation strengthened alloys, only normal after-effects should be expected.

6.5 CHAPTER SUMMARY

In this chapter it is shown that internal stress can be a major contributor to dimensional instability. When internal stresses are present, the possibility always exists that they will not remain constant during the service life of the component in which they reside. Gradual relaxation (thermally activated), material removal (corrosion, abrasion, etc.), shock or vibration, irradiation, or thermal cycling may diminish or redistribute these internal stresses. The net result will be a change in shape and dimensions.

Both long-range (LR) and short-range (SR) internal stresses are described and their origins noted. Although they are analyzed separately, it is pointed out that their effects are additive.

Internal stresses are particularly insidious in precision components because there is no visual evidence of their presence. A component prepared in such a manner as to contain large internal stresses may look identical to one in which internal stress levels are low. Furthermore, conventional mechanical properties such as hardness, Young's modulus, tensile strength, and ductility rarely are good indicators of internal stress levels. Only through relatively complex experimental procedures can the magnitude and distribution of internal stresses be revealed.

There are no general rules that can be applied to indicate what levels of internal stress can be tolerated. This depends entirely on the service conditions, the material employed, and the stability tolerances imposed by the application. Hardened steel gage blocks, for example, may exhibit excellent dimensional stability over several decades if they are handled carefully and maintained at constant temperature and low humidity, even though they contain high levels of residual stress. A beryllium gyroscope rotor, on the other hand, operating under conditions of varying stress and temperature, and for which dimensional instabilities of perhaps 1 part in 10^8 could be troublesome, would likely not tolerate large residual stresses.

Although they are generally viewed with disfavor in precision applications, internal stresses can sometimes be used to advantage. For example, the microcreep rate of a component that is subjected to a steady stress for a long period of time can be reduced by "precreeping" at a slightly higher stress. This prestrain produces SR internal stresses that oppose the applied stress, thereby diminishing the rate of subsequent creep.

The fact that so few rules exist makes it especially important that designers, materials engineers, and manufacturing engineers have a basic appreciation of the sources of internal stresses and the ways in which internal stresses can be controlled. This appreciation, coupled with an empirical approach tailored to a specific application, will provide the groundwork for solving many troublesome dimensional stability problems.

REFERENCES

Abel, A. and Muir, H. 1973a. The effect of cyclic loading on subsequent yielding, *Acta Met.* 21, 93.

Abel, A. and Muir, H. 1973b. The nature of microyielding, *Acta Met.* 21, 99.

Aitchison, I., Honeycombe, R. W. K. and Johnson, R. H. 1962. The thermal
    cycling of uranium, *Prop. Reactor Mater. Eff. Radiat. Damage, Proc. Int.
    Conf.*, D. J. Littler, editor, Butterworths, London, 430.

Andrews, K. W. and Brooksbank, D. 1972. Stresses associated with inclusions
    in steel: a photoelastic analogue and the effects of inclusions in proxi-
    mity, *J. Iron Steel Inst.* 210, 765.

Baldwin, W. M., Jr. 1949. Residual stresses in metals, *Amer. Soc. Test.
    Mater. Proc.* 49, 539.

Barker, R. S. and Sutton, J. C. 1967. Stress relieving and stress control,
    Chapter 10 in *Aluminum, Vol. III, Fabrication and Finishing*, K. R. Van
    Horn, editor, American Society for Metals, Metals Park, Ohio.

Barrett, C. S. 1953. An abnormal after-effect in metals, *Acta Met.* 1, 2.

Bellows, G. 1973. Surface integrity of nontraditional machining processes,
    Proc. Int. Conf. Surface Tech., Society of Manufacturing Engineers, Dearborn,
    Michigan, 469.

Bettman, M., Brown, G. W. and Frankel, J. P. 1953. Dimensional instability
    of uranium and of clad plates subjected to thermal cycling, U.S. Atomic
    Energy Commission Report MTA-36.

Blaha, F. and Langenecker, B. 1959. Ultrasonic investigation of the plasti-
    city of metal crystals, *Acta Met.* 1, 93.

Boas, W. and Honeycombe, R. W. K. 1947. The anisotropy of thermal expansion
    as a cause of deformations in metals and alloys, *Proc. Royal Soc., Ser. A*,
    188, 427.

Bochvar, A. A., Gulkova, A. A., Lolobneva, L. I., Sergeev, G. I. and Tomson,
    G. I. 1958. Effect of thermal cycling on dimensional and structural sta-
    bility of various metals and alloys, *Proc. U.N. Int. Conf. Peaceful Uses
    At. Energy, 2nd, Vol. 5: Properties of Reactor Materials*, United Nations,
    Geneva, 288.

Boggs, B. D. and Byrne, J. G. 1973. Fatigue stability of residual stress in
    shot peened alloys, *Met. Trans.* 4, 2153.

Brooksbank, D. and Andrews, K. W. 1968. Thermal expansion of some inclusions
    found in steels and relation to tessellated stresses, *J. Iron Steel Inst.*
    206, 595.

Brooksbank, D. and Andrews, K. W. 1972. Stress fields around inclusions and
    their relation to mechanical properties, *J. Iron Steel Inst.* 210, 246.

Burke, J. E. and Turkalo, A. M. 1958. The growth of uranium upon thermal
    cycling, *Trans. Amer. Soc. Metals*, 50, 943.

Campbell, J. E. 1971. Shot peening for improved fatigue properties and
    stress-corrosion resistance, Metals and Ceramics Information Center Report
    71-02.

Cullity, B. D. and Allen, C. W. 1965. Accelerated stress relaxation caused
    by an alternating magnetic field, *Acta Met.* 13, 933.

Damask, A. C., Dienes, G. J., Herman, H. and Katz, L. E. 1969. Enhanced dif-
    fusion in alpha-brass during cyclic straining, *Phil. Mag.* 20, 67.

Damask, A. C., Dienes, G. J., Herman, H. and Koczak, M. J. 1973. Enhanced
    diffusion in alpha-brass during cyclic straining; frequency effects, *Phil.
    Mag.* 27, 329.

Davidenkov, N. N., Likhachev, V. A. and Malygin, G. A. 1960. Investigation
    of the irreversible thermal change in shape of zinc, *Phys. Metals Metallogr.
    (USSR)*, 10, 95.

Davidenkov, N. N., Likhachev, V. A. and Ivanov, V. G. 1961. Scale effect in
    irreversible thermal expansion, *Fiz. Metal. Metalloved.* 12, 541.

Decneut, A. and Peters, J. 1973. Continuous measurement of residual stress
    in thin cylindrical pieces using deflection-etching techniques, *Proc. Int.
    Conf. Surface Tech.* Society of Manufacturing Engineers, Dearborn, Michigan,
    262.

Dieter, G. E., Jr. 1961. *Mechanical Metallurgy*, McGraw-Hill, New York.

Digiacomo, G. 1969. Residual stresses in high-strength steel weldments and their dimensional stability during welding and stress relieving, *Mater. Sci. Eng.* 4, 133.

Doroshek, S. I. and Tseytlin, A. M. 1963. Relaxation stability of certain iron-nickel alloys, *Trans. of the 3rd All-Union Scientific Conference on Relaxation Phenomena in Metals and Alloys*, Moscow, Metallurgizdat, 326.

Dubrovin, K. P., Konobeyevskiy, S. T., Levitskiy, B. M., Panteleyev, L. D., Platonov, P. A. and Pravdyuk, N. F. 1962. The relaxation of elastic stresses under neutron irradiation, *Prop. Reactor Mater. Eff. Radiat. Damage, Proc. Int. Conf.*, D. J. Littler, editor, Butterworths, London, 233.

Edelson, B. I. and Robertson, W. D. 1954. The effect of a surface oxide film on torsional relaxation, *Acta Met.* 2, 583.

Eggert, G. L. 1964. The influence of neutron irradiation on the dimensional stability of steel, *Materials Science and Technology for Advanced Applications, Vol. II*, Golden Gate Chapter, American Society for Metals, San Francisco, 62.

Field, M. and Koster, W. P. 1968. Surface integrity in conventional machining chip removal processes, Paper No. EM 68-516, Society of Manufacturing Engineers, Dearborn, Michigan.

Finnie, I. and Heller, W. R. 1959. *Creep of Engineering Materials*, McGraw-Hill, New York.

Forest, J. D. 1974. Advanced composite missile and space design data, General Dynamics Report No. GDCA-CHB72-OO1-6 to U.S. Air Force Materials Laboratory on Contract F33615-72-C-1388.

Forrest, G. 1947. Residual stresses in beams after bending, *Symposium on Internal Stresses in Metals and Alloys*, Institute of Metals, London, 153.

Fox, A. 1971. The effect of extreme cold rolling on the stress relaxation characteristics of CDA copper alloy 510 strip, *J. Mater.* 6, 422.

Freeman, J. W. and Voorhees, H. R. 1956. Relaxation properties of steels and super-strength alloys at elevated temperatures, *Amer. Soc. Test. Mater. Spec. Tech. Publ.* 187.

Freudenthal, A. M. 1958. Stress relaxation in structural materials, A. D. Little, Inc., Report WADC 58-168 to U.S. Air Force.

Goldhoff, R. M. 1971. Creep recovery in heat resistant steels, Chap. 6 of *Advances in Creep Design* (A. I. Smith and A. M. Nicholson, editors), Halsted Press, Wiley, New York.

Graham, W. and Rubenstein, C. 1966. An investigation into the degree and depth of work-hardening produced at the surface of a workpiece by turning, *Proceedings, 7th International Machine Tool Design and Research Conference*, Pergamon Press, Oxford.

Hatch, A. J., Partridge, J. M. and Broadwell, R. G. 1967. Room-temperature creep and fatigue properties of titanium alloys, *J. Mater.* 2, 111.

Hebel, A. G., Jr. and Hebel, A. G., III. 1973. Method for stress relieving metal, U.S. Patent No. 3 741 820.

Henson, F. M. and Inouye, F. 1966. Dimensional instability of aluminum alloys for extreme low temperature cycling application, Aerojet General Corp., Report CR 54829 on Contract NAS 3-2555.

Jakowluk, A. 1969a. The effect of the stress level and the stress amplitude coefficient on the process of vibrational creep, *Polish Society of Theoretical and Applied Mechanics*, 7, 485.

Jakowluk, A. 1969b. The effect of vibration frequency on the vibrational creep process, *Polish Society of Theoretical and Applied Mechanics*, 7, 507.

Johnson, R. H. and Honeycombe, R. W. K. 1962. Some microstructural observations during the thermal cycling of zirconium, *J. Less-Common Metals*, 4, 226.

Kahles, J. F. and Field, M. 1971. Surface integrity guidelines for machining -
    1971, Paper No. IQ71-240, Society of Manufacturing Engineers, Dearborn,
    Michigan.

Kennedy, A. J. 1956. Effect of fatigue stresses on creep and recovery, *Proc.
    Int. Conf. Fatigue Metals*, Inst. Mech. Engrs., London, 401.

Kennedy, A. 1958. The dependence of microcreep properties on the development
    of fatigue in lead, *J. Inst. Metals*, 87, 145.

Kennedy, A. 1963. *Processes of Creep and Fatigue in Metals*, Wiley, New York.

Kempf, L. W. and Van Horn, K. R. 1942. Relief of residual stress in some
    aluminum alloys, *Trans. AIME*, 147, 250.

Khenkin, M. L. 1967. Relationship between changes in fine structure and
    resistance to microplastic deformation in metals and alloys, *Russ. Met.*,
    No. 4, 41.

Khenkin, M. L., Lokshin, I. Kh., Levina, N. K. and Simeonov, S. L. 1966.
    Stabilization of silumin by thermal cycling, *Russ. Cast. Prod.*, February,
    62.

Koiwa, M. and Hasiguti, R. R. 1965. Abnormal plastic after-effect in twisted
    copper, *Acta Met.* 13, 673.

Koster, W. P. 1973. Surface integrity of traditional machining processes,
    Proc. Int. Conf. on Surface Tech. Society of Manufacturing Engineers,
    Dearborn, Michigan, 442.

Koster, W. P., Field, M., Fritz, L. J., Gatto, L. R. and Kahles, J. R. 1970.
    Surface integrity of machined structural components, Metcut Research
    Associates, Inc., Report AFML-TR-70-11 to U.S. Air Force Materials
    Laboratory

Kushner, J. B. 1962. Stress in electroplated metals, *Metal Progr.* No. 2, 88.

Laszlo, F. 1943. Tessellated stresses. Part 1, *J. Iron Steel Inst.* 147,
    173.

Lement, B. S., Averbach, B. L. and Cohen, M. 1951. The dimensional behavior
    of Invar, *Trans. Amer. Soc. Metals*, 43, 1072.

Lessells, J. M. and Broderick, R. F. 1956. Shot-peening as protection of
    surface damaged propeller blade materials, *Proc. Int. Conf. on Fatigue of
    Metals*, Inst. Mech. Engrs., London, 617.

Likhachev, V. A. 1961. Microstructural strains due to thermal anisotropy,
    Sov. Phys. Solid State, 3, 1330.

Likhachev, V.A., Malygin, G.A., Nikiforov, A.V., and Vladimirov, G.V. 1963.
    Creep of Zinc as a result of thermal cycling, *Fiz. Metal. Metallored.*, 16,
    No.6, 908-917.

Lokshin, I. Kh. 1965. Vibration treatment and dimensional stabilization of
    castings, *Russ. Cast. Prod.* No. 10, 454.

Lokshin, I. Kh. 1970. Heat treatment to reduce internal stresses in beryl-
    lium, *Metal Sci. Heat Treat. (USSR)*, May, 426.

Lubahn, J. D. and Felgar, R. P. 1961. *Plasticity and Creep of Metals*, Wiley,
    New York.

Manjoine, M. J. 1971. Measuring stress relaxation by a compliance method,
    J. Mater. 6, 253.

Maringer, R. E., Cho, M. M. and Holden, F. C. 1968. Stability of structural
    materials for spacecraft applications, Battelle Mem. Inst., Final Report
    on Contract NAS 5-10267.

Marschall, C. W., Hoskins, M. E. and Maringer, R. E. 1969. Stability of
    structural materials for spacecraft applications for the orbiting astro-
    nomical observatory project, Battelle Mem. Inst., Final Report on Contract
    NAS 5-11195.

Marschall, C. W. and Maringer, R. E. 1971. Stress relaxation as a source of
    dimensional instability, *J. Mater.* 6, 374.

Masubuchi, K. 1965. Nondestructive measurement of residual stresses in metals and metal structures, Battelle Mem. Inst., Report RSIC-410 to U.S. Army Missile Command, Redstone Arsenal, Alabama.

Maykuth, D. J. 1968. Residual stresses, stress relief, and annealing of titanium and titanium alloys, Defense Metals Information Center Report 2-23.

McClintock, F. A. and Argon, A. S. 1966. *Mechanical Behavior of Materials*, Addison-Wesley, Reading, Massachusetts.

McGoldrick, R. T. and Saunders, H. E. 1943. Some experiments in stress-relieving castings and welded structures by vibration, *J. Amer. Soc. Nav. Eng.* 55, 589.

Morrow, J. and Sinclair, G. M. 1958. Cycle-dependent stress relaxation, *Symp. on Basic Mechanisms of Fatigue*, Amer. Soc. Test. Mater. Spec. Tech. Publ. 237, 83.

Mueller, J. and Marschall, C. W. 1973. Unpublished data.

Nadai, A. 1963. *Theory of Flow and Fracture*, Vol. 2, McGraw-Hill, New York.

Nevill, G. E., Jr. and Brotzen, F. R. 1957. The effect of vibration on the static yield strength of a low carbon steel, *Amer. Soc. Test. Mater. Proc.* 57, 751.

Nordstrom, T.V. and Rohde, R.W. 1972. Mechanical propertier and stress relaxation of Be-Cu Alloy 25, Sandia Laboratory Report SC-DR-770471.

Odegard, B. C. and Maringer, R. E. 1971. The room temperature creep behavior of wrought and as-welded Ti-6Al-4V, Sandia Laboratories Report SCL-DR-710069.

Odegard, B. C. and Thompson, A. W. 1973. Low temperature creep of Ti-6Al-4V, Sandia Laboratories Report SLL-73-5289 (submitted to *Met. Trans.*).

Okazaki, K. and Conrad, H. 1973. Stress relaxation of Ti-N alloys at 300 and 400 K, *Trans. Jap. Inst. Metals*, 14, 368.

Orner, G. M. and Kulin, S. A. 1964. Development of stress relief treatments for high strength aluminum alloys, ManLabs Annual Report on Contract NAS 8-11091.

Orner, G. M. and Kulin, S. A. 1965. Development of stress relief treatments for high strength aluminum alloys, ManLabs Annual Report on Contract NAS 8-11091.

Pattinson, E. J. and Dugdale, D. S. 1962. Fading of residual stresses due to repeated loading, *Metallurgia*, Nov., 228.

Pugh, S. F. 1957. The mechanism of growth of uranium on thermal cycling in the alpha range, *J. Inst. Metals*, 86, 497.

Reimann, W. H. 1971. Room temperature creep in Ti-6Al-4V, *J. Mater.* 6, 926.

Roberts, A. C. 1960. Thermal cycling creep of alpha-uranium, *Acta Met.* 8, 817.

Samuels, L. E. 1956. The nature of mechanically polished metal surfaces: The surface deformation produced by the abrasion and polishing of 70:30 brass, *J. Inst. Metals*, 85, 51.

Terminasov, Yu. S. and Yakhontov, A. G. 1959. Distortion of lattice structure of metals by grinding, *Metalloved. Term. Obrab. Metal.* No. 5, 19.

Turley, D. M. 1968. Deformed layers produced by machining 70/30 brass, *J. Inst. Metals*, 96, 82.

Turley, D. M. 1971. Dislocation substructures and strain distributions beneath machined surfaces of 70/30 brass, *J. Inst. Metals*, 99, 271.

Turley, D. M. 1973. Deformed layers produced by machining, *Proc. Int. Conf. on Surface Technology*, Society of Manufacturing Engineers, Dearborn, Michigan, 39.

Turley, D. M. and Samuels, L. E. 1972. Dislocation substructures and strain distributions beneath abraded surfaces of 70/30 brass, *J. Aust. Inst. Metals*, 17, 114.

Van Horn, K. R.   1953.   Residual stresses introduced during metal fabrication, *Trans. AIME*, 197, 405.

Vasil'yev, D. M. and Pal'mova, N. I.   1966.   Irreversible change in size (after-effect) of specimens heated after plastic deformation, *Fiz. Metal. Metalloved.* 21, 242.

Wick, C. H.   1971.   Mechanical stress relieving by vibrating, *Manu. Engrg. and Management,* 67, 20.

Wood, W. A.   1967.   Instability of titanium and Ti-6Al-4V alloy at room temperature, Columbia University Report No. 45 on Contract NONR 266 (91).

Wozney, G. P. and Crawmer, G. R.   1968.   An investigation of vibrational stress relief in steel, *Weld. Res. (New York),* 47, 411s.

Zeyfang, R., Martin, R. and Conrad, H.   1971.   Low temperature creep of titanium, *Mater. Sci. Eng.* 8, 134.

Zlatin, N., Field, M. and Gould, J. V.   1963.   Machining of refractory materials, Metcut Research Associates, Inc., Final report ASD-TDR 63-581 to U.S. Air Force.

Chapter 7.

Microstructural Effects: Dimensional Instability from Microstructural Changes

An awareness of the role of microstructural changes in causing dimensional instability originated prior to the advent of space age technology. In the United States, considerable attention was given to this problem during World War II, in connection with materials for precision instruments. The most intensive efforts were carried out in the Instrumentation Laboratory of the Massachusetts Institute of Technology. Much of the early work at MIT focused on the dimensional stability of steel. Results of this work were summarized in a series of papers spanning four years (Fletcher *et al.*, 1945, 1948; Averbach *et al.*, 1948; and Lement *et al.*, 1949). Later investigations at MIT covered literally dozens of materials, both ferrous and nonferrous, and material conditions in attempts to develop treatments that would provide optimum dimensional stability (Lement and Averbach, 1955; Schetky, 1957).

The procedures employed for measuring small dimensional changes in the MIT studies are described in Chapter 8. They permitted measurement of unit length changes with a precision of about ± 1 to 5×10^{-6}. Accordingly, in reporting dimensional stability data, unit length changes were reported to the nearest 5×10^{-6}. This information was extremely useful at the time it was generated and remains useful even today. However, in several respects it is not adequate for many current needs. First, it does not include data for recently developed materials. Second, it is of limited help in selecting materials for applications where unit dimensional changes as small as 10^{-7} or 10^{-8} are of significance.

Examination of the literature and government reports concerning dimensional stability in the absence of external stress reveals little new information. The majority of long term stability work reported on newer engineering materials, such as beryllium and titanium alloys, have employed measurement accuracies similar to those reported in the MIT studies. One exception is the work on development of ultrastable gage blocks conducted by the U.S. National Bureau of Standards (Meyerson *et al.*, 1960, 1961, 1964, 1968, 1969). Using advanced metrology techniques and extremely careful control of environment and testing conditions, Meyerson and his coworkers were able to detect changes in unit length approaching 10^{-7}. An example of length measurement data obtained over a period of nearly 8 years on hardened and stabilized AISI 52100 steel gage blocks is shown in Fig. 7.1. Both the stability of the material and the precision of the measurements are impressive.

Other methods for assessing long term dimensional stability on a finer scale than possible heretofore are currently undergoing development, as discussed in Chapter 8. For example, laser interferometry is being used in an attempt to detect unit dimensional changes of as little as 5×10^{-8} in low-expansion mirror materials over periods of several years (Kurtz, 1973).

The dearth of reported data in this area reflects in part the experimental difficulties in obtaining meaningful results and the large investments of both time and money required. It may also reflect the fact that information of this type is regarded as highly proprietary by industrial laboratories that have developed it on their own.

The major emphasis in this chapter is on the various mechanisms by which

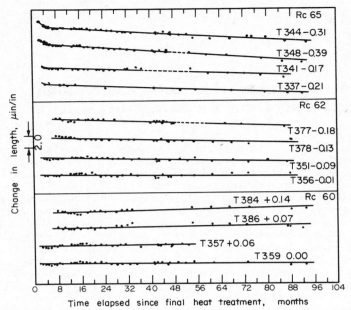

The upper two curves in each section are plots of specimens which were directly quenched
from the austenitizing temperature. The lower two curves in each section are plots of specimens
which were martempered at 300 F and then quenched to room temperature. The values appearing to
the right of the block identification number are the computed stability in microinch per inch
per year. A negative sign indicates a contraction, while a positive sign indicates an expansion.

Fig. 7.1.   Dimensional stability of hardened, stabilized,
             modified Type 52100 steel gage blocks tempered
             to different levels of hardness (Meyerson
             et al., 1968, copyright ASTM, used by
             permission).

materials may undergo dimensional changes as a result of internal changes at
the microstructural level. For simplicity, most of the discussion is divorced
from stress effects, either external or internal, even though it is recognized
that external stress can influence microstructural changes and that internal
stress can gradually change with time and produce dimensional changes that are
difficult to differentiate from those due to microstructural changes.

Of the various external factors that may influence microstructure, temperature is of greatest interest and will receive most of the attention here. It
is important from two standpoints: (1) in a given material, the equilibrium
or stable microstructure is a function of the temperature; and (2) the rate at
which a nonequilibrium microstructure changes toward a stable structure depends
on the temperature; i.e., the change is thermally activated. Thus, in hardenable steel, for example, the stable structure at room temperature is a mixture
of body-centered-cubic iron and iron-carbide. However, after hardening, neither
of these phases is present. Instead, the microstructure consists of martensite
(a body-centered-tetragonal arrangement of iron atoms supersaturated with
carbon) and, in some cases, retained austenite (face-centered-cubic iron).
Although this is not the stable microstructure at room temperature, the rate
at which it transforms to a stable structure may be very slow at this temperature. In principle then, any temperature change should produce some microstructural change and an accompanying volume change but, in fact, the change
may occur so slowly that it can safely be ignored.

The types of microstructural changes considered here are:

- changes in vacancy concentration
- atom rearrangement and ordering
- precipitation and re-solution of a second phase
- phase changes

Specific examples of each are given and the magnitude of the associated dimensional change estimated. Brief discussions of dimensional changes associated with external fields other than temperature are also presented in this chapter. These include magnetic and electrical fields and irradiation.

One additional point should be emphasized. The discussion here points out the very real potential for dimensional instabilities of significant magnitude to arise from microstructural changes. However, when instabilities are actually observed, it can require an extremely complicated piece of detective work to pin down the mechanism of the instability. The instability may result from any of the several microstructural effects described here or it may be from alteration of internal stress, as discussed in Chapter 6.

7.1 CHANGES IN VACANCY CONCENTRATION

A vacancy in a crystal is simply an unoccupied lattice site. As pointed out by McClintock and Argon (1966), thermodynamic equilibrium requires the presence of some vacancies, with the number being a function of the temperature and of the material.

If the temperature of a material is raised and lowered slowly, the number of vacancies is expected to gradually increase on heating and decrease on cooling. However, if a material were held at a high temperature until the required number of vacancies were present and then rapidly quenched to room temperature, it is likely that many of the vacancies formed at the high temperature would remain trapped in the lattice at room temperature. Return to the equilibrium number of vacancies would now be slow, depending on the thermal activation parameters. Obviously, the volume of a material is increased by the presence of vacancies. If the number of vacancies gradually diminishes with time, dimensions will also change.

The magnitude of the effect of vacancies on dimensions can be estimated by examining the expression for the fractional concentration of vacancies, c_v, as a function of absolute temperature, T:

$$c_v = \frac{n}{N} = e^{-h_f/kT}$$

where N is the total number of atoms, n is the number of vacant sites, h_f is the activation energy to form a vacancy, and k is the Boltzman constant (McClintock and Argon, 1966). Consider the case of copper, for which h_f is approximately 0.9 eV.

At room temperature ($\sim$300K) where kT is approximately 1/40 eV, the calculated vacancy concentration is very small.

$$c_v \text{ (at 300K)} = e^{-36} \cong 2.3 \times 10^{-16}.$$

Expressed another way, for every one million occupied lattice sites, there are approximately 2×10^{-10} vacancies. At temperatures approaching the melting point, c_v is much greater. For T = 1200K,

$$c_v \text{ (at 1200K)} = e^{-9} \cong 1.3 \times 10^{-4}$$

or, for every one million occupied lattice sites, there are approximately 130
vacancies. If all of these vacancies formed at 1200K remain in the lattice
at room temperature, the volume at room temperature will be 0.013% greater
than it was originally. Linear dimensions will have increased by about one-
third this amount, or 0.004%. In terms of unit lengths, this is a growth of
40×10^{-6}. If the vacancies then gradually diminish in number by thermally
activated diffusion, the dimensions will gradually shrink toward their original
values.

Whether vacancies actually are an important source of dimensional instability by the mechanism described here is unknown. Rapid cooling from elevated
temperature produces other side effects, such as residual stresses, retention
of unstable phases, and supersaturation, whose gradual change with time can
overshadow vacancy effects. Nonetheless, the calculations indicate that changes
in vacancy concentration have the potential for producing significant dimensional changes.

7.2 ATOM REARRANGEMENT AND ORDERING

In solid solutions of one element dissolved in another, the atom arrangements
are considered to be random. Under certain conditions, a tendency may exist
for the formation of a somewhat more orderly structure. The extreme case
would be when all of the solute atoms occupy certain predictable lattice sites
and have predictable nearest neighbors. Since an ordered structure has a
different specific volume than a random structure, gradual ordering can produce
dimensional changes. As noted earlier, both the true equilibrium structure
and the rate at which equilibrium is achieved are governed by the temperature.
Thus, at any particular time, a material that is susceptible to an ordering
reaction may be in an intermediate state of partially ordered and partially
random atom arrangements.

An example of dimensional instability attributed to a subtle rearrangement
of atoms was reported by Lement *et al.* (1951). In Invar, carbon atoms in
solution in the Fe-36Ni lattice are believed to gradually rearrange themselves
into a more stable configuration with time at room temperature or slightly
above, leading to a gradual length increase (see Fig. 9.11). The rate at which
the change occurs reaches a maximum near $95°C$ ($200°F$). Heating above about
$205°C$ ($400°F$) reverses the trend and destroys the more stable configuration of
carbon atoms developed at somewhat lower temperatures.

7.3 PRECIPITATION AND RE-SOLUTION OF A SECOND PHASE

Many metals are strengthened by deliberate introduction of alloying elements
that have limited solid solubility in the host lattice. By proper selection
of composition and processing, a structure can be developed consisting of
finely dispersed particles of a second phase, usually an intermetallic compound, in a matrix consisting primarily of the host metal. An example given
in Chapter 3 described copper additions to aluminum. Other commercially
important examples include beryllium added to copper, carbon added to iron,
aluminum added to magnesium, and titanium added to nickel.

Dimensional stability considerations again center around the question of
what constitutes a thermodynamically stable microstructure at the service
temperature. Clearly, the solution annealing treatment and rapid quench
usually given to such alloys prior to the precipitation or aging treatment
does not produce a stable structure. The alloy in this condition is highly
supersaturated with dissolved alloying element. If thermal activation is

sufficient to permit atom diffusion, a second phase will gradually precipitate from the matrix and the dimensions will change. Figure 7.2 illustrates the dimensional change observed in several aluminum alloys held at room temperature for several weeks after solution annealing. Notice that the dimensions are still changing after 1000 hr. The magnitude of this change is obviously significant for many precision applications. However, these alloys would seldom be used in the solution annealed condition. Instead, they would be heated to a moderate temperature, ranging from 120°C (250°F) for 7075 aluminum to 190°C (375°F) for 2024 aluminum, to hasten the precipitation reaction and the accompanying dimensional changes. This would accomplish the dual purpose of hardening the material and approaching microstructural stability in a short time. Notice the use of the word approaching. In the strictest sense, the thermodynamically stable structure at room temperature is not identical to that at the precipitation hardening temperature. Generally, the lower the temperature, the greater the equilibrium quantity of precipitate. Thus, in theory at least, further small dimensional changes could occur with time at room temperature following precipitation hardening. Conversely, if an equilibrium two-phase structure has been established at a given temperature, any increase in temperature will tend to cause some of the precipitate to dissolve. This, too, will cause dimensional changes.

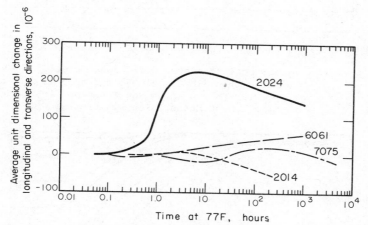

Fig. 7.2.   Average unit dimensional changes during room-
             temperature aging of cold-water-quenched
             sheet for various aluminum alloys (Hunsicker,
             1967; copyright ASM, used by permission).

Another example of the magnitude of the dimensional change that can accompany precipitation is shown in Fig. 7.3 for a 0.2% carbon steel. Initially, this steel was heated to 710°C (1310°F) to dissolve a small fraction (~1/10) of the total carbon present in the bcc iron matrix. The remaining carbon was present as dispersed particles of iron-carbide. The steel was then water-quenched to room temperature, trapping the dissolved carbon in solution. As illustrated in Fig. 7.3, significant dimensional changes occurred as a function of time after quenching, associated with the gradual precipitation of iron-carbide from the bcc iron matrix. Notice the pronounced effect of temperature on the rate of the reaction. That the observed changes are indeed due to carbon is shown in Fig. 7.4 The magnitude of the dimensional change is a function of the solution annealing temperature—in other words, as the temperature is raised, more carbon is dissolved in the bcc iron and more precipitate forms on subsequent aging. However, when the carbon is removed almost entirely from the steel by a decarburization operation, virtually no dimensional change is observed after quenching from 710° C.

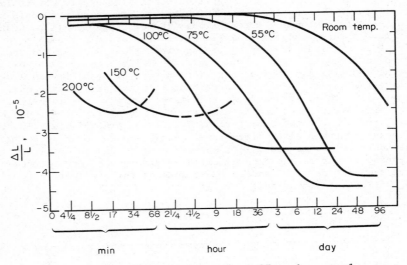

Fig. 7.3.   Dimensional change in 0.2% carbon steel, quenched from 710°C (1310°F) and aged at temperature shown (Bohm and Schumann, 1965; copyright Neue Hutte, used by permission).

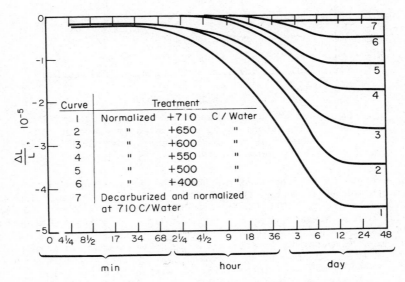

Fig. 7.4.   Dimensional change in 0.2% carbon steel, aged at 75°C after quenching from various temperatures (Bohm and Schumann, 1965; copyright Neue Hutte, used by permission).

It is possible to compute the dimensional change associated with precipitation of an intermetallic phase from the matrix if the crystal structure and interatomic spacing are known for both phases. Consider, for example, a magnesium alloy, AZ31B, containing 3% aluminum and 1% zinc by weight. For Mg–Al–Zn alloys containing more than 2% aluminum, an intermetallic compound, $Mg_{17}Al_{12}$, will be present at equilibrium at room temperature, in addition to the solid solution matrix phase. The Mg–Al–Zn solid solution phase has a hexagonal crystal structure whose lattice parameter depends on the amount of Al and Zn in solution. The intermetallic compound has a cubic structure. As the compound precipitates from the matrix, the volume and linear dimensions increase.

The magnitude of the dimensional increase is calculated below, based on the following facts and assumptions (Dow Chemical Co., 1957).

Composition

Element	Weight %	Atomic %
Mg	96	96.9
Al	3	2.73
Zn	1	0.37

Initial Structure: Solution annealed condition
- Single phase (close–packed hexagonal) with all Al and Zn present in solid solution
- Lattice parameters of cph phase
  a (base of hexagon) = 3.1906Å
  c (height of hexagon) = 5.1841Å
- Volume of cph unit cell = 3/2 a^2c
- Atoms per cph unit cell = 2

Structure after Precipitation of $Mg_{17}Al_{12}$
- Assume matrix phase (cph) containing half of the total Al and all of the Zn, plus $Mg_{17}Al_{12}$ (cubic) containing half of the total Al
- Lattice parameters of cph phase
  a = 3.1957Å
  c = 5.1900Å
- Lattice parameter of cubic phase
  a = 10.54Å
- Volume of cubic unit cell = a^3
- Atoms per cubic unit cell = 58

For the initial structure, the volume occupied by 100 atoms is

$$V_1 = \frac{\text{Number of atoms in cph phase}}{\text{atoms/cph unit cell}} \text{ x volume of cph unit cell}$$

$$= \frac{100}{2} \times \frac{\sqrt{3}}{2} (3.1906\text{Å})^2 (5.1841\text{Å})$$

$$= 2285.17\text{Å}^3$$

For the structure following precipitation of $Mg_{17}Al_{12}$, the volume occupied by 100 atoms is:

$$V_2 = V_{cph} + V_{cubic}$$

$$V_{cph} = \frac{96.701}{2} \times \frac{\sqrt{3}}{2} (3.1957\text{Å})^2 (5.1900\text{Å})$$

$$= 2219.38\text{Å}^3.$$

$$V_{cubic} = \frac{\text{number of atoms in cubic phase}}{\text{atoms/cubic unit cell}} \text{ x volume of cubic unit cell}$$

$$= \frac{3.299}{58} (10.54\text{Å})^3 = 66.60 \text{ Å}^3$$

$$V_2 = 2219.38\text{Å}^3 + 66.60\text{Å}^3 = 2285.98\text{Å}^3.$$

The percentage change in volume accompanying this amount of precipitation is

$$\frac{V_2-V_1}{V_1} \times 100 = \frac{0.81}{2285.17} \times 100 = 0.0354\%.$$

Linear dimensional changes would be about one-third this amount or 0.012%. Expressed in another way, this represents a unit length change of 120×10^{-6}. Similar calculations for a Mg-Al-Zn alloy containing 9% Al (AZ91C) indicate dimensional changes of about 10 times this magnitude.

Although in principle all precipitation reactions produce volume and length changes, there are specific cases where these changes are comparatively small, because the specific volume of the precipitate is nearly the same as that of the matrix. In 6061 aluminum alloys, for example, the length change accompanying precipitation hardening is only a small fraction of that observed for several other aluminum alloys, as indicated in Fig. 7.5.

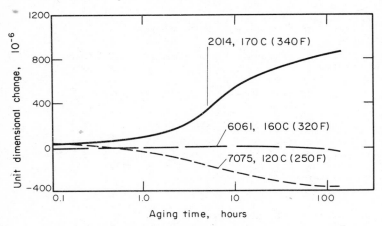

Fig. 7.5.   Dimensional change as a function of time at the precipitation heat treating temperature employed to produce the T6 temper of three aluminum alloys (Hunsicker, 1967; copyright ASM, used by permission).

7.4   PHASE CHANGES

Any material that undergoes a transformation from one crystal structure to another or from a noncrystalline to a crystalline structure will change in volume. The magnitude of the volume change will depend on the efficiency of the atom-packing in the specific structures involved. Because of its great importance in engineering applications, steel is used here to illustrate the importance of this phenomenon with respect to the dimensional stability of materials that are susceptible to phase changes. Steel can also undergo each of the types of microstructural changes already described in Sections 7.1 through 7.3. One of these, precipitation of a second phase (carbide) is of great importance in dimensional stability of steel and is also discussed in this section.

The subject of phase changes and dimensional changes in steel is discussed quantitatively and in detail by Lement (1959). In view of his coverage, the discussion here is relatively brief. Its principal purpose is to illustrate how phase changes and other changes that occur in steel can produce dimensional changes of significant magnitude and to emphasize the importance of recognizing and controlling them.

Plain-carbon Steels, Low-alloy Steels, and Tool Steels

With the exception of maraging steels and certain stainless steels, to be described later, steels are basically iron alloyed with a small amount of carbon, often less than 1%. At high temperatures, the stable structure is referred to as austenite as shown in the iron-carbon phase diagram, Fig. 7.6. Austenite is a face-centered-cubic arrangement of iron atoms, with dissolved carbon atoms located in random interstitial positions. Below about 720° C (1330° F), the stable structure is a mixture of ferrite (bcc iron with interstitial carbon) and iron-carbide. If small percentages of alloying elements are present, for example, Cr, Ni, Mo, etc., they will tend to dissolve in the austenite at high temperatures and to distribute themselves among the ferrite and carbide phases at low temperatures. Nonetheless, the *stable* structure at low temperatures will still be ferrite plus carbide.

The fact that steels can undergo a phase change as the temperature is raised or lowered is the basis of the heat treatability of steels. At the same time, this can be the source of many dimensional instability problems.

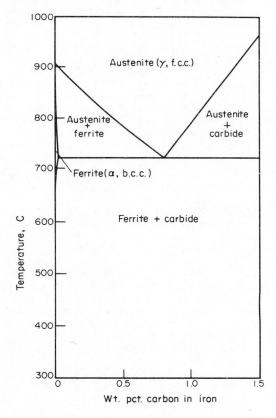

Fig. 7.6.   A portion of the iron - carbon phase diagram.

The latter result from the fact that austenite is more dense than ferrite; when compared at the same temperature, the specific volume of austenite is about 4.9% less than that of ferrite.* Stated in another way, when a steel transforms totally from austenite to ferrite, it will show an increase in linear dimensions of approximately one-third this amount, or 1.6%. This poses significant problems in heat treatment, as discussed in Chapter 6. It can also cause dimensional stability problems if some of the austenite remains untransformed after the heat treatment, which is a common occurrence.

The hardening treatment for steel begins with heating to a temperature at which the ferrite phase rapidly transforms to austenite. Typically this temperature is in the range of 815° to 870°C (1500° to 1600°F). The carbide particles that were dispersed in the ferrite rapidly go into solid solution in the austenite. If the austenite is now cooled very slowly back to room temperature, the process is reversed, i.e., the austenite transforms to a stable mixture of ferrite—plus—carbide by a diffusion controlled process. To get greater hardness and strength, however, the austenite is cooled at a rate sufficient to prevent this diffusion controlled transformation. In its place, a diffusionless transformation occurs in which the austenite transforms to a ferrite—like phase that is supersaturated with dissolved carbon. This phase is relatively hard and strong and is referred to as martensite. The excess carbon in the martensite distorts the normal bcc crystal structure of ferrite to a body—centered—tetragonal structure. The austenite—to—martensite transformation takes place over a range of temperatures designated M_{start} to M_{finish} and may, or may not, be completed when the steel reaches room temperature. In many steels, notably those containing high carbon and/or high alloy content, some austenite may remain untransformed at room temperature. This is referred to as retained austenite.

The effect of the above sequence of events on the linear dimensions of a steel specimen is shown schematically in Fig. 7.7. Point A denotes a stable ferrite—plus—carbide structure at room temperature. Heating causes a length increase along curve ABC as a result of normal thermal expansion. From C to D, the specimen shortens because of the transformation to austenite. Thermal expansion of the austenite is reflected in the DE part of the curve. Cooling the steel back to room temperature from point E can yield several results, as indicated in Fig. 7.7.

(a) Slow cooling will follow the curve $EDFBA$;† the final dimensions will be the same as the original dimensions and the structure will be stable (ferrite plus carbide).

(b) Rapid cooling will follow the curve $EDFGHJ$; the final structure will be martensite and the dimensions will be greater than the original dimensions by the amount JA. For a 1% carbon steel, linear dimensions at point J will exceed those at point A by about 5000×10^{-6}mm/mm (in/in).

(c) If the steel has an M_f temperature below room temperature, rapid cooling will follow the curve $EDFGK$; the final structure will consist of a mixture of martensite and retained austenite and the dimensions will be smaller than the original dimensions by the amount KA. Additional cooling to subzero temperatures can transform the remainder of the austenite to martensite along curve KLN. The return to room temperature will now follow curve NLJ.

---

*This depends on carbon content—the value stated is for carbon—free iron. At 1.0% carbon, the specific volume of austenite is about 4.1% less than that of ferrite.
†If both heating and cooling are extremely slow, the path followed on cooling will be virtually identical to that on heating.

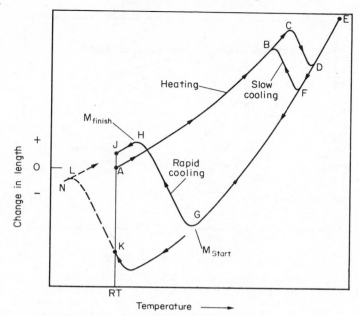

Fig. 7.7.   Schematic depiction of length changes
            accompanying heat treatment of steel. The
            starting point (A) represents a stable
            structure of ferrite plus carbides.

Neither the structure associated with point J nor that associated with point
K is stable. The martensite at point J is supersaturated with carbon and will
tend to gradually precipitate carbide particles and the bct martensite will
revert to bcc ferrite. As this occurs, the steel will gradually shrink toward
point A. This can be accomplished deliberately by "tempering" the martensite
in the range from about 100° to 650°C (200° to 1200°F) to precipitate carbides
more rapidly. This tempered martensite is merely a ferrite–plus–carbide mixture which, as noted earlier, is the stable structure for steels. The retained
austenite at point K likewise will tend to transform to martensite, especially
if it is exposed to low temperatures, with accompanying expansion toward point
A and beyond to point J. This can be accomplished deliberately by subzero
cooling after the original quench to room temperature.
    It is evident from the foregoing that an as–hardened steel can contain two
phases, each of which is unstable. As they gradually change toward a more
stable state, one will cause contraction and the other expansion. Thus, a
variety of types of dimensional instability might be expected. Several examples
from the work of Lement (1959) will help to illustrate this.
    Two steels were examined, W1 and L3. Both contained approximately 1%
carbon. L3 contained, in addition, about 1.5% chromium and 0.2% vanadium.
After quenching from the austenitizing temperature, the structures were primarily martensite and retained austenite. The percentages of retained austenite were approximately 9 and 7% for W1 and L3, respectively. Subcooling to
-195°C (-320°F) reduced these percentages to 3 and 2, respectively. Both as-
hardened and subcooled specimens were then tempered for 1 hr at various temperatures from 20° to 260°C (68° to 500°F). Following tempering, specimen
length was monitored at room temperature as a function of time. The results
are shown in Figs. 7.8 and 7.9.

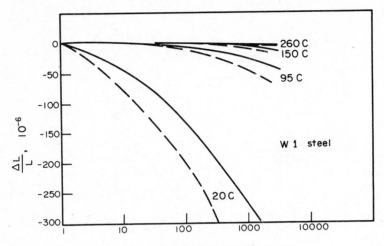

Fig. 7.8.   Length changes at 20°C in the W1 steel
             tempered 1 hr at temperatures indicated.
             Dashed curves are for specimens cold treated
             at -195°C prior to tempering (after Lement,
             1959; copyright ASM, used by permission).

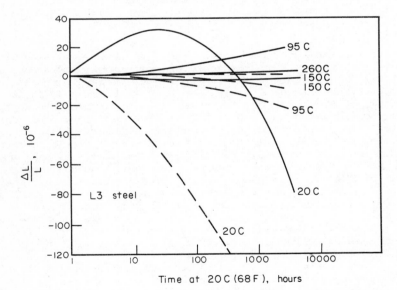

Fig. 7.9.   Length changes at 20°C in L3 steel tempered
             1 hr at temperatures indicated. Dashed curves
             are for specimens cold treated at -195°C
             prior to tempering (after Lement, 1959;
             copyright ASM, used by permission).

The data show clearly that both the martensite and austenite undergo gradual changes with time at room temperature:

austenite → martensite (volume increase)
martensite → ferrite plus carbide (volume decrease)

In the plain carbon steel W1, the martensite tempering reaction is predominant and a net contraction is observed. Cold treatment to reduce the amount of austenite makes the martensite tempering reaction even more predominant, as indicated by the even larger contraction with time. By deliberately tempering the W1 steel at elevated temperatures, to accelerate precipitation of carbides, the stability of the structure at room temperature is vastly improved. Notice that after tempering at 260°C (500°F) the steel shows virtually no length change in several thousand hours.

In the as-hardened low alloy steel L3, on the other hand, the transformation of austenite to martensite predominates initially and a net expansion is observed. This probably reflects the slower rate of the martensite tempering reaction in this steel, due to the presence of chromium. Since the chromium tends to concentrate in the carbide phase, it must diffuse to precipitation sites. This requires more time than diffusion of carbon alone. If most of the retained austenite is eliminated by a cold treatment, the L3 steel exhibits a contraction with time at room temperature, just as does the W1 steel. However, the expansion rate remains noticeably slower for the L3 than for the W1, for reasons already noted.

These data provide a striking example of the potentially damaging effects of microstructural changes on dimensional stability. At the same time, it is encouraging to note that, through knowledge of material behavior and through experiments of the type conducted by Fletcher, Averbach, Lement, and others, a material with a potentially large instability at room temperature can perform in the manner shown in Fig. 7.1.

Maraging Steels and Precipitation Hardening Stainless Steels

Maraging steels and precipitation hardening stainless steels differ markedly in composition from the plain-carbon and low-alloy steels already described. They contain substantially more alloying elements and little, if any, carbon. For example, a typical low alloy steel, such as AISI 4340, contains approximately 3% alloying elements (Ni, Cr, and Mo) and 0.40% carbon. A Type 18 Ni 250 maraging steel, on the other hand, contains over 30% alloying elements (Ni, Co, and Mo) and less than 0.03% carbon. As will be evident from the discussion to follow, the more highly alloyed, lower carbon materials possess several advantages with respect to dimensional stability.

In maraging steels and martensitic precipitation hardening stainless steels, for example 17-4 PH, the stable structure over a wide range of temperatures from absolute zero to about 900°F consists of a mixture of ferrite and small quantities of various complex compounds of the alloying elements. The ferrite is similar to that described earlier, except that it contains a relatively large percentage of dissolved alloying elements. Such a stable microstructure, in combination with desirable mechanical properties, is obtained by means of a heat treating sequence that is remarkably similar to that employed in the steels described earlier.

The steel is heated into the austenite range, cooled to room temperature at a rate sufficient to prevent diffusion-controlled transformation of the austenite and thus, to form a martensitic structure, and then "tempered" (actually, precipitation hardened or aged) by reheating to a moderately elevated temperature to precipitate the complex compounds in the ferrite matrix.

From a dimensional stability point of view, several points are noteworthy:

1. The sluggishness with which diffusion controlled transformations occur in maraging steels and martensitic precipitation hardening steels allows the use of very slow cooling rates during heat treatment. This helps to minimize quenching stresses and warpage during heat treatment.

2. The low carbon content of maraging steels and martensitic precipitation hardening steels reduces the likelihood that retained austenite will be present after cooling to room temperature from the austenitizing temperature. As noted earlier, retained austenite and its gradual transformation in service can lead to significant dimensional instability in steels. The low carbon content should also minimize the relative importance of carbon reactions at room temperature, such as carbide precipitation or migration to preferred lattice sites.

3. The martensite, being supersaturated with respect to certain alloying elements rather than with respect to carbon, has a specific volume that differs less from that of the stable structure than is the case for the steels described previously. With reference to Fig. 7.7, it will be recalled that in going from point J to point A during tempering, a 1% carbon steel will undergo a unit contraction of about 5000×10^{-6}. A maraging steel, on the other hand, will contract on the order of only 400 to 800×10^{-6} during the aging treatment that causes complex compounds to precipitate, thereby relieving the supersaturation of the martensite (International Nickel Company, 1964).

4. The martensite, being relatively soft and ductile, can be formed or machined to final shape and then subjected to the precipitation hardening treatment, consisting of several hours at a temperature near $480°C$ ($900°F$). This treatment should also be effective in relieving residual stresses introduced by forming and machining.

The above-mentioned features of maraging steels and martensitic precipitation hardening steels can be advantageous in many precision applications, both with respect to making a part to close tolerances and to maintaining dimensional stability in service. Meyerson *et al.* (1968, 1969) demonstrated the excellent dimensional stability of gage blocks prepared from both types of steel. Disadvantages of these materials include (a) high cost relative to lower alloy steels, and (b) difficulty in achieving homogeneous distribution of alloying elements; these inhomogeneities, exemplified by banding and segregation, can lead to shape distortion during heat treatment (Hall and Kluz, 1968).

7.5   IRRADIATION EFFECTS

Dimensional change as a result of irradiation is a phenomenon of interest primarily in the development of nuclear reactors, and is discussed only briefly here. As the fuel elements in a reactor are irradiated with neutrons during the fission process, serious dimensional changes can occur, causing adverse effects on heat transfer, cladding integrity, and perhaps on criticality (Kittel and Paine, 1958).

Two types of dimensional change can be distinguished—growth, which is anisotropic and occurs without volume change, and swelling, which is an isotropic volume change caused by accumulated fission products (Buckley, 1962). Growth appears to manifest itself only in crystalline materials that exhibit thermal expansion anisotropy, i.e., in materials with non-cubic crystal structures. It has been observed in both fissile (uranium, for example) and non-fissile (titanium and zirconium, for example) materials. According to Buckley, it arises as a result of the generation of defects such as vacancies and interstitial atoms

which aggregate into plates on different crystal planes, thus producing the anisotropic shape change. The interstitual clusters cause elongation in the direction
of minimum thermal expansion while the vacancy clusters cause contraction in the
direction of maximum thermal expansion. The magnitude of the effect can be very
large. For example, unrestrained uranium rods may more than double their length
during prolonged exposure to irradiation, depending on the manner in which they
were processed and on the temperature.

Although the gross effects of irradiation on dimensional changes of nuclear
fuel materials have been under scrutiny for some time, the authors are unaware
of any comprehensive investigation of the effect of irradiation on the dimensional stability of materials for precision devices. Even in cubic materials,
however, significant effects of irradiation might be anticipated where dimensional changes of a few parts per million are deleterious to performance.
Eggert (1964) showed, for example, that neutron irradiation of hardened SAE
52100 steel gage blocks caused reduction of residual stress, transformation of
some of the retained austenite, and relatively large dimensional changes. Additionally, as discussed in greater detail in Section 6.4, Dubrovin *et al.*
(1962) have shown that stress relaxation in a variety of materials, both fissile and non-fissile and both cubic and non-cubic, is accelerated by neutron
irradiation.

7.6 EFFECTS OF MAGNETIC AND ELECTRICAL FIELDS

In the strictest sense, magnetic and electrical fields produce no microstructural changes in materials. They can, however, give rise to dimensional
changes, depending on the nature of the material with which they interact.
Basically, the interaction modifies the forces that exist between atoms and
causes the atom spacing to be changed slightly.

Electrostriction is the deformation that occurs when a dielectric solid is
subjected to an electric field (Rosenthal, 1964). Consider a small portion of
a sodium chloride crystal, for example, as shown in Fig. 7.10. When an electric field is applied in the direction shown, the central sodium ion will
move upward. To maintain proper spacing with the displaced sodium ion, the
four outside chloride ions (*C*, *D*, *E*, and *F*) will move inward, which in turn
causes the chloride ions at *A* and *B* to move away from the center. The net
result is a deformation of the unit cell and of the crystal as a whole.

Magnetostriction is the deformation that occurs when a ferromagnetic material
is subjected to a magnetic field. It is the result of magnetic domains, originally oriented in several directions, realigning themselves in the direction
of the applied field. As this takes place, the atom spacing is altered
slightly and the external dimensions of a crystal will be changed. As discussed
in Chapters 9 and 10, these ferromagnetic effects have played important roles
in the development of alloys with near-zero thermal expansion coefficients
and near-zero thermoelastic coefficients. The magnitude of the magnetostrictive effect is shown for several ferromagnetic materials as a function of
magnetic field strength in Fig. 10.4.

7.7 CHAPTER SUMMARY

The major emphasis of this chapter is to demonstrate the very real possibility
that significant dimensional instability can occur in all materials as a result
of microstructural changes. Such changes occur because materials rarely possess
their equilibrium microstructure when placed in service. Under the influence
of time and thermal activation, the material will gradually move in the direction of achieving microstructural equilibrium. Furthermore, the equilibrium

microstructure depends on temperature. Thus, every temperature change, in
addition to causing thermal expansion effects, has the potential for producing
a gradual dimensional change as the material seeks a new equilibrium microstructure.

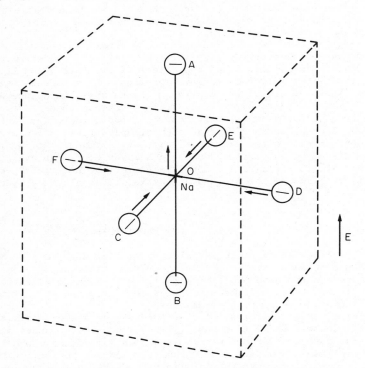

Fig. 7.10. Electrostriction in an Na^+Cl^- type crystal
(Rosenthal, 1971; copyright Litton
Educational Publishing, Inc., used by
permission).

   The types of microstructural changes discussed in this chapter include
changes in vacancy concentration, atom rearrangement and ordering, precipitation and re-solution of a second phase, and phase changes. Specific examples
of each are given and the magnitude of the associated dimensional changes are
estimated.
   From a practical viewpoint, it is often possible to keep these microstructural changes under control. It was shown, for example, that hardened steel
can exhibit relatively large dimensional changes in short times at room temperature as a result of microstructural changes. When properly treated, however, the same material will show virtually no dimensional changes over periods
of many years and can be used successfully in gage blocks.
   Despite the fact that much is known about the dimensional stability of
steels for use in gage blocks operating at room temperature without external
loading, relatively little is known about other materials operating under similar conditions or under more severe conditions. This is particularly true for
applications calling for *extreme* dimensional stability. If dimensional changes
of 1 part in 10^7 or 10^8 over long periods of time are all that can be tolerated

in a design, a designer or manufacturing engineer has little basis for deciding (a) whether it is possible to achieve this degree of stability in a particular material, (b) whether certain materials are more stable than others, or (c) what treatments produce maximum stability? Certain optical and gyroscope applications are causing these questions to be asked and it can be anticipated that they will be asked increasingly in the future.

The chapter concludes with a brief discussion of the effects of various external fields, other than temperature, on dimensions. Neutron irradiation is shown to be capable of producing gross dimensional changes in materials, both fissile and non-fissile, that exhibit anisotropic thermal expansion, by introducing crystalline defects. This is a continuing problem in development of nuclear fuels for economic power generation. Effects of less severe irradiation on the small scale dimensional stability of materials for precision devices have received little attention.

Magnetic and electric fields, while incapable of producing true microstructural changes, can cause changes in atom spacing in ferromagnetic and dielectric materials, respectively. The deformation associated with magnetic fields is termed magnetostriction and that associated with electric fields is termed electrostriction.

REFERENCES

Averbach, B. L., Cohen, M. and Fletcher, S. G. 1948. The dimensional stability of steel. Part III: Decomposition of martensite and austenite at room temperature, *Trans. Amer. Soc. Metals*, 40, 728.

Bohm, D. and Schumann, H. 1965. Dimensional changes during the quench aging of structural carbon steels, *Neue Hutte*, 10, 472.

Buckley, S. N. 1962. Irradiation growth, *Prop. Reactor Mater. Eff. Radiat. Damage*, Proc. Int. Conf., D. J. Littler, editor, Butterworths, London, 413.

Dow Chemical Company. 1957. Growth of Mg-Al-Zn alloys due to precipitation, Technical pamphlet Code 2.50, Dow Chemical Company, Midland, Michigan.

Dubrovin, K. P., Konobeyevskiy, S. T., Levitskiy, B. M., Panteleyev, L. D., Platonov, P. A. and Pravdyuk, N. F. 1962. The relaxation of elastic stresses under neutron irradiation, *Prop. Reactor Mater. Eff. Radiat. Damage*, Proc. Int. Conf., D. J. Littler, editor, Butterworths, London, 233.

Eggert, G. L. 1964. The influence of neutron irradiation on the dimensional stability of steel, *Materials Science and Technology for Advanced Applications*, Vol. II, Golden Gate Chapter, American Society for Metals, San Francisco, 62.

Fletcher, S. G., Averbach, B. L. and Cohen, M. 1948. The dimensional stability of steel. Part II: Further experiments on subatmospheric transformation, *Trans. Amer. Soc. Metals*, 40, 703.

Fletcher, S. G. and Cohen, M. 1945. The dimensional stability of steel. Part I: Subatmospheric transformation of retained austenite, *Trans. Amer. Soc. Metals*, 34, 216.

Hall, R. C. and Kluz, S. 1968. Shape distortion in maraging steel, Paper presented at ASM Materials Engineering Exposition and Congress, Detroit, Michigan.

Hunsicker, H. Y. 1967. The metallurgy of heat treatment, Chapter 5 in *Aluminum*, Vol. I: *Properties, Physical Metallurgy, and Phase Diagrams*, K. R. Van Horn, editor, American Society for Metals, Metals Park, Ohio.

International Nickel Company. 1964. 18 percent nickel maraging steels, International Nickel Company data bulletin.

Kittel, J. H. and Paine, S. H. 1958. Effect of irradiation on fuel materials, *Proc. U.N. Int. Conf. Peaceful Uses At. Energy, 2nd*, Vol. 5: *Properties of Reactor Materials*, United Nations, Geneva, 500.

Kurtz, R. L.   1973.   Private communication: letter dated November 1.

Lement, B. S.   1959.   *Distortion in Tool Steels*, American Society for Metals, Metals Park, Ohio.

Lement, B. S. and Averbach, B. L.   1955.   Measurement and control of the dimensional behavior of metals, Report R-95, Massachusetts Institute of Technology.

Lement, B. S., Averbach, B. L. and Cohen, M.   1949.   The dimensional stability of steel.   Part IV: Tool steels, *Trans. Amer. Soc. Metals*, 41, 1061.

Lement, B. S., Averbach, B. L. and Cohen, M.   1951.   The dimensional behavior of Invar, *Trans. Amer. Soc. Metals*, 43, 1072.

McClintock, F. A. and Argon, A. S.   1966.   *Mechanical Behavior of Materials*, Addison-Wesley, Reading, Massachusetts.

Meyerson, M. R., Young, T. R. and Ney, W. R.   1960.   Gage blocks of superior stability: initial developments in materials and measurement, *J. Res. Nat. Bur. Stand.* 64C, 3.

Meyerson, M. R., Young, T. R. and Ney, W. R.   1961.   The development of more stable gage blocks, *Mater. Res. Stand.* 1, 5.

Meyerson, M. R. and Pennington, W. A.   1964.   Gage blocks of superior stability II: Fully hardened steels, *Trans. Amer. Soc. Metals*, 57, 3.

Meyerson, M. R. and Sola, M. C.   1964.   Gage blocks of superior stability III: The attainment of ultrastability, *Trans. Amer. Soc. Metals*, 57, 164.

Meyerson, M. R., Giles, P. M. and Newfield, P. F.   1968.   Dimensional stability of gage block materials, *J. Mater.* 3, 727.

Meyerson, M. R., Friedman, L. and Giles, P. M.   1969.   Dimensional stability of a 12% nickel maraging steel at ambient temperatures, *Trans. Amer. Soc. Metals*, 62, 809.

Rosenthal, D.   1971.   *Introduction to Properties of Materials*, 2nd edition, Van Nostrand Reinhold, New York.

Schetky, L. M.   1957.   The properties of metals and alloys of particular interest in precision instrument construction, Report R-137, Massachusetts Institute of Technology.

Chapter 8.

Experimental Methods for Measurement of Microplastic Strain and Dimensional Instability

In Chapter 3, various sources of dimensional instability were described briefly. Each was discussed in greater detail in Chapters 4 through 7.

This chapter describes various methods that have been employed to detect small dimensional or shape changes. A major portion of this chapter deals with the measurement of dimensional instabilities that arise from applied stresses. Following this, methods are described for measurement of long-term dimensional instabilities in the absence of applied stresses.

8.1 MEASUREMENT OF ANELASTIC BEHAVIOR

As discussed in Chapter 4, when a material exhibits both elastic and anelastic strain in response to an applied stress, it is likely also that it will display mechanical hysteresis; that is, the stress - strain curve on unloading will differ from that on loading. This will depend on the rate of loading and unloading relative to the rate at which the particular anelastic effect takes place.

Nowick and Berry (1972) have reviewed experimental methods for the investigation of anelastic phenomena, covering a range of frequencies from about 10^5 Hz to 100 GHz. The low frequency end of the range, 1 kHz and below, is generally used to investigate phenomena which involve atomic migration. Where the phenomena exhibit extremely short relaxation times, i.e., they take place at a very rapid rate, high frequency techniques (> 1 MHz) are required for their analysis. The methods described by Nowick and Berry, in order of ascending frequency range, are: (1) quasi-static methods; (2) subresonance methods; (3) resonance methods; and (4) high-frequency wave propagation methods.

Quasi-static methods were employed well over 100 years ago to investigate the time dependent behavior of galvanometer suspension wires. By applying a twist to a wire specimen and detecting the twist angle with light beams and mirrors, it is possible to achieve a high strain sensitivity. Because of the high sensitivity, the technique became popular for studying materials other than those used in galvanometer suspensions.

The arrangement of the galvanometer apparatus is shown schematically in Fig. 8.1. A coil, wound with several hundred turns of fine wire, is suspended between the poles of a magnet by a wire specimen. A regulated current is passed through the coil to induce a torque in the specimen. Several types of experiments can be conducted with this arrangement. In one type, the specimen is suddenly twisted to a preselected strain by application of a suitable current. Thereafter, the current is adjusted by means of the variable resistance to maintain the strain constant with time. If anelastic or plastic strains are slowly occurring, the torque required to maintain constant strain will gradually decrease. Such tests define the stress relaxation response.

In another type of test, the specimen is suddenly twisted to a preselected strain by application of a suitable current. Thereafter, the current is held constant and the strain is monitored as a function of time. If anelastic or plastic strains are occurring, the strains will gradually increase with time.

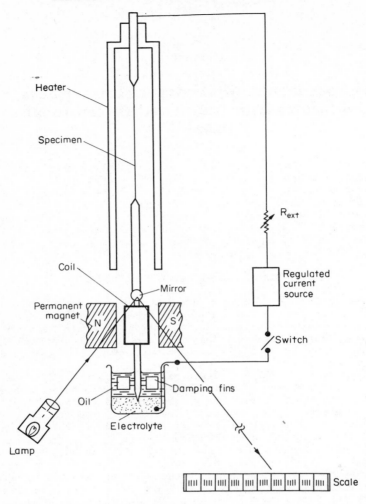

Heater

Specimen

R_{ext}

Regulated
current
source

Coil

Mirror

Permanent
magnet

N   S

Switch

Damping fins

Oil

Electrolyte

Lamp

|||| |||| |||| |||| |||| |||| |||| |||| |||| |||| |||| Scale

Fig. 8.1.   Galvanometer apparatus for quasi-static
             measurements (Nowick and Berry, 1972;
             copyright Academic Press, used by permission).

Such tests define the creep response. Upon completion of either of the above
tests, the current can be turned off and strain recovery can be accurately
monitored as a function of time. These techniques can be employed to investigate plastic, as well as an anelastic, phenomena. However, because of the
torsional state of stress, stress varies from a maximum value at the outer surface to zero at the center of the wire. Thus, when plastic flow occurs, it
will begin at the surface. Thereafter, the stress distribution becomes a
complex function of material properties, rather than a simple linear distribution. This problem can be greatly minimized by employing a thin-walled tubular
specimen in place of the wire.

Many of the methods to be described in Section 8.2 could be termed quasi-
static and, where strain sensitivity is sufficient, could be used to study

relatively slow anelastic phenomena in a manner similar to that employing the galvanometer apparatus. However, they are generally aimed at investigating plastic phenomena as well. Thus, they usually employ stress states in which the stress can be related directly to the load and area, e.g., tension and compression, regardless of whether the strain is elastic, anelastic, or plastic. Generally speaking, more sophisticated strain sensing devices are required for such tests to achieve strain sensitivities comparable to those achieved in the galvanometer apparatus.

Of the other methods described by Nowick and Berry for investigating anelastic behavior, only the resonance methods will be discussed briefly here. These include torsion pendulums which operate at low frequency and resonating bars which operate at intermediate frequency. Use of these methods permits detection of anelastic effects with exquisite sensitivity at extremely low stress amplitudes. This is accomplished by measurement of the energy loss in each cycle as a specimen goes through repeated stress reversals. Unlike methods described in Section 8.2, these methods do not allow direct observation of hysteresis loops in stress - strain curves. The energy loss, arising from anelastic strain lagging behind the stress, leads to a decay of amplitude in a freely oscillating pendulum or a freely vibrating bar. Measurement of the rate of amplitude decay permits calculation of the internal friction or damping capacity. These can be defined in terms of the logarithmic decrement δ, which is the logarithm of the ratio of successive amplitudes:

$$\delta = \ln \frac{A_n}{A_{n+1}}$$

A torsion pendulum apparatus that has been employed in the authors' laboratory is shown schematically in Fig. 8.2. To excite the pendulum, current is pulsed through magnets attracting small pieces of iron at the ends of the inertia bar. Once a certain amplitude is reached, the pendulum is allowed to oscillate freely. The oscillations are observed optically using a narrow beam of light from a projector, reflected by a mirror mounted on the inertia bar clamp to either a ground-glass scale or a recorder. The recorder follows the oscillating light beam and a pen traces the movement on a strip chart. Examples of damping traces are shown in Fig. 8.3.

Theoretical treatment of internal friction arising from anelastic strains is well advanced (Nowick and Berry, 1972) and will not be reviewed here. Specific anelastic effects can be identified by the dependence of internal friction on frequency, temperature, and amplitude.

8.2 MEASUREMENT OF SMALL PLASTIC STRAINS RESULTING FROM APPLIED STRESS

Prior to discussing specific methods that have been employed to investigate small plastic strains, it is useful to discuss general considerations that apply to such investigations.

General Considerations in Microstrain Testing

The importance of paying close attention to all test details in microstrain studies cannot be overemphasized. Specimen preparation techniques, specimen quality and surface condition, test fixture quality, calibration of load-weighing and strain measuring systems, alignment, and ambient temperature must be kept under constant surveillance. The greater the strain sensitivity, the more important these details become. For example, assume a strain sensitivity of 10^{-6}. If the test specimen has a thermal expansion coefficient of 10 x $10^{-6}/°C$, then an increase in temperature of only 0.1°C during a test will indicate an apparent strain of 1 x 10^{-6}. For a strain sensitivity of 10^{-8}, a temperature change of only 0.001°C will indicate an apparent strain of detectable magnitude.

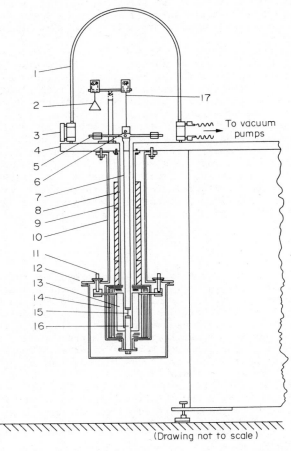

To vacuum pumps

(Drawing not to scale)

1. Bell jar	7. Upper grip	12. High temperature furnace chamber
2. Support system	8. Inner tube	13. Heating element
3. Multiport collar	9. Furnace (low temperature)	14. Heat shields
4. Base plate		15. Specimen
5. Inertia bar	10. Outer tube	16. Lower grip
6. Mirror	11. Power connectors	17. Torsionally weak suspension wire

Fig. 8.2.   Internal friction machine.

The situation with respect to temperature control can be improved somewhat by material selection for the strain measuring device, to match the expansion coefficient of the test specimen. Nonetheless, the temperature problem becomes progressively worse as strain sensitivity increases. Furthermore, it goes beyond simply controlling the temperature of the environment. As pointed

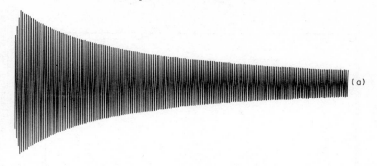

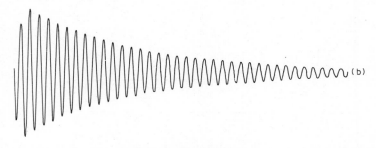

Fig. 8.3.   Typical damping trace recordings.
             (a) Q-1 = 3.7 x 10^{-3} at 2 cm amplitude
             (b) Q-1 = 26.3 x 10^{-3} at 2 cm amplitude

out in Chapter 5, elastic strains induce temperature changes in the specimens themselves. Figure 8.4 shows the temperature change induced in a steel specimen loaded within its nominally elastic range to 345 MN/m^2 (50 ksi) at several rates. Tensile loading caused the temperature to decrease, as expected. The magnitude of the change increased with loading rates, because less time was available for heat to flow into the specimen from the outside. As the stress was maintained, the specimen temperature gradually rose until it reached that of the surroundings. Removal of the stress then produced a reverse effect, as shown in Fig. 8.4.

Suppose this had been a creep test at 345 MN/m^2 and this adiabatic effect had not been recognized. Assuming a value of 11.3 x 10^{-6}/°C for the coefficient of linear thermal expansion of steel, an apparent creep strain of about 2.5 x 10^{-6} would have been observed during the first 30 min of the test. Clearly, effects such as these must be recognized and proper account taken of them in planning and conducting precision tests.

In connection with temperature effects, one additional point should be made. As noted in the preceding paragraphs, when strains are elastic, tension produces a decrease and compression produces an increase in temperature. When plastic strain occurs, heat is generated, regardless of the sign of the strain. This is illustrated in Fig. 8.5. Prior to the test, the steel specimen had been loaded to 760 MN/m^2 (110 ksi), to produce a plastic strain of 830 x 10^{-6}. After waiting for the temperature of the unloaded specimen to equilibrate, the specimen was then reloaded to 890 MN/m^2 (129 ksi), followed by immediate unloading. As expected, the specimen cooled during the initial stages of reloading (up to 760 MN/m^2). As the stress increased from 760 to 890 MN/m^2, an

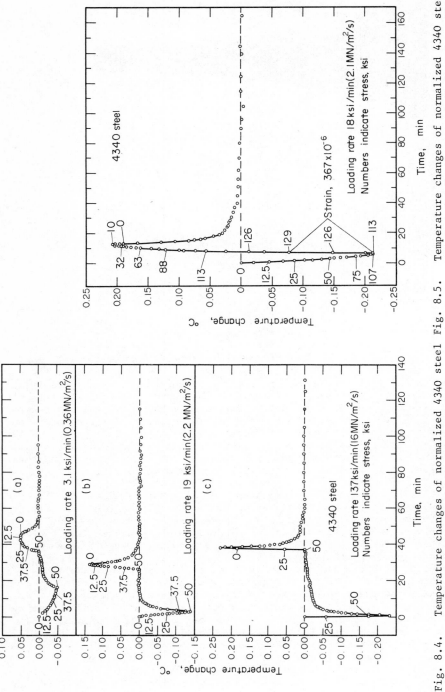

Fig. 8.4. Temperature changes of normalized 4340 steel specimen during test. Specimen loaded within its elastic range at selected rates to 345 MN/m² (50 ksi) with delayed unloading at same rate (Geil and Feinberg, 1969).

Fig. 8.5. Temperature changes of normalized 4340 steel specimen during loading into its microplastic range. Specimen loaded at a rate of 19 ksi/min (2.2 MN/m²/s) to 129 ksi (890 MN/m²) and immediately unloaded at the same rate (Geil and Feinberg, 1969).

additional plastic strain of 369 x 10^{-6} occurred, accompanied by heating of
the specimen.

Another area in which great caution is required in precision testing pertains to the surface condition of test specimens. In Chapter 6, it was shown
that all mechanical methods of metal removal including machining, grinding, and
polishing, alter the test specimen in at least two ways. First, they produce
a plastically deformed layer to a depth that depends on the particular metal
removal operation; and second, they produce residual stresses in the specimen.
Both are undesirable if it is material behavior, rather than specimen behavior,
that is of interest. The smaller the specimen cross-section, the more likely
it is that thin, disturbed surface layers will have an effect on the test
results. Consider, for example, the oversimplified situation illustrated
schematically in Fig. 8.6, in which machining has introduced residual compressive stresses at the surface, balanced by tensile stresses in the specimen
interior. For a specimen with large cross-section, these tensile stresses will
be relatively small. For thin specimens, on the other hand, the tensile
stresses may be quite large. If a tension test is now run on the two specimens
to measure microplastic properties, it is possible that the thin specimen will
exhibit a given amount of plastic strain at a lower applied stress than will
the thick specimen, as illustrated in Fig. 8.6. Such results would indicate
an effect of size on microplastic properties, when, in fact, the difference
may be due entirely to specimen preparation.

In axial load tests, it is extremely important also to avoid parasitic
stresses that may be present during a test. These are usually the result of a
bending moment superimposed on the axial load. Such bending moments can arise

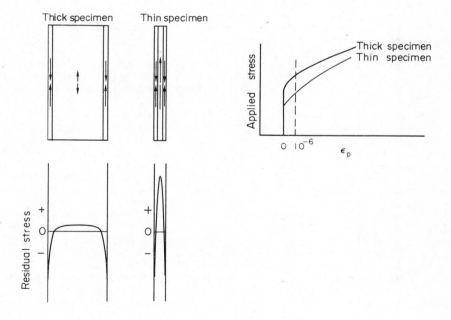

Fig. 8.6.   Possible effect of residual compressive
             stresses introduced by machining on the micro-
             yield behavior of thick or thin specimens.

from several sources, including inadequate alignment of the loading train, improperly designed or machined specimens, and bent specimens.* Ideally, the loading axis will pass through the centroid of each cross-section. Rarely is this condition realized, but every effort should be made to approach it as closely as possible.

Christ (1973) has discussed the role of misalignment in microyield studies. He emphasizes the importance of measuring the local strains at three or four locations around the circumference of a cylindrical specimen to assess the separate contributions of intentionally-applied uniaxial stress and inadvertently-applied bending stress. The most accurate average stress – average microplastic strain relationships are generated when dissimilarities among local strains approach zero. Accordingly, results obtained with strain sensing devices that do not permit extreme strains to be measured may be misleading.

In connection with loading train alignment, a typical loading train employed in the authors' laboratory for investigating microplastic properties in tension is illustrated in Fig. 8.7. The uppermost fitting in the exploded view is a ball-thrust-bearing system to minimize any torque on the specimen. This device supports a universal joint in which the specimen is gripped. The particular specimen shown in the illustration is a flat, pin-loaded specimen. The lower end of the specimen is gripped in another universal joint attached to the moving platen of the tensile machine by a pinned joint. In this lower joint, the pin is confined to a slotted hole to prevent accidental loading in compression.

The type of stress-state employed, e.g., tension, compression, torsion, or bend, can also have some effect on the results of microstrain testing. Tension and compression tests are the easiest to analyze, since stress is simply load divided by area. In bending or in torsion, it is necessary to assume a linear strain distribution in the cross-section and then compute the stress on the basis of the known or assumed stress – strain relationships. While this is relatively simple and straightforward in the nominally elastic region, it becomes very cumbersome when plastic strains are occurring. Furthermore, when plastic strains do occur, they will tend to occur at the outer surface long before they occur in the interior, because of the higher stresses there. This is fundamentally different from the behavior in tension or compression tests and should be recognized as such.

The effects noted for bend and torsion tests can be minimized by modifying the test specimens. For example, if a torsion test is conducted on a thin-walled cylinder of appropriate diameter, rather than on a solid bar, the stress in the wall will be nearly uniform. A similar effect can be obtained in bending by use of a trussed-beam specimen (Fig. 8.8).

Little agreement presently exists on exactly what are the best specimen geometries for microstrain tests. An example of the diversity of approaches taken by various investigators in studying tensile properties is shown in Fig. 8.9. Often, it is necessary to design the specimen around available forms of the test material.

It was pointed out in Chapter 4 that some investigators routinely record stress – strain curves obtained in load – unload tests to obtain σ_E values and to study hysteresis behavior, while others measure only residual strain after unloading. While the first technique obviously gives more information, it frequently requires some sacrifice in strain sensitivity. This is particularly true when tests are conducted on high strength alloys, rather than on high

---

*Mechanical straightening of bent specimens is not a suitable procedure in accurate measurement of microplastic properties, because mechanical straightening requires plastic flow and introduces residual stresses, as discussed in Chapter 6.

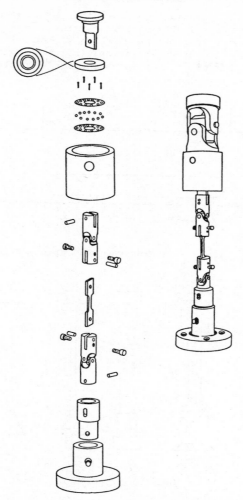

Fig. 8.7.   Exploded view of load train for microstrain
             testing.

purity metals. Consider, for example, a stress – strain curve for a material
with an elastic modulus of 70 GN/m^2 (10 x 10^6 ksi) subjected to a load –
unload cycle to 70 MN/m^2 (10 ksi). If one small division of the recorder chart
equaled a strain of 10^{-7}, the strain record would require 10 000 chart divisions
just to record the elastic strain. If each chart division equaled, say, 1 mm,
the strain record would require 10 m of chart paper. A much more reasonable
size of strain record would be obtained if the strain sensitivity were reduced
to 10^{-5} per chart division, but this would compromise the test objectives.
Clearly, the problem is less severe with low strength, high modulus materials,
and greater with high strength, low modulus materials.

An additional difference between the above two methods of studying microstrain relates to cumulative strains. When load – unload stress – strain
curves are recorded, any plastic strain below the resolution of the strain
record will not be detected. Each cycle of loading will then be compared

against the endpoint of the previous cycle, rather than against the first cycle. Thus, any small unresolvable plastic strains that may occur on each cycle will not be evident. In tests in which only residual strain is measured, however, all readings are compared with an original reading to compute residual strain. Thus, strains that may take place below the resolution of the strain sensing system will be integrated into the results.

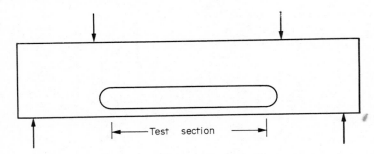

Fig. 8.8.   Trussed-beam bend specimen.

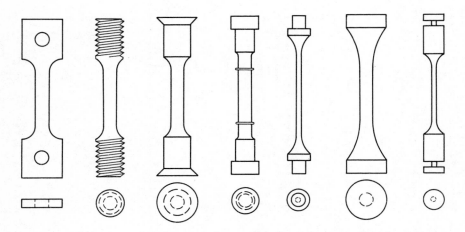

Fig. 8.9.   Typical specimen configurations employed to study tensile microstrain behavior.

Special problems arise when microstrain tests are extended over long time periods, as, for example, in microcreep tests. One problem is the stability of the electronics employed in measuring strain. Unless this can be verified in some independent manner, the degree of uncertainty in long time tests is greater than that in short time tests. Secondly, it is often impractical to employ a high sensitivity strain measuring device for long periods of time on a single test specimen. These problems probably explain, in part, the general scarcity of microcreep data at high strain sensitivity.

One final consideration that has practical implications in microstrain testing pertains to nonhomogeneous test specimens. There is growing interest in the microplastic behavior of various types of joints, including welds, rivets, and adhesives. Test procedures for specimens containing such joints clearly have to be specially tailored. Strain measurement, in particular, requires careful interpretation because of the nonhomogeneous nature of the specimen. Capacitance gages have shown good potential for testing nonhomogeneous specimens.

Strain Sensing Methods

Various methods have been employed successfully to study microstrain behavior at strain sensitivities of 10^{-5} or better. At the present time, no single method appears to be universally accepted. Those receiving most attention include capacitance extensometers, bonded electrical resistance strain gages, Tuckerman optical gages, and linear variable differential transformers. Some attention also has been given to measuring microstrain by laser interferometry and holography. These and other strain sensing systems are reviewed in subsequent paragraphs. Where possible, the major strengths and limitations of each system are emphasized.

A. *Capacitance Gage*

A capacitance gage consists of two parallel plates, usually copper, in close proximity. The capacitance of the system (C) depends on the area of the smaller plate (A), the distance between the plates (D), and the dielectric constant of the medium between the plates (K), according to the relation

$$C = \frac{\alpha . KA}{D} \tag{8.1}$$

where α is a constant. From eq. (8.1) it is evident that changing the plate spacing will produce a change in capacitance. The change in capacitance can be monitored with a capacitance bridge network and the output of the bridge recorded. Thus, changes in length of a specimen test section can be detected from observation of the bridge output. Figure 8.10 is a schematic diagram of a typical arrangement for recording stress - strain data in a tension test.

The sensitivity of the device depends primarily on the sensitivity with which capacitance changes can be detected and on the initial plate spacing. Differentiating eq. (8.1) gives

$$\frac{dC}{dD} = \frac{-\alpha\ KA}{D^2}$$

or   (8.2)

$$dD = \frac{D^2 dC}{-\alpha\ KA}.$$

If a capacitance change (dC) of $\sim 10^{-5}$ picofarads can be detected and the initial plate spacing (D) is $\sim 10^{-2}$ cm, then displacements (dD) of $\sim 10^{-9}$ cm can be detected if α, K, and A are taken to be of the order of unity. Strain sensitivity depends on the particular gage length employed. This sensitivity can obviously be increased significantly by decreasing the initial plate spacing.

Capacitance gages were employed in microstrain studies as early as 1960 (Roberts and Brown, 1960) and since then have been used by many investigators (Hughes and Rutherford, 1969; Wilson and Teghtsoonian, 1970; Brown and Ekvall, 1962; Solomon and McMahon, 1971; Argon and East, 1967; Brown and Lukens, 1961; Rutherford *et al.*, 1968; Roberts and Hartmann, 1964; Geil and Feinberg, 1969; and Lyons *et al.*, 1970). Several desirable features have been cited for the capacitance gage: (a) it has both a high theoretical and high practical strain sensitivity which is maintained even after large prestrains; (b) since it can be mounted on the specimen, it does not record deformations in the tensile assembly; (c) it measures bulk behavior and does not modify or alter the surface of the specimen; and (d) it is readily adapted to test temperatures both

below and above room temperature. Furthermore, Rutherford *et al.* (1968) have
demonstrated the usefulness of the capacitance gage in studying the microstrain behavior of nonhomogeneous specimens, particularly of adhesive-bonded
joints. They used various modifications of the gage to study adhesive behavior
in tension, compression, and shear for bond-line thicknesses as small as
13×10^{-4} cm (5×10^{-4} in). In some of the tests, the metal adherends were
used as the capacitor plates and the adhesive material served as the dielectric
medium. This confined the strain measurement to the adhesive itself.

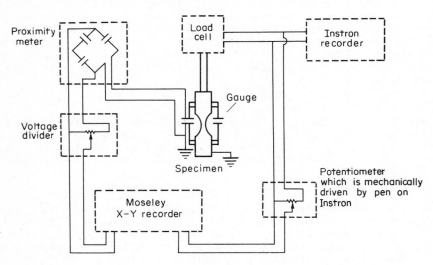

Fig. 8.10.   Schematic diagram of equipment used with the
             capacitance gage (Brown, 1968; copyright
             Wiley, used by permission).

    In some of the early investigations, the capacitance gage was attached to
the specimen shoulders. Thus, recorded strains included strains in the fillet
sections, making specification of an exact gage length difficult. Attachment
of the gage to the reduced section overcomes this difficulty but, particularly
in soft materials or thin specimens, the bulkiness of the gage may require that
special precautions be taken to avoid damage to the specimen. Geil and
Feinberg (1969) designed a ribbed tensile specimen and a capacitance gage
holder to permit measurement of strain over a known gage length without damage
to the specimen. The specimen is shown in Fig. 8.11 and the test assembly
in Fig. 8.12. Note that three capacitance gages, spaced at 120 degree inter-
vals, are used to obtain average strain readings, thereby minimizing the
effects of any bending that may take place.
    Lyons *et al.* (1970) have made use of the capacitance gage to develop a
computerized microstrain measuring system. The system utilizes an automatic
absolute-capacitance measuring bridge with either an analog or hybrid computer
to measure plastic deformations of 10^{-6} or less, hysteresis loop areas, low-
and medium-frequency high-amplitude internal friction, unidirectional fatigue
behavior, and nonelastic strain recovery. In an additional adaptation, Lyons
(1970) has used capacitance gages for investigation of long-term microcreep
behavior. The system is said to have high strain sensitivity, long-time
stability, and to be capable of measuring and recording on a printer as many
as 100 microcreep measurements on a consecutive sampling basis. The capacitance gage itself is made primarily of stabilized Invar having a low thermal

expansion coefficient (see Chapter 9) and the capacitance plates are bonded
to the gage base with a thin layer of low expansion dielectric adhesive. The
specimen, capacitance gage, grips, and three-terminal wiring connection are
shown in Fig. 8.13a. A simplified block diagram of the entire measuring
system is shown in Fig. 8.13b. Microcreep results obtained by Lyons on I400
beryllium and 440C stainless steel are shown in Figs. 8.14 and 8.15, respectively.

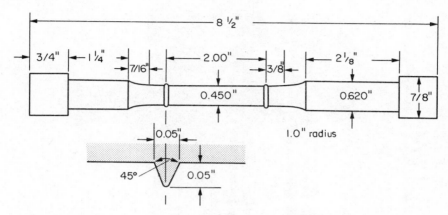

Fig. 8.11. Microplasticity specimen for use with capacitance gage (Geil and Feinberg, 1969).

Despite its wide use and attractive characteristics, the capacitance gage
has some limitations. It is inherently nonlinear, as indicated by eq. (8.1),
and, except at very small deformations, the nonlinearity must be calibrated
and corrections applied to the data. Lyons *et al.* (1970) make this correction
mathematically in a computer, thereby linearizing the parabolic-shaped capacitance versus gap curve. Christ (1973), on the other hand, employs an induc-
tively-coupled capacitance bridge circuit, rather than the usual parallel-
plate capacitance bridge, to obtain a linear output with displacement. Temperature changes can also be troublesome, particularly at very high strain
sensitivities. Rutherford and Swain (1966) point out that temperature fluctuations as large as $0.1°C$ do not influence the results if the thermal expansion coefficients of the gage and specimen are matched. This, of course,
does not get around the problem of adiabatic heating and cooling discussed
earlier in this section. As noted by Geil and Feinberg (1969), at strain
sensitivities of 10^{-7}, adiabatic heating and cooling effects must be taken into
account when using capacitance gages.

B. *Bonded Electrical Resistance Strain Gages*

Another method often used to sense small nonelastic strains employs thin
metal foils, on a paper or plastic backing, bonded to a test specimen. These
foils, usually referred to simply as strain gages, have the characteristic of
changing their electrical resistance in direct proportion to strain. Resistance changes can be detected in a suitable bridge circuit and the values
converted to strain from knowledge of the strain sensitivity factor of the
particular gage employed. The assumption is made that, if the combined stiffness of the gage, backing and cement is several orders of magnitude less than
that of the test section, it will have negligible effect on the behavior of
the test specimen and will experience strain identical to that experienced by
the specimen. In this regard, present day foil gages of 0.005 mm (0.0002 in)
thickness are far superior to the older wire gages. A typical foil gage is
pictured in Fig. 8.16.

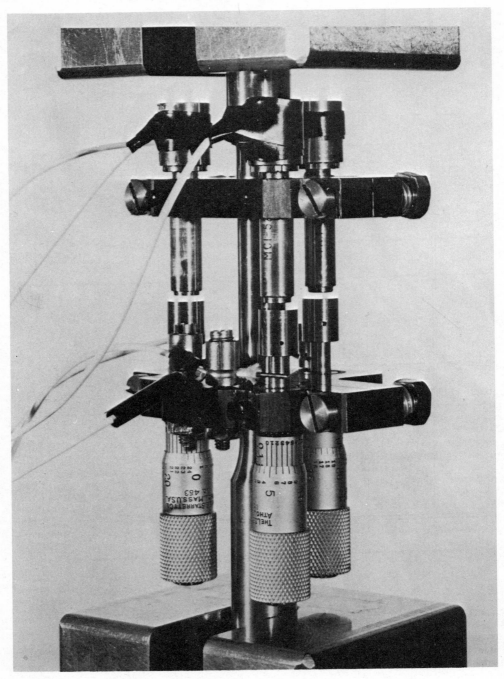

Fig. 8.12.   Capacitance gages installed on specimen (Geil and Feinberg, 1969).

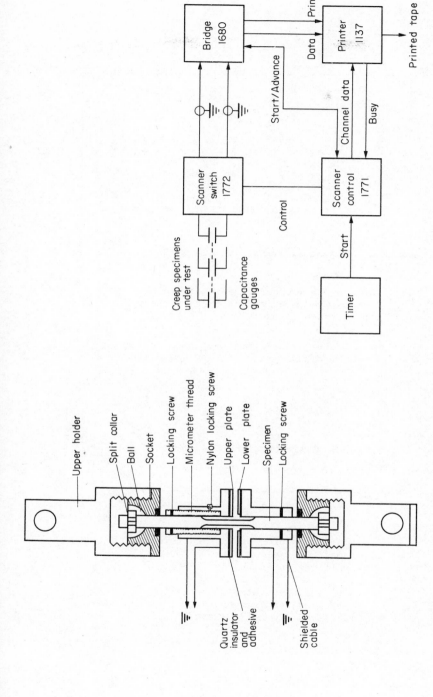

Fig. 8.13a. Capacitance gage assembly showing specimen, three terminal wiring connection and grips for automatic microcreep measuring system (Lyons, 1970).

Fig. 8.13b. Simplified block diagram of automatic micro-creep measuring system (Lyons, 1970).

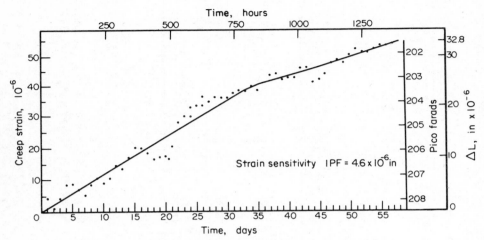

Fig. 8.14.   Example of microcreep data obtained with
automatic measuring system that employs
capacitance gages (Lyons, 1970).

I400 beryllium, tested at room temperature
at an applied stress of 69 MN/m^2 (10 ksi)

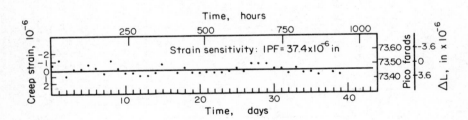

Fig. 8.15.   Example of microcreep data obtained with
automatic measuring system that employs
capacitance gages (Lyons, 1970).

440C stainless steel (Rockwell C 62),
tested at room temperature at an applied
stress of 520 MN/m^2 (82 ksi)

Dimensions

      Gage length   6.35 mm (0.25 inch)

      Overall length   9.53 mm (0.375 inch)

      Grid width   3.18 mm (0.125 inch)

Grid Alloy

      Constantan foil with temperature compensation for 0, 3, 6,
      13 or 15 x 10^{-6}/°F

Backing

      0.023 mm (0.0009 inch) epoxy resin film

Electrical Properties

      Gage resistance   120 ohms

      Gage factor   2.095

      Linearity better than 0.05 percent

Fig. 8.16.   Typical foil gage used in microplastic strain
measurement.   Type MA-XX-250BG-120.

Bonded foil strain gages are used routinely to sense strains of $\sim 10^{-5}$.
However, by careful attention to details, both in gage application procedures
and in instrumentation, sensitivities of about 10^{-7} can be achieved.   These
details include: (1) strict adherence to manufacturer's specifications in
surface cleaning and neutralization procedures, choice of bonding agent, curing
conditions, and protective coating; (2) special care in wiring and instrumentation to avoid variable contact resistance and resistance changes arising
from temperature gradients in the lead wires; (3) specially ordered, or factory
modified, resistance bridges (strain indicators); and (4) good control of
temperature and humidity.

A number of investigators have employed bonded gages to study the microstrain behavior of various materials (Hughel, 1960; Carnahan and White, 1964;
Maringer *et al.*, 1968; Marschall *et al.*, 1972).   Among the advantages cited
for this method are:

- Gage application is relatively simple; no mechanical attachments are
required; gage is of small size and weight and can be used easily on
small specimens.
- Response to strain is essentially linear over fairly wide ranges of
strain.
- Temperature compensation can be achieved by proper matching of temperature coefficient of resistivity of gage element to thermal expansion
coefficient of test specimen; furthermore, small gage mass and intimate
contact with specimen ensure rapid temperature equilibration.   This is
an extremely important consideration where high strain sensitivities are
employed and adiabatic heating and cooling produces significant temperature changes in the specimen.

- Multiple gages can readily be employed to minimize bending errors in tension or compression, and simultaneously improve sensitivity.
- Relatively low cost.

In an investigation conducted by Maringer *et al.* (1968), foil strain gages were employed in the following manner to study the microplastic properties of various alloys. Strain gages were applied to both faces of a flat, pin-loaded tensile specimen. A second (dummy) specimen was identically gaged to provide temperature compensation in the circuitry in addition to the temperature compensation built into each gage by the manufacturer. Active and dummy gages were connected in a full-bridge circuit (Fig. 8.17) with four conductor wires, shielded in pairs. Bridge connections were made to a terminal strip attached rigidly to the loading frame, as pictured in Fig. 8.18. All the gages were kept close together physically to minimize temperature differences. Strain measurements were made with a BLH Model 120 strain indicator, factory modified to increase its strain sensitivity by a factor of 5. The indicator was further modified by using a highly sensitive galvanometer in place of the normal null indicator. Tests were conducted in a room whose temperature was maintained at $20° \pm 0.15°C$ ($68° \pm 1/4°F$).

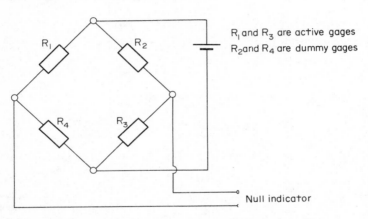

Fig. 8.17. Full bridge measuring circuit.

At the start of a test, a small load was applied and the active gages were read individually in a half-bridge circuit. Any observed eccentricities were minimized by slight adjustments in the loading train. Following this, the full-bridge circuit was used and a small load was applied and removed several times to assure stability of the strain gage readings at the reference load of about 225 g (1/2 lb). The actual test then consisted of repeated loading and unloading, to a higher stress on each succeeding cycle. The strain indicator was read at the reference load after each cycle to sense any permanent strain resulting from the loading. This procedure was followed until permanent strain of the order of 50×10^{-6} was achieved. Examples of the type of data obtained by this technique have already been given in Chapter 5 (Fig. 5.5). Additional examples are shown in Fig. 8.19, where the data are plotted on logarithmic coordinates.

Several investigators have employed bonded strain gages to measure microstrain behavior under continuous loading (Lee, 1965; Carnahan *et al.*, 1967; Holt, 1973), rather than by load - unload techniques. This method requires that the plastic strain be computed by subtracting the elastic strain, estimated from modulus data, from the measured total strain.

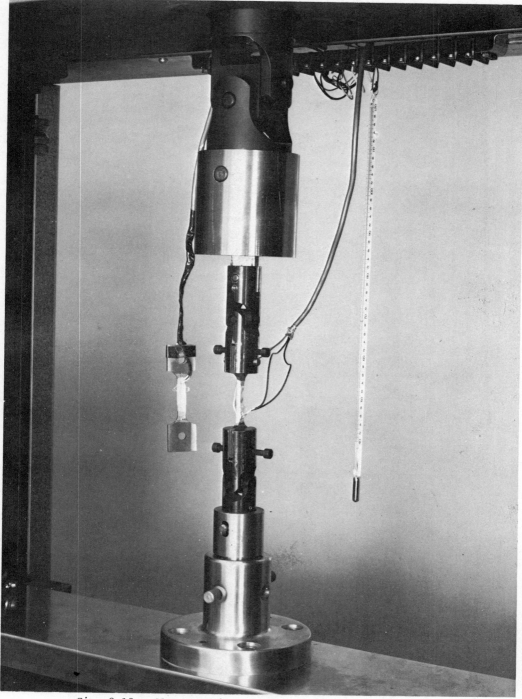

Fig. 8.18.   Close-up view of load train for microstrain
             testing.

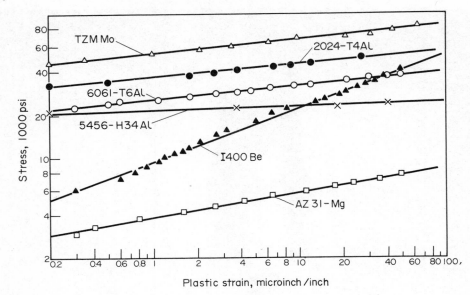

Fig. 8.19. Stress versus plastic strain for several
            alloys; data obtained with bonded foil
            strain gages.

Maringer *et al.* (1968) have also employed bonded foil strain gages to study
microcreep behavior for times up to 1400 hr. Procedures were essentially the
same as those described above for measurement of microplastic properties,
except that loading was accomplished in creep frames and a special low-resis-
tance switching arrangement was incorporated into the circuitry to permit
reading of creep strain in numerous specimens. Results of a stability check,
shown in Fig. 8.20, indicate that the measuring system is stable within about
1×10^{-6} for several hundred hours.

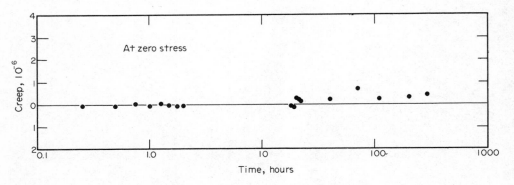

Fig. 8.20. Strain-time relationship obtained from a test
            for microcreep of AZ 31 Mg at zero load
            (Maringer *et al.*, 1968).

As with other strain sensing systems, bonded strain gages have a number of limitations. In some instances, the chemical cleaning and use of a thin layer of cement may produce undesirable effects on the specimen surface. Additionally, the total strain limits of strain gages, which depend on the specific gage system, are generally less than those of other systems used to study microstrain. At elevated temperatures, gage performance for precision measurements is generally unsatisfactory, due to use of polymeric cements and backings. In long-term tests, moisture may have adverse effects on the cements unless exceptionally good protective coatings are used. Furthermore, the resistance of the gage element and of the circuitry may change significantly during a long-term test. Special techniques can minimize errors that arise from the resistance changes but cannot eliminate them entirely.

C. *The Tuckerman Optical Gage*

An optical extensometer, known as the Tuckerman optical strain gage, has been used by Bonfield (1966) for the measurement of microstrain. The system, used in conjunction with an autocollimating telescope (Fig. 8.21), provides an optical lever in which, as shown in Fig. 8.22, a light beam emanating from the autocollimator is reflected by the lozenge to the fixed mirror (a 90° roof prism) and back to the autocollimator to form an image on a vernier reticule.

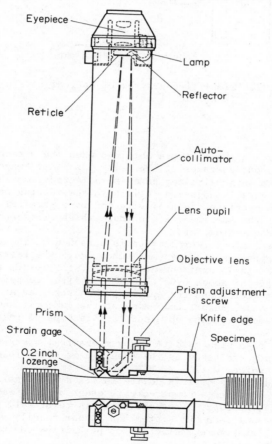

Fig. 8.21. Tuckerman optical strain gage and autocollimator (Bonfield, 1966).

The gage makes contact with the specimen by the lozenge, which is free to rotate, and by a fixed knife edge. As the specimen is extended, the rotation of the lozenge produces a deflection of the light beam and a corresponding movement of the image on the graduated scale in the autocollimator. One division on the vernier scale corresponds to a strain of 2×10^{-6}.

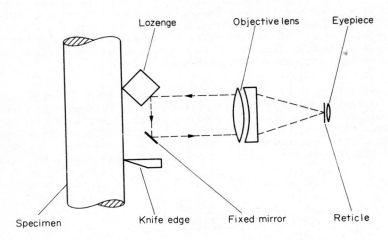

Fig. 8.22.   Simplified diagram of optical strain gage (Bonfield, 1966).

The reading image on the scale is obtained when the autocollimator beam is reflected from the roof surfaces of the prism. Another image (the flash image) is obtained when the autocollimator beam is reflected from the flash surface of the prism. The gage is adjusted until both the reading and the flash images are received in the eyepiece field of view and move parallel to one another during extension of the specimen, a procedure which eliminates errors due to the rotation of the extensometer.

In an actual test, the Tuckerman gage is attached to the test specimen with a spring-loaded support and the assembly mounted in the testing machine. The specimen is enclosed in an environmental chamber to provide the necessary temperature stability of about $0.05°C$ ($0.1°F$). After temperature equilibration, the zero point is measured and the specimen then strained at a constant rate to a given stress value, before unloading at the same rate and remeasuring the zero point. An example of data obtained in this type of test is shown in Fig. 8.23.

This system has also been used to study mechanical hysteresis. In this case, strain readings are taken as the specimen is being loaded and unloaded. Figure 8.24 shows typical data.

The principal advantage of this device is its directness and simplicity. No electronics are involved. Compared with other systems employed in micro-strain measurements, however, its sensitivity is relatively poor. In addition, it is strictly a manual read-out system and is not readily adaptable to automatic recording. Furthermore, detection of parasitic bending stresses is cumbersome with this device.

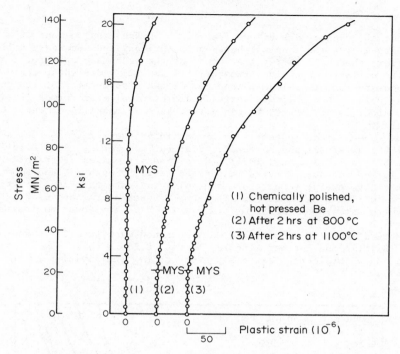

Fig. 8.23. Example of microstrain data obtained with
            Tuckerman optical gage (Bonfield, 1966).

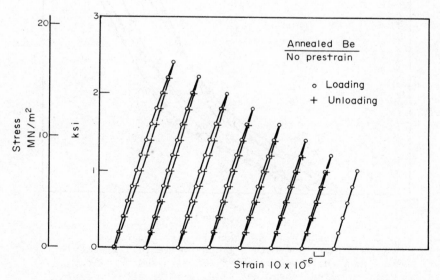

Fig. 8.24. Example of mechanical hysteresis data obtained
            with Tuckerman optical gage (Bonfield, 1966).

D. *Differential Transformer Gages*

Differential transformers are devices in which the displacement of a ferromagnetic core alters the coupling between the primary and secondary coils of a transformer, thereby altering the output of the secondary coil for a given input to the primary. When the transformer is suitably designed, the output of the secondary coil is a linear function of the displacement of the core. Such a device is termed a linear variable differential transformer (LVDT). Since the output is a voltage, it can either be read manually or recorded automatically.

LVDTs have been used frequently in microstrain investigations (Lukas and Klesnil, 1965; Brentnall and Rostoker, 1965; Eul and Woods, 1969; Bilello and Metzger, 1969; Parikh and Hay, 1971). Strain sensitivities of the order of 10^{-7} are claimed for such devices in tension tests, though Eul and Woods, by use of extremely careful temperature controls, indicate that sensitivities approaching 10^{-8} can be achieved. The same investigators have used LVDTs to obtain reported strain sensitivities of 5×10^{-10} in torsional microstrain experiments.

An example of tensile microstrain data obtained with an LVDT is shown in Fig. 8.25 (Bilello and Metzger, 1969). In Fig. 8.26 is shown some torsional microstrain data obtained at extremely high strain sensitivities by Eul and Woods (1969).

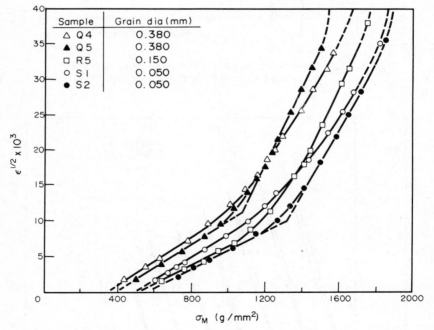

Sample	Grain dia (mm)
△ Q 4	0.380
▲ Q 5	0.380
□ R 5	0.150
○ S 1	0.050
● S 2	0.050

Fig. 8.25.   Example of microstrain data for copper obtained with a linear variable differential transformer (Bilello and Metzger, 1969; copyright AIME, used by permission).

   In addition to their relatively high strain sensitivities, LVDTs possess
several other desirable features. They are mechanically simple and durable
and can be used over a relatively wide range of temperatures. No bridge
circuitry is required. They can be used to measure strain on nonhomogeneous
specimens, as described earlier for capacitance gages.

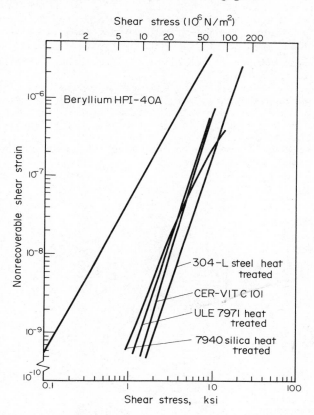

   Fig. 8.26. Example of torsional microstrain data obtained
               with LVDT (Eul and Woods, 1969).

   Temperature control is extremely important when employing LVDTs at high
sensitivity. For example, Eul and Woods (1969) employ a test chamber whose
temperature is held within ± 0.005°C. Precautions noted earlier in connection
with adiabatic heating and cooling of test specimens when subjected to stress
must also be observed when employing LVDTs for measurement of microstrains.

E. Interferometric Techniques

   In principle, it should be possible to investigate microplastic behavior by
interferometric techniques. For example, if bend specimens were prepared with
one surface of optical quality such that interference fringes would be returned
when viewed in an interferometer, then load – unload procedures in bending
should reveal the stress at which permanent deformation begins.
   Paquin and Goggins (1971) employed such a technique to investigate the
microstrain behavior of several nonmetallic materials for mirror applications.

Specimens were mirror disks, nominally 10 cm (4 in) in diameter and 0.625 cm (1/4 in) thick. They were polished on both faces to better than $\lambda/5$ (λ = wavelength of light = 0.6328 μm = 24.9 x 10^{-6} in) and one surface was coated with a thermally evaporated aluminum coating, approximately 500 Å thick. A Fizeau laser interferometer was set up, as shown schematically in Fig. 8.27, so that the mirror disks could be subjected to load without being removed from the interferometer. The mirror disks were uniformly supported near their edge and were centrally loaded. The procedure employed in a microyield strength test was as follows:

1. The sample was clamped in the fixture and an initial interferogram was obtained at zero stress.
2. The load necessary to produce a maximum surface stress of 3.5 MN/m^2 (0.5 ksi) was applied for 10 min and then removed.
3. Ten minutes after load removal, a matching interferogram was obtained.
4. Steps 2 and 3 were repeated for increasingly higher stresses, in increments of 3.5 MN/m^2.
5. Interferograms were compared to determine any permanent changes in the flatness of the mirror specimens.

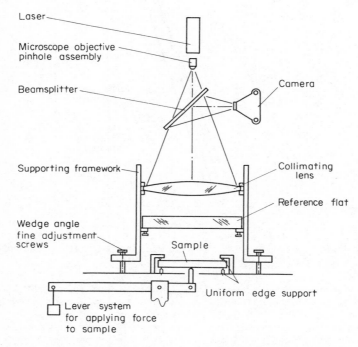

Fig. 8.27.   Schematic diagram of Fizeau interferometer for anelasticity measurements (Paquin and Goggin, 1971).

The reported measurement sensitivity was ± 0.02 λ. For the sample size employed, this is equivalent to a surface strain sensitivity of about ± 0.04 x 10^{-6}.

The same device was employed to investigate anelastic behavior and microcreep of nonmetallic mirror materials. Anelastic behavior was studied by taking interferograms as a function of time after removal of load. Microcreep

was investigated by maintaining a constant load on the specimen for times up
to 500 hr, after which the load was removed and an interferogram obtained.

   In subsequent studies on graphite/epoxy composites, the technology for
producing smooth optical surfaces in a manner which would not upset the intrinsic stability of the materials had not yet been developed. To overcome this
problem, a holographic test technique was devised that does not require highly
precise, extremely smooth optical surfaces. This technique can detect shape
changes regardless of the shape or geometrical accuracy of the part under test
(Freund and Goggin, 1972).

   Figure 8.28 is a schematic diagram of the holographic test facility.
Instrument components are rigidly attached to a granite base located in a
temperature controlled room. In making a hologram, the object is recorded on
the photographic plate using two coherent beams—an object beam and a reference
beam. The photographic plate is then developed and replaced in the holder.

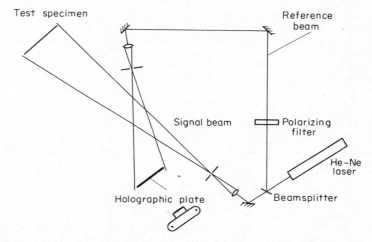

Fig. 8.28.   Optical schematic of holographic test facility
             (Freund and Goggin, 1972).

The object and reconstruction of the object are viewed through the plate and
are seen as superimposed. When the superpositioning is accurate, the result
is a bright object, i.e., a null. If the object is now subjected to bending
while in the holographic facility, a fringe pattern will develop that is indicative of the strain. If, upon removal of the load, the strain reverses
completely, the fringe pattern will disappear and a null condition will again
develop. If, on the other hand, some residual strain remains after unloading,
a fringe pattern will still be evident. The nature of the fringe pattern will
be indicative of the residual strain magnitude.

   The method has certain obvious advantages for studying microstrain behavior,
both microyielding and microcreep, of mirror materials. However, plate exposure periods can be quite long if low-power lasers are employed and this
requires that vibration and temperature fluctuations be kept to an absolute
minimum. The procedure is applicable only to bend tests which, as noted
earlier, are often difficult to correlate with tension results because of stress
gradients. Finally, strain sensitivity levels achievable in practice have not
yet been well defined.

F. *Etch Pitting Techniques*

Each of the methods described heretofore for sensing strain is based on integration of the deformation over a known gage length. Actually, on a microscopic scale, plastic strain is often very heterogeneous. If this could be observed in some way, it might be possible to detect localized plastic deformation at stresses below those at which general plastic deformation can be first detected. Etch pitting techniques can sometimes be used to accomplish this.

The rest position of individual dislocations can be observed by etching the surface of a specimen with a suitable chemical solution. The disturbed lattice about the dislocation intersection with the surface permits accelerated etching at this point and leads to the formation of an etch pit. An etch pit will generally have a geometric shape such as a triangle or a square that depends on the crystal structure of the material and the orientation of the particular crystalline grain relative to the etched surface.

Suppose that an annealed specimen is etch-pitted to reveal dislocation positions and that a known stress is then applied and removed. Re-etching will then reveal whether any of the dislocations have moved to new positions. If they have, this means that plastic deformation has taken place. Dislocation multiplication can also be detected in this way.

Young (1961) has used etch pitting to observe dislocation motion in copper single crystals at very low stresses. More recently, Vellaikal (1969) used similar techniques to study microplastic deformation in polycrystalline copper. Figure 8.29 shows some of his results.

This technique is not intended to replace those that employ integrating extensometers to detect average strain. Nonetheless, it is capable of detecting the very beginnings of plastic strain and is useful in fundamental investigations.

G. *Precision End-length Measurements*

Because of gage and instrument stability problems in long-term strain measurements such as microcreep tests, procedures have been devised for precise measurement of overall length changes of tension or compression specimens after exposure to known stresses for long times. These procedures require that the specimen ends be carefully prepared to enable total length measurements to be made with reasonable precision. This can be accomplished by lapping to provide flat and parallel end surfaces or by grinding convex end surfaces having a radius equal to half the specimen length.

Lement (1959) has described a length measuring technique employing cylindrical specimens, 10 cm (4 in) in length and 0.94 cm (3/8 in) in diameter, with the ends spherically ground to a 5 cm (2 in) radius. As shown in Fig. 8.30, the specimen is mounted vertically in a V-block fixture and held in position by rubber bands. The jig is placed on the anvil of a comparator which has a sensitivity of about 0.1 μm (4 x 10^{-6} in). In making a measurement, the fixture is moved back and forth so that the bottom end of the specimen slides across the supporting gage block while the upper end contacts the gaging point. A reading corresponding to the highest point of the specimen is taken. Next, a standard gage block of nominally the same length is wrung onto the anvil and a reading taken. The length of the specimen is the algebraic sum of the length of the standard block and the difference in readings obtained by measuring the specimen and standard, after appropriate corrections are made for thermal expansion differences in the two materials. With a sensitivity of about 0.1 μm and a specimen length of 0.1 m, the accuracy in measuring length changes is stated to be about 1 μm/m (1 μin/in).

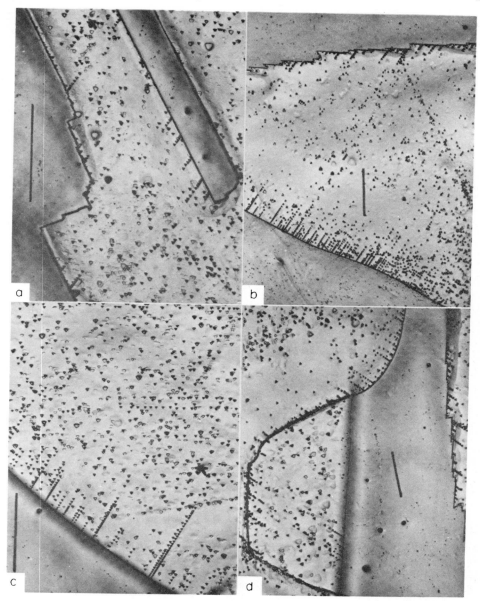

Fig. 8.29. Typical dislocation pileups in copper specimens revealed by etching after application of
a compressive stress of (a) 40 g/mm^2, (b) 50
g/mm^2, (c) 50 g/mm^2, and (d) 55 g/mm^2
(Vellaikal, 1969; copyright Pergamon, used by
permission). Black lines are 0.5 mm in
length and indicate direction of applied
stress.

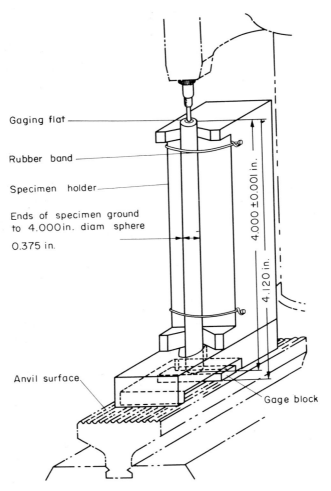

Gaging flat

Rubber band

Specimen holder

Ends of specimen ground
to 4.000 in. diam sphere

0.375 in.

4.000 ±0.001 in.

4.120 in.

Anvil surface

Gage block

Fig. 8.30. Apparatus for precision end-length measurements (Lement, 1959; copyright ASM, used by
permission).

Roberts and Averbach (1950) and Hordon *et al*. (1958) modified the specimen
by introducing threaded end sections to permit investigation of tensile microyield behavior. After the initial length was measured, the specimen was loaded
in tension to a known stress. A second length measurement was made after
removal of the load to detect any permanent length change. These steps were
repeated to ever greater stresses until measurable amounts of strain were observed. This technique would be equally applicable to long-term microcreep
tests.
In the authors' laboratory, microcreep of both tension and compression
specimens has been investigated on specimens with ends lapped flat and parallel.
Cylindrical specimens approximately 7.5 cm (3 in) in length and 1.25 cm (0.5
in) diameter are used. Tension specimens normally are prepared with a reduced

section and threaded ends for gripping. Their initial length is measured by comparison with a wrung stack of laboratory grade chromium carbide gage blocks, employing an electronic gage block comparator with an LVDT sensing head, illustrated schematically in Fig. 8.31. Each small division on the indicator dial represents 0.025 μm (10^{-6} in).

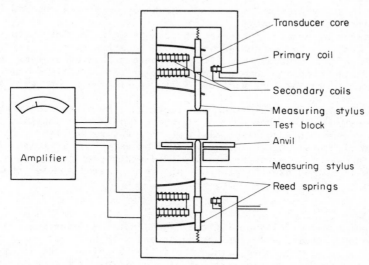

Fig. 8.31.   Schematic illustration of electromechanical comparator employed in precision-end-length measurements (Meyerson *et al.*, 1960).

Measurements are conducted in a constant temperature laboratory, held at $20° \pm 0.15°$C ($68° \pm 1/4°$F). Actual temperatures of the reference block and workpiece are monitored to the nearest $0.01°$C with contacting digital thermometers and corrections for thermal expansion are applied to each measurement. Since the specimen ends are not always perfectly flat and parallel, length readings are taken at three reproducible locations and averaged. These are repeated several times to assess the reproducibility of the measurements. Specimens are removed from the gage block comparator between repeat measurements to more closely duplicate the situations that exist in microcreep testing.

Once an initial length has been established, the specimens are again removed from the comparator and transferred to a creep stand, where loads of known magnitude are applied for predetermined times. After the creep load is removed, the specimens are remeasured relative to the same gage block reference, employing procedures identical to those employed in establishing the initial length. With proper attention to details, it is possible to detect plastic strains of approximately 1 to 2 x 10^{-6} with this method.

Another way to do the type of experiment just described is to measure total length interferometrically rather than by physical contact methods of measurement. In the interference method, the difference in length between a test specimen and a reference standard (laboratory grade gage block) is measured in terms of fractional wave lengths of light, employing an interferometer similar to that employed by the U.S. Bureau of Standards for calibrating gage blocks.

This procedure has the potential to improve strain sensitivity slightly— Meyerson *et al.* (1960) indicate that specially designed optical interference comparators, operating under ideal conditions, can provide precision to approximately 1 or 2 parts in 10 million when used to measure gage blocks. Assuming a 5 cm (2 in) gage block, this represents an ability to detect a length change of 0.005 to 0.01 μm (0.2 to 0.4 μin). Young (1965), on the other hand, feels that a more realistic value for most laboratories employing interferometric techniques is 0.025 μm (1 μin). While this latter figure still represents a slight improvement over physical contact methods of measurement, interferometry requires considerably greater care in end-preparation procedures and is more tedious to carry out.

The advantages of precision end-length measurements in microcreep testing are twofold. First, they are based on a known relatively stable reference standard. Therefore, gradual drift of electronic measuring systems does not cloud the results. Second, they do not tie up precision strain sensing devices and recorders for long periods of time while the creep test is in progress.

The major disadvantage centers on the fact that the creep stress must be removed in order to make a measurement of creep strain. To position the specimen for measurement and allow time for temperature equilibration generally requires several hours. Creep recovery could be occurring in this interval, but there is no way to observe it. A second disadvantage is strain sensitivity. This method will not be useful for studying creep strains smaller than about 10^{-6} to 10^{-7}.

H. *Other Techniques*

In addition to the strain sensing techniques already described, various other methods are available for measuring strain with an accuracy of 10^{-5} or better. The two described here, linear variable capacitors and laser diffraction devices, have not been used extensively in microstrain investigations.

A linear variable capacitor, according to Wolfendale (1968), provides an accurate and sensitive means of measuring displacement over a wide range. Such a device consists of a movable pick-up electrode in the form of a cylinder which can move coaxially within two or more other cylinders which are high-potential electrodes. When used with inductive ratio arms, accuracies to 1 part in 10^6 can be achieved and resolution of better than 1 part in 10^7 is possible. The use of an automatic self-balancing bridge enables measurement to be made rapidly with digital output and data recording. Excellent stability is claimed for these devices.

Strain measurement by laser diffraction is illustrated in Fig. 8.32. The

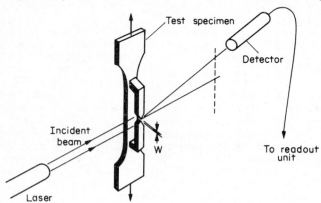

Fig. 8.32. Schematic representation of microstrain
measurement by laser diffraction.

gap, w, formed by two arms attached to a test specimen, is illuminated with light from a laser. The light rays are diffracted by the gap and form a fringe pattern resembling a row of dots. If the gap spacing changes, the fringe pattern will move, resulting in a magnification of several hundred times. If an optical detector is placed in the pattern, the change in gap spacing can be monitored by counting the fringes that sweep past the detector head. Increased resolution can be achieved by an analog fringe splitting technique. Resolutions of better than 1 part in 10^6 are claimed for this method. Stability is also said to be excellent because operation is based on the highly stable wavelength of the laser light (Pryor, 1974).

8.3 MEASUREMENT OF SMALL DIMENSIONAL CHANGES IN THE ABSENCE OF APPLIED STRESS

Precision end-length measurement is the method that has been most commonly employed to observe dimensional changes that may occur over long periods of time or that may accompany thermal cycling. This can be accomplished with either physical contact measurements, employing gage block comparators of the type described earlier, or with an interferometer. Both require specimens with carefully prepared measuring surfaces.

In the authors' laboratory, length changes are measured on cylindrical specimens nominally 7.5 cm (3 in) in length and 1.25 cm (0.5 in) diameter, lapped to produce flat and parallel ends. An electronic gage block comparator, depicted schematically in Fig. 8.31, is used to compare the specimen length with that of a wrung stack of laboratory grade gage blocks. The procedures employed in conducting periodic end-length measurements are similar to those described for measurement of microcreep in the previous section. By proper attention to test details, particularly with regard to temperature compensation, length changes of about 1 part per million can be measured reproducibly in this way.

Meyerson (1960), at the U.S. National Bureau of Standards, has employed gage block comparators extensively in monitoring the dimensions of gage blocks over long periods of time. His experience indicates that this technique is capable of obtaining a precision approaching 2 parts in 10 million. According to Meyerson, similar, or slightly better, precisions are attainable through interferometric measurement of length. However, attainment of thermal equilibrium is extremely important in the interferometric technique. This leads to long hold times in the interferometer—for example, a minimum of 4 hr is considered necessary for a 2 in gage block. Thus, the test capacity of an interferometer is appreciably smaller than that of a mechanical comparator.

Detection of dimensional changes of even less than 1 part in 10^7 is receiving a great deal of attention in connection with planned large earth-orbiting telescopes. Dimensional changes of 5 parts in 10^8 can seriously impair the performance of a diffraction-limited mirror. Accordingly, the U.S. National Aeronautics and Space Administration is conducting programs aimed at defining the dimensional stability over a period of several years of various mirror-substrate materials to an accuracy of about 5 parts in 10^8 (Kurtz, 1973). One approach, being conducted jointly by Corning Glass works and the U.S. National Bureau of Standards, employs Fizeau interferometry. The light source is an iodine stabilized HeNe laser developed by the NBS. Monolithic etalons of the type illustrated in Fig. 8.33 are prepared from the candidate materials with surfaces A and B plane parallel to $\lambda/20$. A modified Dyson polarization interferometer is expected to be used in conducting periodic measurements on the etalons to detect changes in dimension L with an accuracy of about 1 part in 10^8.

A second approach, being followed at the University of Arizona, employs a Fabry - Perot interferometer. A frequency-stable laser beam is shone through an optical resonator whose mirrors are spaced by the test specimen. By moni-

toring cavity resonance as a function of time, length changes can be measured
with an accuracy of approximately 1 part in 10^9 (Jacobs *et al.*, 1973).

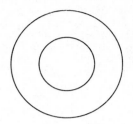

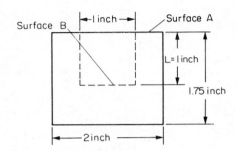

Fig. 8.33. Monolithic etalons for dimensional stability
measurement in Fizeau interferometer.

   Methods involving various types of extensometers attached to the test specimens have also been employed for measuring long-term stability. Problems can
arise from several sources, however, including drift in the electronics with
time, changes in contact resistance and lead wire resistance with time,
instabilities in the extensometer itself, and temperature changes. If
proper accounting and control of these sources of difficulty can be maintained,
any of various extensometer methods can be employed.
   Lyons (1970) has employed capacitance gages to measure long-time (1000 hr)
dimensional changes in free-machining Invar, Ti-6Al-4V, and Ferrotic MS-5,
a sintered carbide material. His techniques were described earlier in connection with microcreep testing. Figure 8.34 shows some of his results.

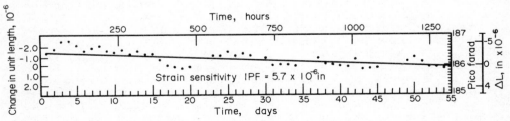

Fig. 8.34. Example of dimensional stability measurements made with a capacitance gage;
material investigated is free-cutting Invar,
water quenched from 815°C (1500°F) and held
1 hr at 315°C (600°F) (Lyons, 1970).

   In this chapter, experimental methods are described for measurement of small
strains--10^{-5} and less--that can arise from several sources. Strains associated with applied stresses are considered first and include anelastic strain
and microplastic strain. Following this, measurement of strains that can occur
in the absence of applied stress, for example, as a result of gradual microstructural changes or gradual alteration of internal stresses, is discussed.
   Techniques for measurement of anelastic strain are described only briefly
and include mainly low- and intermediate-frequency ranges. A comprehensive
review of experimental methods for investigation of anelastic phenomena has
been prepared by Nowick and Berry (1972), to which the reader seeking additional details is referred.
   Various methods are described for measuring microplastic strains resulting
from applied stresses. These include capacitance gages, electrical resistance
strain gages, Tuckerman optical gages, linear variable differential transformers, interferometry, holography, etch pitting, electromechanical comparators, linear variable capacitors, and laser diffraction devices. Each has
advantages and disadvantages that depend strongly upon the desired strain
sensitivity and accuracy, as well as upon the form of the material to be investigated. A system suitable for detecting plastic strains of 10^{-5} might not
be the best system for detecting strains of 10^{-7}, and vice versa.
   It is also emphasized that even the best system for measuring microplastic
strain will not ensure valid results if insufficient attention is paid to test
details. The necessity for paying close attention to specimen preparation
techniques, specimen quality and surface condition, test fixture quality,
calibration of load-weighing and strain measuring systems, alignment, and
temperature is discussed in detail. Temperature is extremely critical because
of thermal expansion and becomes increasingly critical as strain sensitivity
increases. At strain sensitivities of 10^{-6} and beyond, adiabatic heating and
cooling of the specimen associated with elastic strains can lead to serious
strain measurement errors, even if the environmental temperature is perfectly
constant.
   Techniques for measuring microplastic strains over periods of weeks, months,
and years are shown to be lagging behind those for short duration tests. Instability of the electronics employed in measuring strain, dimensional instability of the strain-measuring device, and imperfect temperature control are
contributors to uncertainties associated with microcreep testing. For example,
when optimum techniques are employed, bonded electrical resistance strain
gages can detect plastic strains with an accuracy approaching 1×10^{-7} in
short duration tests. However, in 1000 hr creep tests, the accuracy diminishes
by a factor of perhaps 10 to 50, because of the factors cited above. Precision
end-length measurements, in which microcreep specimens are periodically unloaded and measured relative to a stable reference gage block, have been
employed with considerable success to avoid the above uncertainties. However,
during the period following unloading and prior to measurement, an unknown
amount of creep recovery may be occurring.
   With respect to measurement of small dimensional changes in the absence of
applied stress, methods described include electromechanical end-length measurements, interferometry and various types of attached extensometers. Since
dimensional stability is normally measured over long periods of time, the
attached extensometers suffer from the same uncertainties described earlier in
connection with microcreep testing. Thus, accuracies better than a few parts
in 10^6 appear difficult to achieve with such techniques. Under ideal conditions, electromechanical end-length measurements are capable of detecting
dimensional changes approaching 1 part in 10^7 over periods of several years.

However, a more realistic figure appears to be about 1 part in 10^6. For greater sensitivity and accuracy, methods that employ interference of light waves offer the greatest promise. These techniques are undergoing rapid development in attempts to measure dimensional changes approaching 1 part in 10^8 or 10^9 over periods of several years. Two of these techniques, Fizeau interferometry and Fabry - Perot interferometry, employ highly stable laser light and are described briefly in this chapter.

REFERENCES

Argon, A. S. and East, G. 1967. The microyield strength of beryllium—iron alloys, *Trans. TMS-AIME*, 239, 598.

Bilello, J. C. and Metzger, M. 1969. Microyielding in polycrystalline copper, *Trans. TMS-AIME*, 245, 2279.

Bonfield, W. 1966. Microplasticity—measurement, interpretation, and application in precision instruments, Honeywell, Inc., Report HR-66-271:5-7.

Brentnall, W. D. and Rostoker, W. 1965. Some observations on microyielding, *Acta Met*. 13, 187.

Brown, N. 1968. Observations of microplasticity, *Microplasticity*, C. J. McMahon, Jr., editor, Interscience, New York, 45.

Brown, N. and Ekvall, R. A. 1962. Temperature dependence of the yield points in iron, *Acta Met*. 10, 1101.

Brown, N. and Lukens, K. F., Jr. 1961. Microstrain in polycrystalline metals, *Acta Met*. 9, 106.

Carnahan, R. D. and White, J. E. 1964. Some comments on strain-gage techniques for determining microstrain, *Trans. TMS-AIME*, 230, 249.

Carnahan, R. D., Arsenault, R. J. and Stone, G. A. 1967. Effect of purity and temperature on dynamic microstrain of niobium, *Trans. TMS-AIME*, 239, 1193.

Christ, B. W. 1973. Effects of misalignment on the pre-macroyield region of the uniaxial stress-strain curve, *Met. Trans.* 4, 1961.

Eul, W. A. and Woods, W. W. 1969. Shear strength properties to 10^{-10} of selected optical materials, Boeing Co., NASA Report CR-1257.

Freund, N. P. and Goggin, W. R. 1972. Advanced composite missile and space design data, Perkin-Elmer Corp., Interim Technical Report on U.S. Air Force Materials Laboratory Contract No. F33615-C-2033.

Geil, G. W. and Feinberg, I. J. 1969. Microplasticity I. Measurement of small microstrains at ambient temperature, U.S. National Bureau of Standards Report NBS 9996.

Holt, R. T. 1973. The flow stress dependence on grain size during microstrain tests, *Met. Trans.* 4, 875.

Hordon, M. J., Lement, B. S. and Averbach, B. L. 1958. Influence of plastic deformation on expansivity and elastic modulus of aluminum, *Acta Met*. 6, 446.

Hughel, J. 1960. An investigation of the precision mechanical properties of several types of beryllium, General Motors Corp., Report MR-120.

Hughes, E. J. and Rutherford, J. L. 1969. Microstrain in continuously reinforced tungsten-copper composites, *Amer. Soc. Test. Mat. Spec. Tech. Publ.* 460.

Jacobs, S. F., Norton, M. A. and Berthold, J. W., III. 1973. Dimensional stability of fused silica and several ultralow expansion materials, Paper presented at International Symposium of Thermal Expansion of Solids, Nov. 7 to 9, Lake of the Ozarks, Missouri.

Kurtz, R. L. 1973. Private communication; letter dated Nov. 1.

Lee, W. L. 1965. Determination of proportional limit by direct readout, *Mater. Res. Stand.* November, 571.

Lement, B. S.   1959.   *Distortion in Tool Steels*, American Society for Metals, Metals Park, Ohio.

Lukas, P. and Klesnil, M.   1965.   Hysteresis loops in the microstrain region, *Phys. Status Solidi*, 11, 127.

Lyons, J. W.   1970.   Absolute capacitance microcreep and dimensional stability measuring system, NASA Electronics Research Center Report C-134.

Lyons, J. W., Pambookian, H. C., Krawiec, J. P. and Curran, T. P.   1970. Computerized microstrain test system, NASA Electronics Research Center Report C-110.

Maringer, R. E., Cho, M. M. and Holden, F. C.   1968.   Stability of structural materials for spacecraft application, Battelle Columbus Laboratories, Final Report on NASA Contract No. NAS5-10267.

Marschall, C. W., Maringer, R. E. and Cepollina, F. J.   1972.   Dimensional stability and micromechanical properties of materials for use in an orbiting astronomical observatory, AIAA Paper 72-325.

Meyerson, M. R., Young, T. R. and Ney, W. R.   1960.   Gage blocks of superior stability: initial developments in materials and measurement, *J. Res. Nat. Bur. Stand.* 64C, 3.

Nowick, A. S. and Berry, B. S.   1972.   *Anelastic Relaxation in Crystalline Solids*, Academic Press, New York.

Parikh, P. D. and Hay, D. R.   1971.   Effect of plastic prestrain on microstrain behavior of <100> and <110> tantalum single crystals, *Scr. Met.* 5, 1039.

Paquin, R. A. and Goggin, W. R.   1971.   Micromechanical and environmental tests of mirror materials, Perkin-Elmer Corp., Final Report on NASA Contract NAS5-11327.

Pryor, T. R.   1974.   Private communication; letter dated Jan. 28.

Roberts, C. S. and Averbach, B. L.   1950.   Yielding in plain carbon steels, *J. Metals*, 188, 1211.

Roberts, J. M. and Brown, N.   1960.   Microstrain in zinc single crystals, *Trans. TMS-AIME*, 218, 454.

Roberts, J. M. and Hartman, D. E.   1964.   The temperature dependence of the microyield points in prestrained magnesium single crystals, *Acta Met.* 10, 430.

Rutherford, J. L., Bossler, F. C. and Hughes, E. J.   1968.   Capacitance methods for measuring properties of adhesives in bonded joints, *Rev. Sci. Instrum.* 39, 666.

Rutherford, J. L. and Swain, W. B.   1966.   Research on materials for gas-lubricated gyro bearings, The Singer Company-Kearfott Division, First technical summary report on NASA Contract NAS 12-90.

Solomon, H. D. and McMahon, C. J., Jr.   1971.   Solute effects in micro- and macro-yielding of iron at low temperatures, *Acta Met.* 19, 291.

Vellaikal, G.   1969.   Some observations on microyielding in copper polycrystals, *Acta Met.* 17, 1145.

Wilson, F. G. and Teghtsoonian, E.   1970.   Interpretation of microflow measurements in niobium crystals, *Phil. Mag.* 22, 815.

Wolfendale, P. C. F.   1968.   Capacitive displacement transducers with high accuracy and resolution, *J. Sci. Instrum.* Ser. 2, 1, 817.

Young, A. W.   1965.   Industrial interferometers, Instrument Society of America, Preprint No. 33.1-5-65.

Young, F. W., Jr.   1961.   Elastic-plastic transition in copper crystals as determined by an etch-pit technique, *J. Appl. Phys.* 32, 1815.

Chapter 9.

Low Thermal-Expansion Materials

Most solid materials expand when their temperature is increased. This expansion
is attributed to the increased thermal vibration of individual atoms associated
with the temperature rise. As each atom vibrates with increased amplitude, it
occupies additional space. To accommodate this, the atom spacing is increased
slightly, i.e., the solid expands. This is expressed quantitatively through
the coefficient of thermal expansion, α.

$$\alpha = \frac{1}{L_0} \frac{dL}{dT}$$

where L_0 is the length at a reference temperature (usually $0°$ C or $20°$ C) and
dL/dT is the slope of a graph of length L versus temperature T. Frequently,
an average expansion coefficient, $\bar{\alpha}$, over a range of temperature, T_0 to T_1, is
employed.

$$\bar{\alpha} = \frac{1}{L_0} \frac{(L_1 - L_0)}{(T_1 - T_0)}.$$

Actual values of the expansion coefficient are related to the strength of the
atomic bonds, which in turn influence both the elastic moduli and the melting
point. As indicated in Table 9.1, high expansion coefficients are associated
with low modulus values and low melting points.

Thermal expansion is normally considered to be a reversible phenomenon—
dimensional changes associated with a temperature change will disappear when
the temperature is returned to the original value.* Nonetheless, even though
reversible, thermal expansion can frequently prove to be extremely troublesome
in various precision applications. For example, in an orbiting telescope
subjected to alternating periods of sunlight and darkness, temperature changes
and temperature gradients would occur in the telescope support structure. This
could cause shifts in both the axis and the spacing of the secondary mirror
relative to the primary mirror, thereby impairing the quality of the image.
To minimize this problem, methods could be devised to hold the entire telescope
assembly within narrow limits of temperature. Generally speaking, however,
this would increase both the cost and the weight of the assembly. It would be
much simpler to employ materials in the structure that did not change dimensions
with changing temperature. Such materials are available and it is not surprising that they are of great interest to designers of precision devices.

------

*In the strictest sense, this is not true. Likhachev and Malygin (1963)
describe several "temperature after-effects", i.e., internal changes that occur
not instantaneously when temperature is changed. Each produces a dimensional
change in addition to that associated with thermal expansion. They include:
(1) variation in vacancy concentration (see Section 7.1), (2) variation in
solute atom clustering around lattice imperfections, (3) variation in the
stable positions of dislocations, (4) phase transformation and change in degree
of ordering, (5) diffusion of solute atoms to preferential lattice sites, and
(6) variation in internal stresses.

Table 9.1 Relationship of Thermal Expansion Coefficient to
Melting Point and Young's Modulus for selected Metals (ASM, 1961)

Element	Melting point, $^\circ$C	Young's Modulus of elasticity		Approximate thermal expansion coefficient near 20°C, 10^{-6}/$^\circ$C
		GN/m^2	10^6psi	
Tin	232	41	6	23[a]
Lead	327	14	2	29.3[b]
Magnesium	650	44	6.4	26[c]
Aluminum	660	62	9	23.6[b]
Silver	961	76	11	19.7[a]
Copper	1083	110	16	16.5 ·
Nickel	1453	207	30	13.3[a]
Cobalt	1495	207	30	13.8
Iron	1537	197	28.5	11.8
Chromium	1875	248	36	6.2
Molybdenum	2610	324	47	4.9
Tungsten	3410	345	50	4.6[b]

[a] measured from 0° to 100°C.

[b] measured from 20° to 100°C.

[c] average of measurements along a-axis and c-axis.

In this chapter, several types of materials that exhibit anomalous thermal expansion characteristics are discussed, including metals, glasses, ceramics, and composites. In discussing each type, characteristics of importance in design of dimensionally critical components are emphasized.

9.1   METALLIC MATERIALS

If a metal is to overcome its usual expansion tendencies with increasing temperature, a change must occur concurrently within the metal to produce an offsetting volume decrease. Metallurgical changes provide one possibility. For example, precipitates may form within the matrix phase or a phase transformation may take place as temperature is raised. If the precipitates or the new phase have a smaller specific volume than the original matrix material, the net thermal expansion will be diminished as a result of the metallurgical change. This, however, is not a practical method for achieving low thermal expansion, for several reasons.

First, such metallurgical changes usually depend both on time and temperature. Hence, the expansion characteristics would be a function of the rate of temperature change. Secondly, precipitation reactions are irreversible, i.e., precipitates formed during heating do not redissolve during cooling. Likewise, phase transformations, though usually reversible, display temperature hysteresis effects whose magnitude depends on the rate of change of temperature. For these reasons, no low expansion alloys have been developed that depend on metallurgical reaction or phase changes to offset the normal thermal expansion.

The difficulty in achieving very low expansion coefficients in metallic materials is apparent from the above examples. To achieve near-zero thermal expansion, the internal change opposing the normal thermal expansion due to atom vibration must be approximately equal in magnitude to the thermal expansion associated with atom vibration, be reversible and independent of time, and occur over a temperature range that includes room temperature. With these restrictions, it is perhaps not surprising that only one class of low expansion alloys has been developed commercially. This class, appropriately referred to as Invar (invariant) was discovered by Guillaume, a French physicist, in the early 1900's.* It is based on the composition Fe-36Ni and includes also alloys in which cobalt is substituted for some of the nickel and iron.

From a somewhat oversimplified view, Invar alloys achieve near zero expansion coefficients by virtue of their ferromagnetic behavior near room temperature. As the temperature is raised, the ferromagnetic nature of the material diminishes in intensity, reaching zero at the Curie temperature. The diminution of the ferromagnetic characteristics is accompanied by a slight modification of the forces that exist between atoms. This, in turn, results in a slight decrease in atom spacing which nearly compensates for the increased spacing associated with greater atom vibrations. Since this effect is reversible and not dependent on time, the net result is that, over a certain range of temperature including room temperature, Invar shows virtually no change in dimensions as its temperature is varied.

A typical curve illustrating the thermal expansion characteristics of Fe-36Ni Invar is shown in Fig. 9.1. For comparison, the thermal expansion behavior of several other metallic materials is included.   Note particularly two features. First, the expansion coefficient of Invar (the slope of the curve in Fig. 9.1) while small, does not achieve a value of zero. The minimum value in Fig. 9.1 is near $0.9 \times 10^{-6}/°C$ ($0.5 \times 10^{-6}/°F$). Second, below about $-50°C$ ($-60°F$) and above about $+100°C$ ($+210°F$) the thermal expansion coefficient of this material is appreciably greater than at room temperature. In fact, the expansion tendencies in these regions do not differ appreciably from those of other metallic materials.

Although Fig. 9.1 shows a minimum expansion coefficient of approximately $1 \times 10^{-6}/°C$, it is, in fact, possible to exercise some control over this value, as discussed in the next section.

Factors Influencing Thermal Expansion of Invar

The actual thermal expansion coefficient displayed by Invar alloys depends both on composition and processing. Minimum expansivity occurs at a nickel content very close to 36% by weight, as shown in Fig. 9.2. However, slight deviations from this value within about ±0.5% nickel do not seriously increase the expansion coefficient.

Impurities present in Invar tend to increase the expansivity. Manganese and carbon appear to have the strongest effects. Each 1% of manganese increases the expansion coefficient approximately $1.5 \times 10^{-6}/°C$, while each 0.1% of carbon increases it by about $0.4 \times 10^{-6}/°C$ (International Nickel Co., 1956).

---

*Invar is a registered trademark of Soc. Anon. de Commentry-Fourchambault et Decaziville (Aciéries d'Imphy).

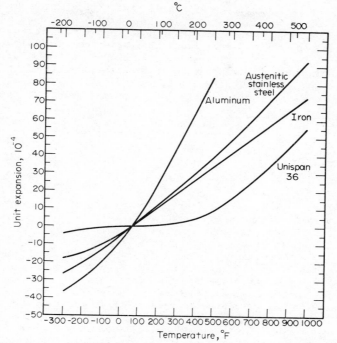

Fig. 9.1.   Unit expansion as a function of temperature
             for several metals, including an Invar alloy,
             Unispan 36 (Copyright Universal Cyclops,
             1968; used by permission).

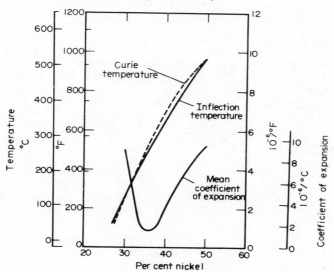

Fig. 9.2.   Effect of nickel on the mean expansivity
             between room temperature and the inflection
             temperature of annealed iron-nickel alloys
             (Copyright International Nickel Company,
             1956; used by permission).

Silicon is reported to have little effect on expansivity but tends to lower the inflection temperature in Fig. 9.1. Thus, free-machining grades of Invar, which generally contain somewhat more manganese (about 0.8% versus about 0.4% for normal grades) as well as small amounts of selenium, exhibit significantly greater α-values than do ordinary grades. To obtain the lowest possible α-values, some manufacturers have produced Invar with lower levels of impurities. Unispan LR35* is one example (Universal Cyclops, 1970). A typical analysis of this material is compared below with that of a normal grade, Unispan 36.

	Unispan 36	Unispan LR35
Manganese	0.35%	0.05% max
Silicon	0.12%	0.05% max
Carbon	0.04%	0.10% max

For reasons related to melting and fabrication practice, it is generally undesirable to remove all of the carbon from Invar.

It has long been known that substitution of small amounts of cobalt for some of the nickel in Invar alloys can further reduce the expansion coefficient (Scott, 1930). Alloys containing 31 to 32% nickel and 4 to 5% cobalt are said to achieve zero expansion coefficients without special processing. The term Super-Invar has been used to describe these materials. Caution must be exercised in use of the cobalt-bearing Invars, however, because of possible partial transformation of austenite to martensite at low temperatures, with an accompanying volume increase of appreciable magnitude. Data from Starr (1973) on two Super-Invar compositions, slowly cooled from 1095°C (2000°F), are tabulated below. The temperature at which transformation to martensite begins is denoted by M_s temperature.

	Fe-31Ni-7Co	Fe-31Ni-8Co
$\bar{\alpha}(30°$ to $100°$ C)	$0.69 \times 10^{-6}/°$ C	$0.98 \times 10^{-6}/°$ C
Curie temperature	$220°$ C	$240°$ C
M_s temperature	$-58°$ C	$-79°$ C

With respect to processing variables, the expansion coefficient of Invar alloys can be reduced substantially by rapid cooling from the annealing temperature and by cold working. The effect of rapid cooling is illustrated below for Unispan 36 (Universal Cyclops, 1968):

	Mean coeff. of linear expansion, $10^{-6}/°$ C	
Temperature range	$830°$ C(1525° F), slow cool	$830°$ C(1525° F), quench
$21°$ to $93°$ C (70° to 200° F)	2.0	0.63

At least a portion of the large effect of rapid cooling on the expansion coefficient is attributed to the presence of carbon in the material. When Invar is slowly cooled from the annealing temperature, any carbon in solution tends to precipitate in the form of tiny graphite particles. These graphite particles act to increase the expansivity. By rapid cooling from the annealing temperature, the carbon in solution is prevented from precipitating and lower expansion coefficients are obtained.

---

*Unispan is a registered trademark of Universal-Cyclops Corporation.

Avoidance of graphite precipitation may not be the entire explanation
quenching effects, however. For example, Unispan LR35 samples that had b
decarburized in moist hydrogen to reduce their carbon content from about 0.07%
to less than 0.01% showed similar effects of quenching. Slow cooling from the
decarburizing temperature (1095° C) produced an $\bar{\alpha}$-value of about 0.45 x $10^{-6}/°$ C.
When these specimens were reannealed at 845° C and rapidly quenched in water,
$\bar{\alpha}$-values of less than 0.05 x $10^{-6}/°$ C were obtained (Marschall, 1973a).

Cold working is yet another way to reduce the expansion coefficient of Invar
alloys. Figure 9.3 reveals that cold drawing annealed-plus-water-quenched
Unispan LR35 to a reduction level of only 15% reduced the expansion coefficient
from its initial value of 0.47 x $10^{-6}/°$ C to zero. Greater reductions caused
the coefficient to be slightly negative. In a companion experiment on a different lot of the same material, a cold reduction of 35% reduced the coefficient
from an initial value of about 0.4 x $10^{-6}/°$ C to about −0.5 x $10^{-6}/°$ C.

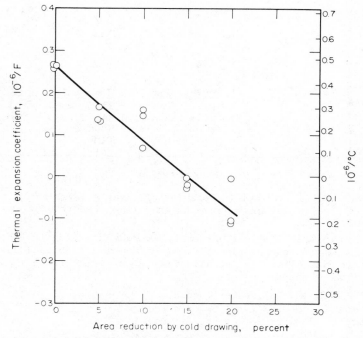

Fig. 9.3.   Thermal expansion coefficient of Unispan LR35
rod as a function of reduction level by
cold drawing (Marschall, 1973a).

It is evident from this discussion that the original Fe-36Ni Invar alloy,
containing residual impurities and slowly cooled from the annealing temperature,
is not a zero thermal expansion material. However, through control of composition and processing, it is possible to significantly reduce Invar's expansivity
and, in some cases, to develop zero or slightly negative expansion coefficients.

Micromechanical Behavior of Invar

The micromechanical behavior of Invar alloys can be varied over wide ranges,
depending primarily on the thermal and mechanical treatments employed in pro-

cessing. As noted earlier, it is common to water quench Invar from the annealing temperature to achieve a low thermal expansion coefficient. In this condition, it exhibits detectable plastic strains at relatively low stresses. As
shown in Fig. 9.4, independent investigations of two different Invars, one with
normal inpurity levels and the other with low residual elements, indicate that
plastic strains of 10^{-6} occur at stresses of about 40 to 60 MN/m^2 (6 to 8.5
ksi). Figure 9.4 shows also that a thermal treatment at 93° C (200° F) for 100
hr following the water quench significantly raises the stress required to
produce small amounts of plastic strain. As the plastic strain level approaches
100 x 10^{-6}, however, the curve for the stabilized specimens merges with that
for the as-quenched material.

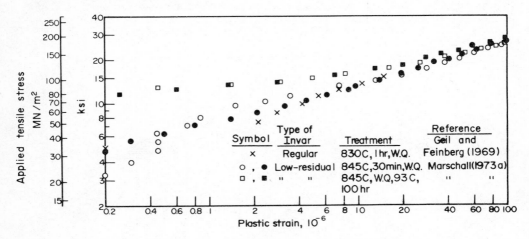

Fig. 9.4. Microyield behavior of Invar alloys, water
quenched from the annealing temperature.

When Invar is used in applications where maximum dimensional stability is
required, it is common to employ a more complex thermal treatment developed by
Lement *et al.* (1951). This is discussed in greater detail in the next section.
It consists of three steps:

1. Anneal 830° C (1525° F), water quench.
2. Heat to 315° C (600° F) for 1 hr, air cool.
3. Heat to 95° C (205° F) for 48 hr, air cool.

Figure 9.5 shows the yield behavior of Unispan 36 specimens that have been
subjected to this treatment. At a plastic strain of 10^{-6}, the yield strength
is seen to vary among the five specimens from about 75 to 115 MN/m^2 (11 to 17
ksi). However, at higher strain levels, the curves merge and all specimens
exhibit essentially identical yield behavior beyond about 100 x 10^{-6} strain.
The 0.2% offset yield strength (2000 x 10^{-6} strain) is approximately 290 MN/m^2
(42 ksi).

It appears from the results shown in Figs. 9.4 and 9.5 that thermal stabilization treatments following water quenching from the annealing temperature
can improve the microyield properties of Invar at plastic strain levels of
about 10^{-5} and smaller. Maringer and Hoskins (1971), however, found $\sigma_y(10^{-6})$
values of only about 40 MN/m^2 (6 ksi) for a cobalt-containing Invar (Fe-32Ni-
5Co) following the three-step treatment already described.

Another stabilization treatment that has been recommended for Invar following water quenching from 830° C (1525° F) is to reheat to 650° C (1200° F) for 1
hr, air cool, heat to 93° C (200° F) for 48 hr, and air cool (Schetky, 1957).

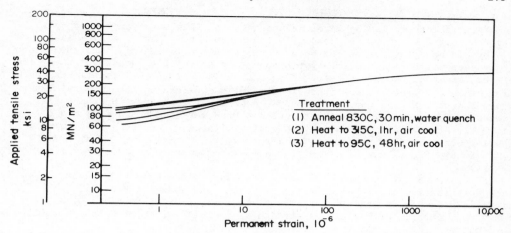

Fig. 9.5.   Stress versus permanent strain for Unispan 36,
             after stabilization treatment (Marschall, 1973b).

In microyield strength tests conducted on free-machining Invar that was
stabilized in this manner, Weihrauch and Hordon (1964) found $\sigma_y(5 \times 10^{-7})$ to
be 182 MN/m^2 (26.4 ksi). This is substantially higher than the values shown
in Fig. 9.5 for Unispan 36 that had received the earlier stabilization treatment. These investigators also found that test temperature had little effect
on the microyield strength, as shown below:

Test temperature	$\sigma_y(5 \times 10^{-7})$	
	MN/m^2	ksi
24° C (75° F)	182	26.4
66° C (150° F)	173	25.0
93° C (200° F)	170	24.7

Another treatment mentioned earlier to reduce the thermal expansivity of
Invar is cold working. Cold working is known to improve the macroyield
behavior of most metals and Fig. 9.6 demonstrates that it also improves the
microyield strength of Invar. In this experiment, Unispan LR35 rods were first
annealed and water quenched and then they were cold drawn in four passes to a
total area reduction of about 35%. This cold drawing increased $\sigma_y(10^{-6})$ from
its initial value of 40 to 60 MN/m^2 (6 to 8.5 ksi) to 310 to 340 MN/m^2 (45 to
49 ksi). Though not shown in Fig. 9.6, subsequent thermal treatment at 93° C
(200° F) for 48 hr had little additional effect on microyield behavior.

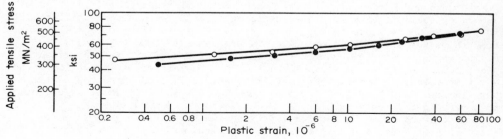

Fig. 9.6.   Microyield behavior of Unispan LR35, cold
             drawn 35% after annealing.

In applications involving sheet material, it is often necessary to flatten the Invar sheets following water quenching. This can be accomplished by stretching the sheets a few percent. This undoubtedly has an effect on the microyield behavior, depending both on the amount of stretching and the nature of subsequent thermal treatments. Although no systematic investigation of the effects of small amounts of stretching and subsequent thermal treatments on microyield behavior has been reported, results obtained in the authors' laboratory indicate a wide range of microyield behavior for Invar sheet material. In tests of 1 mm (0.040 in) sheet that had been water quenched from 845° C (1550° F), stretched a few percent to achieve flatness, and subjected to various thermal stabilization treatments at temperatures varying from 93° to 315° C (200° to 600° F), values of $\sigma_y(10^{-6})$ ranged from about 70 to 170 MN/m^2 (10 to 25 ksi).

In the course of the same investigation, several Unispan LR35 sheet samples were deliberately stretched approximately 20%. As shown in Fig. 9.7, the microyield curves for this material are at levels similar to those of cold drawn rod, particularly when the stretching is followed by a thermal stabilization treatment.

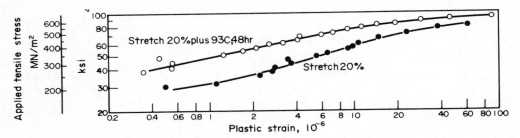

Fig. 9.7.   Microyield behavior of Unispan LR35 sheet, stretched 20% after annealing.

When Invar is subjected to an applied stress for long periods of time, microcreep strains may be observed. Some room temperature data for Unispan 36 which had received the three-step stabilization treatment described earlier (water quench plus 315° C plus 95° C), are shown in Fig. 9.8. As noted in Fig. 9.5, $\sigma_y(10^{-6})$ values for this material ranged from 75 to 115 MN/m^2 (11 to 17 ksi). It is evident from Fig. 9.8 that significant microcreep occurs within the first few minutes of loading at stresses as low as 69 MN/m^2 (10 ksi). At higher stresses, substantially more creep is observed. However, the rate of creep at 138 MN/m^2 (20 ksi) does not differ appreciably from that at 96 MN/m^2 (14 ksi) if the comparison is made on the basis of the amount of creep occurring between 0.1 hr and 1000 hr.

Weihrauch and Hordon (1964) also investigated microcreep of Invar. An example of their results for free machining Invar, quenched and stabilized at 650° C and 95° C, is shown in Fig. 9.9. As mentioned earlier, microyield strength values at a strain of 5×10^{-7} were near 180 MN/m^2 (26 ksi). The results on this material confirm the findings of Fig. 9.8 that significant creep occurs at stress levels below $\sigma_y(10^{-6})$. The effect of stress level and test temperature on total plastic strain observed in a 500 hr creep test are shown in Fig. 9.10. Note that temperature has a relatively small effect on microcreep in this material.

It is evident from the information presented here that a great deal of control can be exercised over the micromechanical properties of Invar alloys.

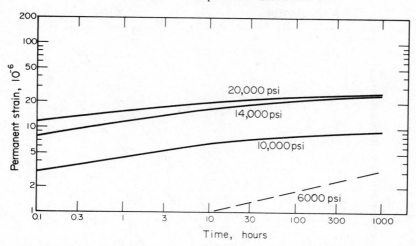

Fig. 9.8.   Average room-temperature microcreep of
Unispan 36 at several stress levels (Marschall,
1973b). Each curve is an average of three
tests.

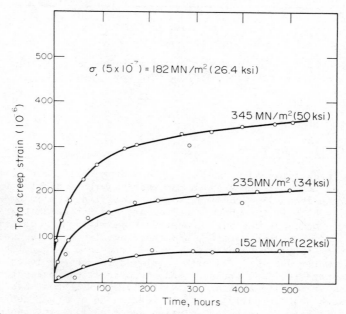

Fig. 9.9.   Total creep curves of free-machining Invar at
30°C (85°F) (Weihrauch and Hordon, 1964).

Cold working and thermal treatments, singly or in combination, can increase
the microyield strength substantially. The effect of such treatments on microcreep and stress relaxation has not yet been determined.

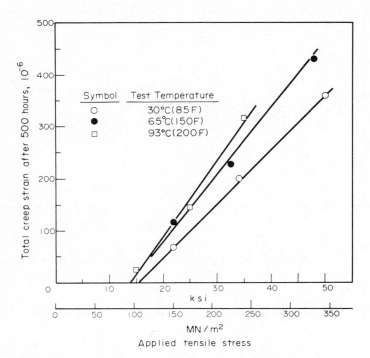

Fig. 9.10. Microcreep of free-machining Invar (after
Weihrauch and Hordon, 1964).

Dimensional Stability of Invar

Dimensional stability can be maximized in most metallic materials by complete removal of residual stresses in combination with a stable microstructure.
Invar is no exception. Slow cooling from the annealing temperature comes very
close to meeting both of the above requirements and, as demonstrated by Lement
et al. (1951), produces excellent dimensional stability at room temperature.
Unfortunately, as indicated earlier, slow cooling also produces thermal expansion coefficients that may be greater than desired in a particular application.
In such cases, water quenching is often employed to achieve lower α-values,
giving rise to high residual stresses and an unstable microstructure. With
passage of time at room temperature or slightly above, Lement has shown that
residual stresses can gradually diminish, leading to a length decrease in rod
specimens, and the carbon in solid solution can gradually rearrange itself
into a more stable configuration, leading to a length increase. The magnitude
of the dimensional change and the effect of temperature is shown in Fig. 9.11.
The potential dimensional instabilities resulting from water quenching to
achieve low α can be greatly minimized by appropriate thermal treatments that
essentially "use up" the dimensional changes prior to placing the material in
service. These thermal treatments are designed for two purposes: (1) to

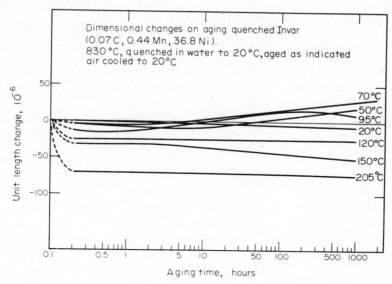

Fig. 9.11. Dimensional instability on aging quenched
Invar (Lement *et al.*, 1951; copyright ASM,
used by permission).

reduce residual stresses sufficiently so that they will no longer be subject
to gradual reduction at the service temperature, and (2) to rearrange the
carbon atoms to a configuration that is not subject to further change at the
service temperature. Residual stress reduction is accomplished most readily
at elevated temperatures (the higher, the better) although, in theory, stress
relaxation can occur even at very low temperatures if sufficient time is
allowed. For nearly complete removal of residual stress in Invar in times of
1 hr or less, temperatures near 560° C (1000° F) would be required. However,
Lement has shown that treatments at 425° to 560° C (800° to 1000° F) result in
large increases in α because of graphite precipitation. The highest allowable
temperature for stress relieving Invar without precipitating graphite appears
to be about 315° C (600° F).

Unlike stress relieving, rearrangement of carbon atoms in the Fe–Ni lattice
is not necessarily favored by high temperatures. Lement has indicated that
this phenomenon takes place primarily between room temperature and about 205° C
(400° F), with the rate of the reaction being maximum near 93° C (200° F). Once
completed, the reaction can be reversed by reheating above about 205° C (400° F)
but below the graphite precipitation temperature, after which it can be made
to reoccur at room temperature to 205° C (400° F).

Based on the above observations, Lement *et al.* recommended a series of
thermal treatments to provide both a low α and good dimensional stability.
These consisted of

1. 830° C (1525° F), 30 min, water quench; this treatment is designed to
    place all of the carbon in solution; rapid cooling will minimize the
    precipitation of graphite particles.

2. 315°C (600°F), 1 hr, air cool; this treatment is designed to reduce residual stresses introduced by quenching.
3. 95°C (205°F), 48 hr, air cool; this treatment is designed to rearrange carbon atoms into a relatively stable configuration.

Steps 2 and 3 must be done in the order indicated. Reversing them would nullify the benefit of the 95°C (205°F) treatment. Temperatures other than those shown in steps 2 and 3 might also be employed in certain situations. For example, step 2 might be done at 205° or 260°C (400° or 500°F) but, presumably, appreciably longer times would be required or less stress relief would be accomplished. Or, step 2 might be eliminated completely and step 3 used to accomplish both stress relief and carbon atom rearrangement.

The effectiveness of this stabilization treatment is shown in Fig. 9.12. Unit length changes in Unispan 36 bars, treated according to the recommendations of Lement *et al.*, were found to be only about 2 to 4 x 10^{-6} over a period of 2 years.

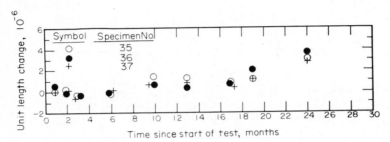

Fig. 9.12.   Length change of Unispan 36 as a function of time at 20°C (68°F) (Marschall, 1973b). Specimens received the three-step treatment described in the text.

In earlier discussions of thermal expansion coefficient and microyield behavior, the benefits of cold working were described. The extent to which cold working may influence long-term dimensional stability is not known with certainty. Limited work done in the authors' laboratory indicates that some instability follows cold working. However, it is likely that this can be minimized by appropriate thermal treatments.

9.2   GLASS AND CERAMIC MATERIALS

Several glass and ceramic materials are known to exhibit near-zero thermal expansion coefficients near room temperature. Examples are shown in Fig. 9.13. Each derives its anomalous expansion behavior in a somewhat different way as described below.

Looking first at glassy materials, the zero expansion-coefficient of fused silica at low temperatures has been attributed to the relative openness of its atomic arrangement permitting additional lateral vibration of the oxygen ions as the temperature is raised without increasing the average ion spacing. The relationship, shown in Fig. 9.13, between expansion coefficient and temperature for fused silica can be altered by the addition of titania. Titanium ions substitute for some of the silicon ions in the random glassy network, thus changing the vibrational characteristics of the ions and shifting the expansion

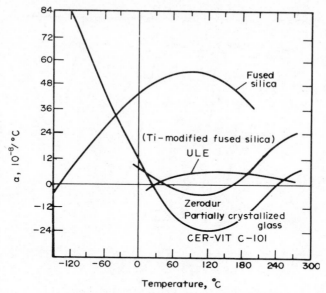

Fig. 9.13.   Thermal expansion behavior of several glass
and ceramic materials (Jacobs *et al.*, 1973).

coefficient to more negative values. This is the basis for ULE fused silica*
whose expansion characteristics are also shown in Fig. 9.13.
Certain other glasses that normally exhibit positive expansion coefficients
of appreciable magnitude can be made to exhibit near-zero coefficients by subjecting them to thermal treatments that cause partial crystallization. Figure
9.14 illustrates this effect for a $Li_2O-Al_2O_3-SiO_2$ composition to which TiO_2
and ZrO_2 were added as nucleating agents. Note that the relatively large
initial α-value can be reduced to zero or even to a slightly negative value as
crystallization proceeds at elevated temperatures. The mechanism by which
this crystallization reduces the expansion is believed to be associated with
the highly anisotropic thermal expansion behavior of the randomly oriented
crystallites. In at least one crystalline direction, the expansion coefficient
is negative and of sufficient magnitude that the net expansion coefficient of
the entire assemblage of crystals, averaged over the entire specimen, is
negative. This acts to reduce the positive expansion of the glassy matrix,
with the effect becoming more pronounced as the amount of crystalline phase
increases. Finally, when a sufficient percentage of crystallinity is achieved,
the expansion coefficient reaches zero. Such materials are often referred to
as glass ceramics. Cer-Vit C-101, whose expansion characteristics are shown
in Fig. 9.13, is one such glass-ceramic material.[†] Another is called Zerodur.[‡]

---

*ULE is a registered trademark of Corning Glass Works.
†Cer-Vit is a registered trademark of Owens-Illinois, Incorporated. Two grades
are produced: C-101 is referred to as mirror grade and C-126 as structural
grade.
‡Zerodur is a registered trademark of Jenaer Glaswerk Schott and Gen.

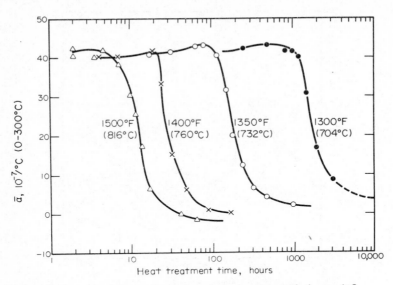

Fig. 9.14.   Average thermal expansion coefficient (0°
to 300°C) versus heat treatment for Li_2O-
Al_2O_3-SiO_2 system (Rapp, 1973; copyright
American Ceramic Society, used by permission).

Certain wholly crystalline ceramics also exhibit near-zero thermal expansion
coefficients over wide temperature ranges. One example is hafnium titanate,
$HfTiO_4$. Simpson (1960) has reported the average coefficient between 20° and
980° C to range from -1 to $+1 \times 10^{-6}/^{\circ}$ C. The anomalous expansion behavior of
this material is believed to result from the highly anisotropic thermal expan-
sion characteristics of individual crystals or grains. In one direction, the
coefficient is negative and of appreciable magnitude. The net effect in specimens with random orientation of crystals is a near zero expansion coefficient.

Micromechanical Behavior and Dimensional Stability
of Glass and Ceramic Materials

Because of their potential usefulness in high precision optical systems,
certain of the zero-expansion glass and ceramic materials have come under
careful scrutiny from the standpoint of micromechanical behavior and dimensional
stability. Moberly (1971) has stated that the strain stability of diffraction-
limited mirrors should be better than 10^{-7}, approaching 10^{-8}. Ideally, the
material would be perfectly elastic.
The low-expansion materials that have received the greatest attention for
mirror substrates in the United States are fused silica, ULE fused silica, and
Cer-Vit C-101. In addition, Cer-Vit C-126 has been investigated for suitability
for use in structural applications where zero-expansion is required. Each of
these materials can be classified as brittle and, hence, incapable of supporting large stresses in tension without fracturing. According to Brock (1972),
the usual maximum design stress for brittle materials is 10 to 15% of the

modulus of rupture of abraded bend specimens. This means that the maximum
design stress for these mirror materials is likely to be 7 to 10 MN/m² (1 to
1.5 ksi) and perhaps slightly larger for the structural material, Cer-Vit
C-126. For this reason, most studies of micromechanical behavior have been
limited to relatively low stresses.

Most evidence indicates that fused silica, ULE fused silica, and the Cer-
Vit materials experience little or no permanent plastic strain when exposed to
short-duration external loading and, hence, have a yield strength that equals
or exceeds the fracture strength (Marschall *et al.*, 1972; Paquin and Goggin,
1971; Moberly, 1971). Exceptions have been reported by Eul and Woods (1969,
1973) and by Woods (1970) who studied microyield behavior in torsion with
exquisite strain sensitivity ($\sim 10^{-9}$). These latter investigators found that
specimens with untreated ground surfaces exhibited detectable permanent strain
of the order of 10^{-8} at relatively low stresses and that the magnitude of the
permanent strain increased with increasing stress. An example of the type of
behavior found for rough ground ULE fused silica is shown in Fig. 9.15. Similar
trends were reported for fused silica and Cer-Vit C-101. When the ground
surfaces were subsequently etched or polished, Eul and Woods were unable to
detect permanent strains.

Some puzzling observations have been reported by Marschall (1972) in connection with microyield testing of ULE fused silica. Testing was done in
compression so that large stresses could be applied without danger of fracture,
thereby increasing the likelihood of detecting microyielding in this material.
Above a certain compressive stress level, near 70 MN/m² (10 ksi) for one
specimen and 140 MN/m² (20 ksi) for two others, strain-gage measurements

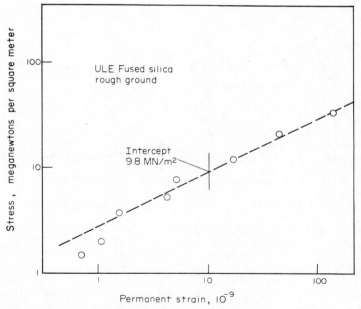

Fig. 9.15. Microyield behavior of rough ground ULE fused
silica (Eul and Woods, 1973).

indicated a <u>positive</u> residual strain of about 1 x 10^{-6} after unloading. As
the stress level was increased, the amount of positive residual strain increased
dramatically. After unloading from 550 MN/m^2 (80 ksi), one specimen showed
about +135 x 10^{-6} residual strain; another showed +250 x 10^{-6} after unloading
from 690 MN/m^2 (100 ksi). A small amount of recovery was observed to occur
in the first few minutes after unloading. As described later, similar unusual
results were obtained in compressive microcreep tests on ULE fused silica.
Identical tests conducted concurrently on Cer-Vit C-101 and C-126 produced no
unexpected results. Thus, there is reason to believe that the effect is real.
From a practical standpoint, however, the stresses at which this effect was
observed are well above those normally employed in designing with these brittle
materials.

Although microyielding appears to be a relatively minor problem in these
materials, several investigators have reported that their behavior is not
perfectly elastic. Anelastic effects are frequently observed. Upon removal
of an applied stress of sufficient magnitude, some residual strain will be
present but this will gradually disappear with time. This is illustrated for
Cer-Vit C-126 in Fig. 9.16. The greater the applied stress, the larger will
be this anelastic strain. Paquin and Goggin (1971), who tested mirror disks
with a centrally applied load and uniform edge support at a reported strain
sensitivity of 4 x 10^{-8}, found that the anelastic behavior is a function of
the material and the stress level. Fused silica and ULE fused silica show
no anelastic or microcreep strain when exposed to a tensile stress of 7 MN/m^2
(1 ksi). At 20 MN/m^2 (3 ksi), some anelastic strain occurs but disappears
within about one minute. At 35 MN/m^2 (5 ksi), approximately 15 min are required
for the anelastic strain to be recovered.

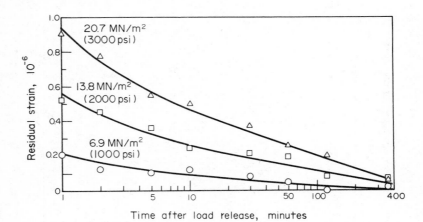

Fig. 9.16. Recovery of anelastic strain in Cer-Vit
C-126 after unloading from the indicated
stress (Paquin and Goggin, 1971).

In tests on Cer-Vit, Paquin and Goggin reported that Cer-Vit C-101 exhibits
no anelastic effects when loaded to 20 MN/m^2 (3 ksi) for periods of up to 16
hr, but does show discernible anelasticity when loaded for 500 hr, even with
the stress level at only 7 MN/m^2 (1 ksi). The reported anelastic strain
amplitudes were 0.09 and 0.33 x 10^{-6} for samples tested at 7 and 35 MN/m^2 (1

and 5 ksi), respectively. Cer-Vit C-126 exhibits much larger anelastic strains than Cer-Vit C-101 or either of the fused silica materials. At 20 MN/m^2 (3 ksi), the anelastic strain was 0.91 x 10^{-6}. Approximately 5 hr were required for this strain to disappear.

Eul and Woods (1973) also have compared the anelastic behavior of ULE fused silica and Cer-Vit C-101. Their tests were conducted in torsion at a somewhat greater strain sensitivity ($\sim$10^{-9}) than that reported by Paquin and Goggin. They report that Cer-Vit C-101, when stressed to 80 MN/m^2 (12 ksi), a stress approaching its ultimate strength, will retain a strain of 4 x 10^{-5} 1 sec after release, 6.7 x 10^{-7} at 1 min, and 1.1 x 10^{-8} at 1 hr. ULE fused silica recovers at a rate significantly greater, by a factor of about 10, than that for Cer-Vit. Accordingly, only a few minutes are required for the anelastic strain to diminish to a level of about 10^{-8}. This anelastic behavior, according to Eul and Woods, is independent of surface preparation procedure.

These anelastic effects are undoubtedly responsible for the reported reduction in apparent modulus with increasing stress for these materials (Marschall, 1972). As shown in Table 9.2, Young's modulus measured over a stress range of 0 to 70 MN/m^2 (0 to 10 ksi) is approximately 2% less than that measured over a stress range of 0 to 14 MN/m^2 (0 to 2 ksi).

Table 9.2
Young's Modulus Values Obtained in Compression for Glass
and Glass-Ceramic Materials as a Function of Stress Level (Marschall, 1972)

Material	Average Young's modulus (10^6 psi) for indicated range				
	0 - 2 ksi	0 - 4 ksi	0 - 6 ksi	0 - 8 ksi	0 - 10 ksi
Cer-Vit C-101	13.19	13.17			13.05
	13.18	13.14			13.03
	13.32	13.16			12.98
Average	13.23	13.16			13.02
Cer-Vit C-126	12.19	12.17	12.08		11.98
	12.16	12.04	11.98		11.90
	12.16	12.08	12.02		11.92
Average	12.17	12.10	12.03		11.93
Corning ULE	10.20	10.00	9.95	9.90	9.89
	10.14	10.01	9.97	9.95	9.94
	9.94	9.86	9.84	9.81	9.80
Average	10.09	9.96	9.92	9.89	9.88

Exposure to stress over long periods of time produces little microcreep in low expansion glasses and glass-ceramics if the stresses are maintained at a relatively low level. Paquin and Goggin (1971), employing centrally loaded edge supported mirror disks, reported that the permanent strain in these materials is no greater than about 0.1×10^{-6} when the maximum fiber stress does not exceed 35 MN/m^2 (5 ksi) and when sufficient time has elapsed after removal of the load to allow recovery to occur. Inelastic strains somewhat larger than the above value are present while the material is under load, but the strain measurement method employed in these tests required that the specimen be unloaded before measurement. Hence, appreciable recovery occurred in the interval between unloading and measurement and continued for several weeks.

Significantly greater amounts of microcreep have been reported for the Cer-Vit materials when stressed in compression at 69 MN/m^2 (10 ksi). As shown in Figs. 9.17 and 9.18, Cer-Vit C-101 averaged approximately 2×10^{-6} creep strain after 1000 hr and C-126 averaged nearly 20×10^{-6}. These measurements were made while the specimens were under load. After the load was removed, the creep strain was observed to recover gradually with time. Identical tests conducted concurrently on ULE fused silica specimens indicated unusual behavior, similar to that noted earlier for microyield tests. <u>Positive</u> creep strains of 12 to 18×10^{-6} were observed over a period of 1900 hr under a compressive stress of 69 MN/m^2 (10 ksi). Little recovery was noted over a period of 5 days after unloading (Marschall, 1972). Subtle microstructural rearrangements might be occurring under stress to produce a volume increase that could account for this unusual behavior. From a practical standpoint, this is an academic question because, as noted earlier, the stress level at which this effect was observed is well above the maximum design stress.

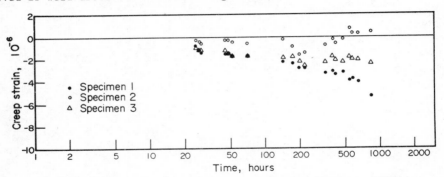

Fig. 9.17. Microcreep curves for CER-VIT C-101 at 20°C (68°F), stressed in compression at 69 MN/m^2 (10 ksi) (Marschall, 1972).

Thus, it appears that stress induced dimensional instabilities in low expansion glasses and glass-ceramics are extremely small, so long as the stresses are kept at low levels. Assuming design stresses no greater than about 15 MN/m^2 (2 ksi) to avoid brittle fracture, these materials may be assumed to exhibit nearly ideal elasticity.

From the standpoint of dimensional stability in the absence of external loading, the behavior of the low-expansion glass and ceramic materials also appears favorable. Mirror samples of fused silica, ULE fused silica, and Cer-Vit C-101 showed no figure change greater than $\lambda/50$ (0.04×10^{-6} strain) as a result of 100 thermal cycles between $-46°$ and $+38°$ C ($-50°$ to $+100°$ F) (Paquin and Goggin, 1971). An identical test on Cer-Vit C-126 resulted in a figure change of $\lambda/25$ (0.08×10^{-6} strain). Somewhat larger figure changes were reported for the interferometric measurements conducted at temperatures other than room temperature. This indicates possible nonuniform expansion coefficients throughout the workpiece or surface effects. Improvements in quality control should help to minimize this. For example, Slomba and Goggin (1972)

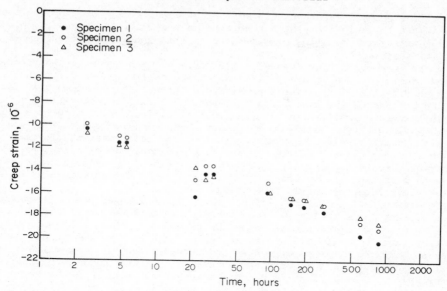

Fig. 9.18.   Long-time microcreep curves for Cer-Vit
             C-126 at 20°C (68°F). Stressed in compression
             at 69 MN/m^2 (10 ksi) (Marschall, 1972).

more recently found that a Cer-Vit C-101 mirror blank showed no figure change
greater than $\lambda/20$ when examined at temperatures as low as $-57°$C $(-71°$F),
compared with room temperature measurements. They reported also that reducing
the weight of the mirror blank to approximately one-third of its initial value
by machining hexagonal cores from the back-side caused only localized and small
figure changes ($\sim \lambda/10$ in magnitude). Subsequent etching of the machined
cavity surfaces produced additional small changes in surface figure, such that
the small irregularities noted earlier became smoothed, leaving a low area
approximately $\lambda/5$ in depth over the same area. The light-weighted mirror also
showed excellent thermal stability.   No change of figure was detected, within
$\lambda/20$, when the temperature was reduced to $-57°$C $(-71°$F) or when the temperature
was cycled 20 times between $20°$ and $-57°$C $(68°$ and $-71°$F).

    Based on the available experimental evidence, low expansion glasses and
glass-ceramics appear to possess attractive properties for use in dimensionally
critical applications. Because of their brittleness, however, applications
are limited to stress-free or low-stress designs. No clear-cut advantages of
one type of material over another (glasses versus glass-ceramics) are yet
evident at the current stages of development. Each possesses suitably small
thermal expansivity and relatively good dimensional stability, though this
latter characteristic is continuing to come under scrutiny for demanding applications. Thus, secondary properties, such as material homogeneity, density,
ease of fabrication, and susceptibility to fracture will govern material selection for specific applications.

9.3   LOW-EXPANSION COMPOSITE MATERIALS

    A material that is receiving increasing attention for use in thermally
stable structures is a composite made of graphite fibers in an epoxy matrix.
Its low expansivity derives from the negative thermal expansivity of the
graphite fibers and the extremely high elastic modulus of the fibers relative

to that of the epoxy. If present in sufficient quantities, the graphite fibers can counteract the relatively high thermal expansivity of the epoxy and produce a material with zero, slightly negative, or slightly positive thermal expansion coefficients.

There are a number of fundamental differences between graphite—epoxy composites and other low expansion materials discussed earlier. One difference pertains to directionality. The expansivity of graphite—epoxy is highly directional and depends on the type of fiber layup. For example, in a unidirectional layup, the expansion coefficient in the fiber direction may be zero or slightly negative, while at right angles to the fiber, the expansion coefficient will have a high positive value, approaching that of the epoxy matrix. Thermal expansion isotropy can be obtained in one plane of the composite by appropriate cross—plying but the expansivity normal to this plane remains high. Cross—plying also increases the necessary volume percentage of fibers to achieve a given low coefficient of thermal expansion.

Another feature of composites is that temperature changes produce internal stresses because the normal dimensional changes of each component are prevented from occurring by the presence of the other component. Thus, as temperature rises, the graphite fibers are placed in tension and the adjacent epoxy in compression. If the epoxy experiences stress relaxation at the particular temperature in question, some strain readjustment will occur in the graphite fibers as well. This can lead to gradual dimensional changes with time.

Finally, the dimensions of graphite—epoxy composites can be strongly influenced by moisture. They tend to expand in the presence of moisture and contract as moisture is removed, as, for example, in a vacuum.

The problems with gradual relaxation of internal stresses and with moisture effects pose obvious problems in using graphite-epoxy composites. Even measurement of thermal expansion coefficients is made difficult by these effects. In a program conducted to ascertain the applicability of advanced graphite—epoxy composite materials to dimensionally stable aerospace systems and structures, Freund (1974) confirmed that these materials are subject to dimensional instability associated with temporal changes, thermal cycling, and humidity changes. He concluded that the feasibility of using eight—ply composite laminates for structures requiring moderately good dimensional stability is marginal; for structures where dimensional stability is critical, additional development studies are required.

In spite of the problems reported by Freund, interest in these materials for thermally stable structures remains high, primarily because of the extremely favorable stiffness—to—density ratios that can be achieved relative to those attainable in Invar alloys or in glasses and glass-ceramics. Furthermore, by selecting different fibers and matrix materials and varying the type of layup, the material can be tailored to suit the application.

9.4 CHAPTER SUMMARY

The need for materials that maintain constant dimensions in the face of changing temperatures has led to a number of interesting developments. As a result of these developments, the designer now has available metallic, glass, ceramic, and composite materials that exhibit zero or near—zero coefficients of thermal expansion.

In dimensionally critical designs, other aspects of material behavior, such as micromechanical properties and long-term dimensional stability, may be equally as important as thermal expansion. In this chapter, each of these properties is discussed insofar as information is available. The discussion indicates that the behavior of each material depends on numerous factors, including composition and processing.

In selecting one of these materials for a particular thermally stable design, the designer must consider the strengths and weaknesses of each type. Some of these are listed in Table 9.3. Though not included in the table, economic factors may be of equal or greater importance.

A basic understanding of the behavior of each material can also be of great benefit. For example, rejection of Invar for a particular component solely because its expansion coefficient is too large would overlook the fact that the coefficient can be made zero or even negative by suitable processing. In fact, some control can be exercised over the expansivity of each of the types of materials described in this chapter, through control of composition and processing. Unfortunately, less is often known about the micromechanical properties and dimensional stability that accompany these compositional and processing variations. Nonetheless, as these materials are used in greater quantities, information of this type will gradually become available and the confidence with which designers can employ these materials in thermally stable structures will increase.

Table 9.3 Comparison of Several Types of Low Thermal
Expansion Materials

Strengths	Weaknesses

Invar Alloys

• Readily fabricated by conventional metal working processes	• High density
	• Achievement of $\alpha = 0$ requires special processing
• Resists fracture (tough)	• Ferromagnetic
• State-of-the-art technology	• Exhibits measurable dimensional instability over long periods of time
	• Limited temperature range

Glass and Glass-Ceramics

• Zero-expansion	• Fractures at low stress in tension (brittle)
• Chemically inert	
• Apparently dimensionally stable	• Low modulus
• Wide temperature range	• Relatively high density

Graphite-Epoxy Composites

• Low density	• Joints and attachments
• High modulus	• Moisture effects
• Properties can be tailored to the application (within limits)	• Stress relaxation in the epoxy
	• Requires special tooling

REFERENCES

ASM. 1961. *Metals Handbook*, Vol. 1, *Properties and Selection of Materials*, American Society for Metals, Metals Park, Ohio.

Brock, T. W. 1972. Private communication, letter dated Feb. 11.

Eul, W. A. and Woods, W. W. 1969. Shear strain properties to 10^{-10} of selected optical materials, Boeing Company, NASA Report CR-1257.

Eul, W. A. and Woods, W. W. 1973. Effects of surface polishing on the microstrain behavior of telescope mirror materials, Boeing Company, NASA Report CR-112217.

Freund, N. P. 1974. Advanced composite missile and space design data, Perkin-Elmer Corp., Air Force Materials Laboratory Report AFML-TR-74-33.

Geil, G. W. and Feinberg, I. J. 1969. Microplasticity II, Microstrain behavior of normalized 4340 steel and annealed Invar, U.S. National Bureau of Standards Report NBS 9997.

International Nickel Company. 1956. Iron-nickel and related alloys of the Invar and Elinvar types (30 to 60% nickel), International Nickel Company, New York (revised 1962).

Jacobs, S. F., Norton, M. A. and Berthold, J. W., III. 1973. Dimensional stability of fused silica and several ultralow expansion materials, Paper presented at International Symposium of Thermal Expansion of Solids, Nov. 7 to 9, Lake of the Ozarks, Missouri.

Lement, B. S., Averbach, B. L. and Cohen, M. 1951. The dimensional behavior of Invar, *Trans. Amer. Soc. Metals*, 43, 1072.

Likhachev, V. A. and Malygin, G. A. 1963. Temperature after-effect in metals, *Fiz. Metal. Metalloved*. 16, 435.

Maringer, R. E. and Hoskins, M. E. 1971. Unpublished data.

Marschall, C. W. 1972. Micromechanical properties and dimensional stability of materials for use in orbiting observatories, Battelle Columbus Laboratories, Final Report on NASA Contract NAS5-11351.

Marschall, C. W., Maringer, R. E. and Cepollina, F. J. 1972. Dimensional stability and micromechanical properties of materials for use in an orbiting astronomical observatory, AIAA Paper No. 72-325.

Marschall, C. W. 1973a. Unpublished data.

Marschall, C. W. 1973b. Investigation of micromechanical behavior and dimensional stability of Unispan 36 low thermal expansion alloy, Battelle Columbus Laboratories, Final Report to Itek Corporation, Lexington, Massachusetts.

Moberly, J. W. 1971. Tensile microstrain properties of telescope mirror materials, Stanford Research Institute, NASA Report CR-111948.

Paquin, R. A. and Goggin, W. R. 1971. Micromechanical and environmental tests of mirror materials, Perkin-Elmer Corp., Final Report on NASA Contract NAS5-11327.

Rapp, J. E. 1973. Thermal expansion coefficient - crystallinity relations in Li_2O-Al_2O_3-SiO_2 glass ceramics, *Ceram. Bull.* 52, 499.

Schetky, L. M. 1957. The properties of metals and alloys of particular interest in precision instrument construction, Massachusetts Institute of Technology, Report R-137.

Scott, H. 1930. Expansion properties of low-expansion Fe-Ni-Co alloys, *Trans. AIME*, 89, 506.

Simpson, F. H. 1960. High temperature structural ceramics, *Mater. Des. Eng.* 52, (4), 16.

Slomba, A. F. and Goggin, W. R. 1972. Research study to determine critical optical/mechanical properties of materials considered for selection as substrates for the primary mirror on a large telescope, Perkin-Elmer Corp., NASA Report CR-130141.

Starr, C. D.   1973.   Private communication, W. B. Driver Co., Newark, N. J.,
    letter dated May 9.
Universal-Cyclops.   1968.   Unispan 36 low thermal expansion alloy (pamphlet).
Universal-Cyclops.   1970.   Unispan LR35 low thermal expansion alloy (pamphlet).
Weihrauch, P. F. and Hordon, M. J.   1964.   The dimensional stability of
    selected alloy systems, Alloyd General Corp., Final Report on U.S. Navy
    Contract N140(131)75098B.
Woods, W. W.   1970.   Microyield properties of telescope materials, Boeing
    Company, NASA Report CR-66886.

Chapter 10.

Materials with Near-Zero Thermoelastic Coefficients

It is normally observed that the elastic moduli of structural materials diminish gradually as temperature is raised. This is an expected consequence of the slightly increased atom spacing associated with the greater atom vibration that accompanies heating. Typically, structural metals decrease in modulus from 1 to 5% of their room temperature value as temperature is increased $100°$ C. This is illustrated in Fig. 10.1 for several commercially available materials.

It is common to express the modulus change with temperature in terms of a thermoelastic coefficient, g. This is defined as

$$g = \frac{1}{E_0} \frac{dE}{dT}$$

where E_0 represents the elastic modulus at some reference temperature (often $0°$ C or $20°$ C) and dE/dT is the slope of the E versus T curve at temperature T. Thus, modulus decreases of 1 to 5% with a temperature increase of $100°$ C correspond to g values of $-100°$ to $-500°$ x $10^{-6}/°$ C. Examples of g values for several metals are shown in Table 2.2.

In certain precision applications it would be highly desirable if spring elements could maintain invariant load – deflection response over a range of temperatures. Since the response of a spring to load is directly proportional to the stiffness, i.e., to the elastic modulus of the material from which the spring is constructed, the material must display a constant stiffness over an appropriate range of temperature if it is to meet the objective of $g = 0$. Fortunately, such materials do exist. In fact, numerous alloys can be prepared that actually increase in stiffness as the temperature is raised within a certain range. Although a positive thermoelastic coefficient is not generally advantageous to a designer, it permits the metallurgist to add additional alloying elements to develop other important spring properties, such as high yield strength and resistance to stress relaxation. The additional alloying generally diminishes the magnitude of the modulus anomaly. Thus, if the thermoelastic coefficient of the basic alloy is positive, it is possible to reduce this to zero by adding other alloying elements in proper proportions.

Anomalous modulus–versus–temperature relationships can accompany certain metallurgical changes, just as was noted in Chapter 9 for anomalous thermal expansion behavior. However, as was also pointed out in Chapter 9, these internal changes must follow certain rules if practical advantages are to be realized: (1) they must produce an effect of appropriate magnitude that opposes the normal reduction of modulus with increasing temperature; (2) they must be reversible and independent of time; and (3) they must occur over the desired range of temperature. To date, the only metallic materials known to meet these requirements are certain ferromagnetic alloys containing primarily iron, nickel, and cobalt.* Masumoto and coworkers (1952, 1954,,

---

*Some ceramic materials also exhibit positive thermoelastic coefficients. For example, ULE titanium silicate displays thermoelastic coefficient values ranging from $+100°$ to $+200°$ x $10^{-6}/°$ C over a rather wide temperature range (DeVoe, 1969). However, since ceramics are not widely used in spring applications, the discussion in this chapter is centered around metallic materials.

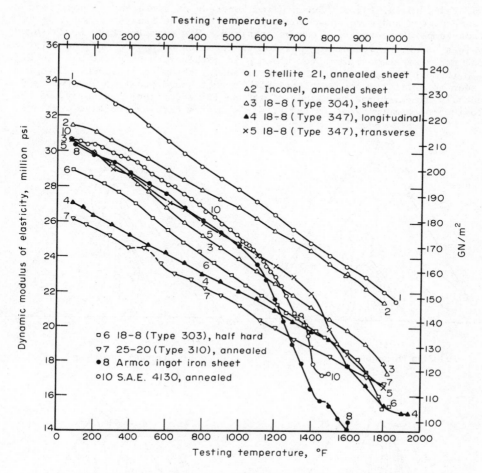

Fig. 10.1. Effect of temperature on the modulus of
elasticity of selected materials (Andrews,
1950; copyright ASM, used by permission).

No.	Alloy	Composition
1	Stellite 21	62 Co, 27 Cr, 6 Mo, 2 Ni, 0.25 C
2	Inconel	80 Ni, 15 Cr, 5 Fe
3	18-8 (Type 304)	19 Cr, 9 Ni, 0.07 C, 2.0 max. Mn
4,5	18-8 (Type 347)	19 Cr, 10 Ni, 0.07 C, 1.0 Cb
6	18-8 (Type 303)	18 Cr, 9 Ni, 0.12 C, 0.1 Se
7	25-20 (Type 310)	25 Cr, 20 Ni, 0.20 C, 2.0 max. Mn
8	Armco Iron	0.012 C, 0.017 Mn, trace Si
10	4130 Steel	0.31 C, 1.0 Cr, 0.21 Mo, 0.44 Mn

1955, 1956, 1957a, 1957b) have studied hundreds of different compositions
of such alloys and have shown conclusively that wide ranges of modulus-
versus-temperature behavior can be attained. For example, Fig. 10.2a shows
g values as a function of composition in Fe–Ni–Co alloys. Note the relatively
broad range of compositions for which g is either zero or positive. In Fig.
10.2b through Fig. 10.2d, the effect of adding chromium is shown. Chromium
is seen to shift the compositional limits that bound positive g-values and to
gradually diminish the magnitude of the positive g-values. Finally, at 15%
chromium, no compositions in the Fe–Ni–Co system display positive g-values.

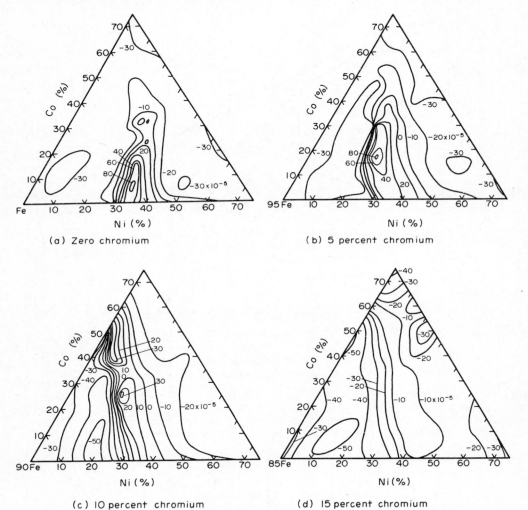

Fig. 10.2.   Temperature coefficient of shear modulus for
              Fe-Ni-Co-Cr alloy containing various
              percentages of chromium (Masumoto *et al.*,
              1956; copyright Tohoku University, used by
              permission).

Several commercial alloys have been developed that exhibit essentially constant modulus as a function of temperature. The first such alloy developed is credited to Guillaume, who also discovered Invar. He named the alloy Elinvar for invariant elasticity. Essentially, this is Invar in which 12% chromium is added in place of some of the iron. Other alloys also have been developed, based on the Elinvar composition. These are known by various trade names, including Isoelastic, Vibraloy, and Ni-Span-C.

10.1 ATTEMPTS TO RATIONALIZE ANOMALOUS MODULUS-TEMPERATURE RELATIONSHIPS

The majority of alloys that exhibit positive or zero thermoelastic coefficients in the absence of microstructural transformations are ferromagnetic. The relationship between modulus and test temperature for a typical ferromagnetic alloy is shown in Fig. 10.3.

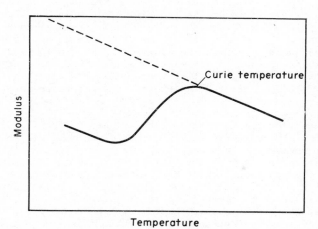

Fig. 10.3.   Schematic representation of modulus versus test temperature for an unmagnetized ferromagnetic alloy.

Above the Curie temperature, i.e., the temperature above which ferromagnetism disappears, ferromagnetic materials are seen to have negative thermoelastic coefficients similar to nonferromagnetic materials. Below the Curie temperature, however, the modulus is lower than would be anticipated from extrapolation of the high temperature data. This gives rise to a positive thermoelastic coefficient at intermediate temperatures.

Although the existence of positive thermoelastic coefficients has been known for many years, debate continues concerning the source of this behavior. Weiss (1963) introduced the concept that ferromagnetic Fe-Ni alloys can exist in two states with identical crystal structures and slightly different atom spacing. At low temperatures, according to this concept, the ferromagnetic high-volume γ_2 state is favored, while at higher temperatures, the antiferromagnetic low-volume γ_1 state is favored. As the temperature is raised, there is a gradual transition from the γ_2 state to the γ_1 state, accompanied by a strengthening of the interatomic forces, i.e., by an increase in the elastic moduli. According to Hausch and Warlimont (1972), this theory suffers from the fact that the existence of two electronic states has never been verified experimentally.

Another explanation for positive thermoelastic coefficients involves the small magnetostrictive strains that accompany reorientation of magnetic domains. Such an explanation is employed in this chapter because it is a relatively simple and straightforward way to rationalize observed phenomena. At the same time, the authors recognize that the explanations based on magnetostriction contain inconsistencies and are incorrect in certain of their details. The interested reader is referred to a paper by Hausch and Warlimont (1972) who quantitatively explain the anomalous modulus − temperature relationship in Fe-Ni alloys in terms of the contribution of the exchange energy.

In attempting to explain the modulus anomaly in terms of magnetostriction, it is helpful to briefly review ferromagnetic domain theory. In a ferromagnetic material below its Curie temperature with no externally applied magnetic field, the magnetic moments of adjacent atoms are held parallel to each other by extremely strong forces. Over a given small region or domain of a crystal, all of these atomic magnets point in the same direction and the material in this region is magnetized to saturation. In another small domain of the same crystal, the atomic magnets will likewise be aligned, but in a different direction. It is easy to see that a specimen made up of numerous crystals can have a net magnetization anywhere from zero to saturation, depending upon the orientation of the individual domains. If randomly oriented in all directions, as they are in the unmagnetized state, the net magnetization will be zero; if all are oriented in the same direction, as they are in the presence of a strong magnetic field, the entire specimen will be magnetized to saturation.

The alignment of domains in a ferromagnetic material can be achieved not only by imposition of a magnetic field but also by mechanical stress. In the latter case, however, no net magnetization results, because half of the domains align themselves in one direction and half in the opposite direction.

Accompanying the reorientation of ferromagnetic domains is an extremely small strain, of the order of a few parts per million, called magnetostriction. This term usually refers to a change in length; more generally, it means any change of dimensions brought about by reorientation of domains, including transverse and volume magnetostriction. The change in length associated with magnetization is shown in Fig. 10.4 for several different ferromagnetic materials. It can be seen that some materials lengthen while others shorten during magnetization.

As already noted, mechanical stress also produces domain realignment and, hence, magnetostrictive strain. The magnetostrictive strain accompanying stressing of a ferromagnetic material is shown in Fig. 10.5 for an Fe-40Co alloy loaded in torsion. The curve on the left was obtained in the presence of a strong magnetic field which reoriented all of the domains prior to stressing, thus eliminating any domain reorientation during stressing. The difference between the two curves at a given stress represents the magnetostrictive strain associated with domain reorientation. It has been demonstrated by Cochardt (1954) that when tests are conducted in torsion, the effect will always be as shown in Fig. 10.5, whether the material exhibits positive magnetostriction (Invar) or negative magnetostriction (nickel).

From this example, it is clear how magnetostriction can lead to a lowering of the apparent elastic modulus. Indeed, it has been shown that the magnitude of the modulus decrease, frequently termed the ΔE effect,* is strongly dependent upon the magnitude of the magnetostriction. This is shown in Fig. 10.6 for iron − nickel alloys. Note the close correspondence between the magnitude of the ΔE effect and the magnitude of the magnetostriction for all compositions

---

*$\Delta E = E$(with field) − E(without field).

studied. For example, alloys containing 28 to 30% nickel exhibit very little magnetostriction and a correspondingly small ΔE effect, while alloys containing 40 to 60% nickel show a sizable ΔE effect resulting from large magnetostriction.

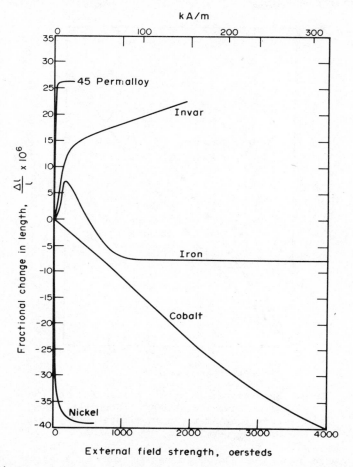

Fig. 10.4.   Magnetostriction of several common materials
             (Bozorth, 1951; copyright Litton Educational
             Publishing, Inc., used by permission).

In addition to its direct dependence upon magnetostriction, the ΔE effect is inversely proportional to the internal stresses existing in an alloy (Bozorth, 1951). For example, fully annealed materials (low internal stresses) exhibit a larger ΔE effect than do cold-worked materials (high internal stresses).

Consider next what happens as the temperature is varied. As temperature is lowered from the Curie point in a given material, the ΔE effect is expected to become larger because the magnetostriction increases, as shown for nickel in Fig. 10.7. While this is generally true at temperatures somewhat below the Curie point, at still lower temperatures the ΔE effect is frequently observed to diminish. This is shown for nickel in Fig. 10.8. Attempts have been made to explain this behavior in terms of internal stress variations with temperature (Köster, 1943a, b) but so far an entirely satisfactory explanation is lacking. It is not known whether all ferromagnetic alloys behave similarly to nickel in this respect.

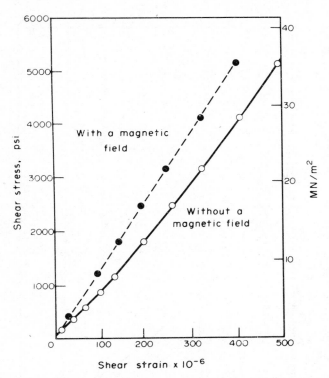

Fig. 10.5.   Shear stress – strain curves for Fe–40% Co
             alloy annealed at 900°C (Cochardt, 1954;
             copyright American Institute of Physics,
             used by permission).

Figure 10.8 also shows that the ΔE effect in nickel can be completely eliminated by superimposing a strong magnetic field. This is to be expected, since this eliminates the magnetostrictive component of strain during stressing. In certain alloys, however, superimposing large magnetic fields does not completely eliminate the ΔE effect, as indicated in Fig. 10.9 for an Fe–42Ni alloy. This is thought to be characteristic of alloys that exhibit large volume magnetostriction in addition to linear magnetostriction. Neither iron nor nickel, when magnetized at room temperature, exhibit an appreciable amount of volume magnetostriction (less than 0.1×10^{-6}) whereas iron alloys containing 20 to 50% nickel have relatively large volume magnetostriction (from 1 to 50 $\times 10^{-6}$) (Bozorth, 1951).

The domain theory of ferromagnetism thus offers an explanation for the origin of positive thermoelastic coefficients exhibited by certain alloys. In addition, it predicts that external magnetic fields and internal stresses, such as result from cold working, can exert a marked influence on the thermoelastic behavior. Likewise, it should be expected that a static stress applied to a specimen will influence its modulus behavior. Referring to

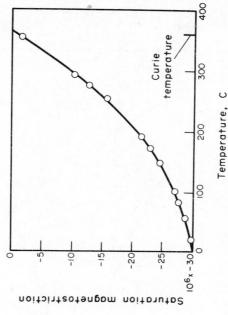

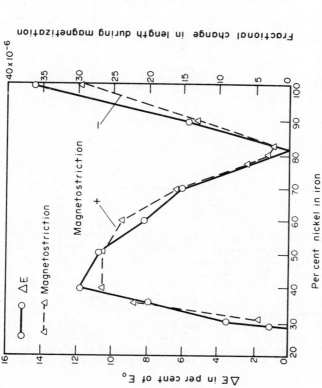

Fig. 10.6. Comparison of magnitude of ΔE effect with magnitude of magnetostriction for iron – nickel alloys annealed at 700°C (Bozorth, 1951; copyright Litton Educational Publishing, Inc., used by permission).

Fig. 10.7. Change of magnetostriction of nickel with temperature (Bozorth, 1951; copyright Litton Educational Publishing, Inc., used by permission).

All measurements at room temperature.

Fig. 10.5, it will be noted that the shear modulus without field measured between 0 and 1000 psi shear stress is considerably lower than the modulus measured between 3000 and 4000 psi. Hence, static stress appears to influence the ΔE effect in much the same way as magnetic fields and internal stress.

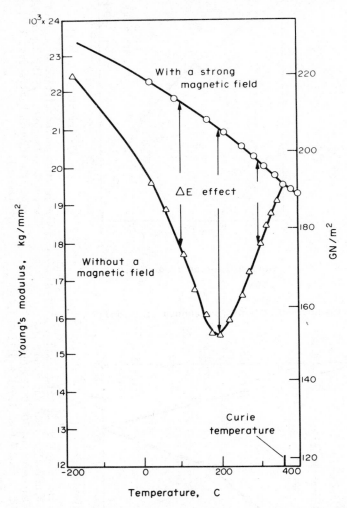

Fig. 10.8. Variation of modulus with temperature for annealed nickel (Bozorth, 1951; copyright Litton Educational Publishing, Inc., used by permission).

This aspect has not yet been extensively investigated, although it is important to the application of these materials in springs. For example, most modulus--versus-temperature measurements are conducted at very low stresses. If a material behaves as shown in Fig. 10.5, thermoelastic coefficients measured at low stresses could differ appreciably from values observed in springs operating at relatively high stresses. Additional attention is given to this matter later in this section.

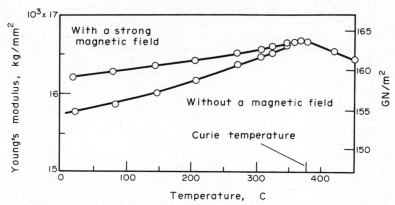

Fig. 10.9. Variation of modulus with temperature for Fe-
42% Ni alloy (Bozorth, 1951; copyright Litton
Educational Publishing, Inc., used by
permission).

Another feature of ferromagnetic alloys associated with domain reorientation
is damping. When a material is subjected to alternating stresses, the domain
reorientation accompanying each cycle absorbs energy and causes vibrations to
be damped.   In accord with previous considerations, it should be expected that
damping of ferromagnetic alloys will be influenced by temperature, magnetic
fields, and stress because of their effects upon domain movement.

To this point, the discussion has centered on materials that display a
positive thermoelastic coefficient over a certain range of temperature.
However, as shown earlier in Fig. 10.2, it is possible to obtain a coefficient
of zero by adjusting the composition of the alloy. For example, in binary
iron – nickel alloys, a zero thermoelastic coefficient is obtained at compositions containing either 27 or 44% nickel, as shown in Fig. 10.10. Early
attempts to manufacture these alloys produced discouraging results, because
the value of the coefficient depended so strongly on the nickel content.
Referring to Fig. 10.10, a variation of only 1% in nickel alters g approximately $100 \times 10^{-6}/°C$. Guillaume discovered, however, that adding 12%
chromium in place of some of the iron altered the g-versus-composition
relationship shown in Fig. 10.10 to that shown in Fig. 10.11. This chromium
addition greatly reduces the sensitivity of g to composition and, as noted
earlier, formed the basis for the first Elinvar alloy.

A possible problem concerning use of constant-modulus ferromagnetic
materials in springs was alluded to earlier. As shown in Figs. 10.8 and
10.9, application of a strong magnetic field reduces the magnitude of the
modulus anomaly; i.e., it moves g in a negative direction.   Similar effects
would be expected from an applied stress, for reasons already advanced. Thus,
if a ferromagnetic spring were to operate in either a magnetic field or under
the influence of a bias stress, its thermoelastic coefficient would depend on
the magnitude of the field or the bias stress, respectively.

From a practical standpoint, it would be preferable for g to be independent
of both magnetic field strength and applied stress magnitude. There is evidence
to suggest that it is possible to precondition alloys of suitable composition
to achieve this latter type of behavior. Such preconditioning, or prealigning
of magnetic domains, might result from internal stresses introduced by cold
working or by judicious alloying element additions. For example, Fine and
Ellis (1951) have shown that both cold work and molybdenum additions in-

fluence the modulus anomaly in Fe-42Ni alloys. This is illustrated schematically in Fig. 10.12. Note that the region of positive slope becomes less
positive when the alloys are cold worked or when the molybdenum content is increased, suggesting that both may be acting to minimize the stress-induced
magnetostriction. Thus, it is likely that both cold working and Mo additions
will decrease the dependency of g on magnetic fields and applied stresses.

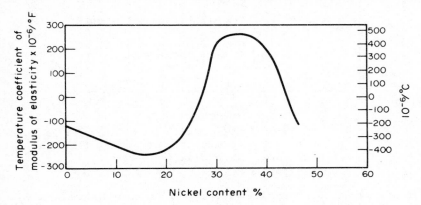

Fig. 10.10. Effect of composition on the temperature coefficient of modulus of elasticity of iron -
nickel alloys (copyright International Nickel
Company, 1963; used by permission).

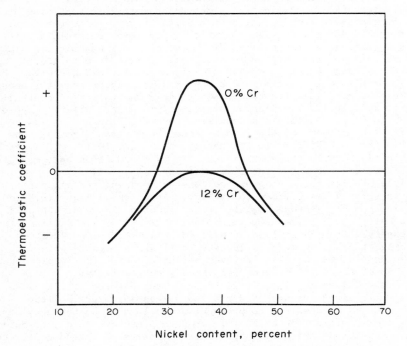

Fig. 10.11. Schematic depiction of effect of chromium on
thermoelastic coefficient of Fe - Ni alloys.

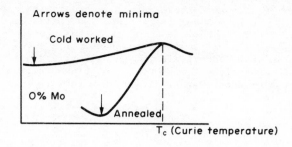

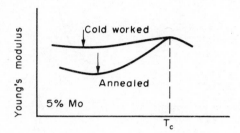

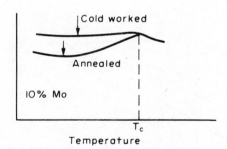

Fig. 10.12. Schematic diagram of modulus versus tempera-
          ture for Fe-42Ni alloys containing varying
          amounts of molybdenum (Fine and Ellis, 1951;
          copyright AIME, used by permission).

Introduction of coherent precipitates that elastically strain the matrix
might be expected to act similarly. In fact, one of the commercially available
constant-modulus alloys, known as Ni-Span-C, contains small amounts of
titanium and aluminum to promote precipitation of complex compounds of Ni,
Ti, and Al. This, in combination with cold working, not only helps to minimize the stress dependency and field dependency of g, but permits significantly
greater strengths to be realized in this alloy than in alloys of the Elinvar
type.

   An attempt has been made in this section to explain the anomalous modulus –
temperature relationships exhibited by ferromagnetic Fe-Ni-Co alloys. An
understanding of why these materials behave as they do and knowledge of the
range of thermoelastic coefficients attainable should broaden the options
available to precision designers. In addition, it should make designers

more aware of the limitations of these materials and of the extreme importance
of processing details on subsequent properties. Not only the chemical
composition of a particular lot of an alloy, but the degree of cold working
and the temperature and time of heat treatment as well, have pronounced
effects on the thermoelastic coefficient. This is illustrated in Fig. 10.13
for Ni-Span-C and demonstrates the importance of close interaction among
designers and materials engineers in specifying materials for precision
applications.

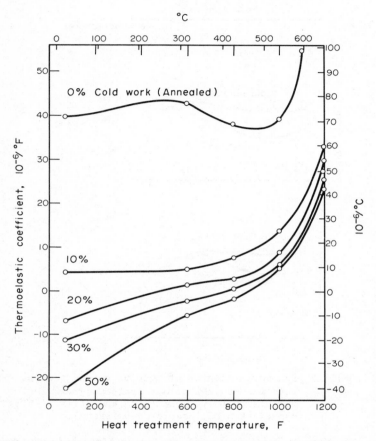

Fig. 10.13. Effect of cold work and 5 hr heat treatment
at temperature shown on thermoelastic coefficient of Ni-Span-C alloy 902 (copyright
International Nickel Company, 1963; used by
permission).

In the next section, attention is given to the micromechanical behavior
and dimensional stability of alloys that exhibit near-zero thermoelastic
coefficients.

10.2 MICROMECHANICAL BEHAVIOR AND DIMENSIONAL STABILITY

Information pertaining to the micromechanical behavior and dimensional
stability of materials that exhibit near-zero thermoelastic coefficients is
limited. Of the properties of importance in precision design, mechanical

hysteresis has received most of the attention while microyield and microcreep
have been largely ignored. To the authors' knowledge, these latter
properties have been investigated only for Ni-Span-C, a precipitation-
hardenable alloy. Accordingly, the discussion centers around this material.

The nominal composition of Ni-Span-C is Fe-42Ni-5Cr-2.5Ti-0.6Al. Solution
treating is accomplished by heating to approximately $1000°C$ ($1830°F$) and
quenching in water. This is followed by a precipitation hardening treatment
at a temperature ranging from about $315°$ to $700°C$ ($600°$ to $1300°F$). The temperature
selected depends both on the desired value of thermoelastic coefficient and on
the amount of cold working, if any, introduced prior to the precipitation
treatment. Empirical formulas for calculating the desired precipitation
hardening temperature are normally furnished by the alloy manufacturer. By
proper selection of processing procedures, the modulus of Ni-Span-C can be
made essentially constant over the range from $-45°$ to $+65°C$ ($-50°$ to $+150°F$).
The temperature range can be widened, from $-70°$ to $+115°C$ ($-90°$ to $+240°F$), by
slight alterations in processing with only small deviations from the constancy-
of-modulus behavior (McCain and Maringer, 1965).*

Mechanical Hysteresis

Ferromagnetic constant-modulus materials typically exhibit mechanical
hysteresis; that is, load - deflection curves observed on loading do not
coincide precisely with those on unloading, even though no permanent strain
occurs. With reference to Fig. 10.14, mechanical hysteresis can be expressed
quantitatively as

$$\text{Percent hysteresis} = \frac{\text{max. width of hysteresis loop (MN)}}{\text{max. deflection (OD)}} \times 100.$$

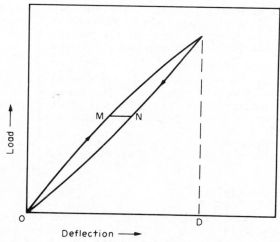

Fig. 10.14. Schematic load - deflection curve to illustrate
mechanical hysteresis. Percent hysteresis =
$\frac{MN}{OD}$ X 100.

---

*An Fe-Ni-Co alloy (Inconel 903) has recently been developed to provide nearly
constant modulus over a much wider temperature range than is possible with
Ni-Span-C (ASM, 1973). Between $-240°$ and $+640°C$ ($-400°$ to $+1200°F$), the
Young's modulus is reported to be 150 ± 3.5 GN/m^2 ($21.8 \pm 0.5 \times 10^6$ psi).

In near-zero thermoelastic coefficient materials, the magnitude of mechanical hysteresis is a complicated function of the processing conditions employed, including cold working and subsequent thermal treatments. In addition, it depends on the stress levels attained, increasing with higher stress levels. Figure 10.15 shows the effect of amount of cold work and subsequent heat

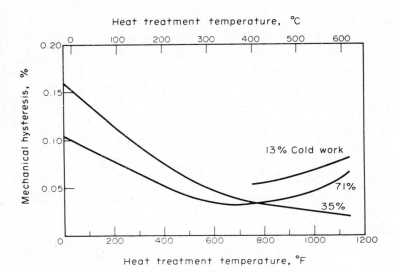

Fig. 10.15. Effect of 5 hr heat treatment of Ni–Span–C at indicated temperature on mechanical hysteresis in torsion (copyright International Nickel Company, 1963; used by permission).

treatment temperature on the mechanical hysteresis of Ni–Span–C, determined at a maximum torsional stress of 172 MN/m^2 (25 ksi). As is evident, the magnitude of the hysteresis can be reduced to very low levels by appropriate processing. The processing must, however, be compatible with thermoelastic coefficient requirements (see Fig. 10.13).

Similar data relating hysteresis to processing details for other zero thermoelastic coefficient materials are generally available from the alloy manufacturers.

Microyield and Microcreep Behavior

Imgram *et al.* (1968) examined the microyield behavior of Ni–Span–C in the solution treated plus aged condition ($1\frac{1}{4}$ hr at 985°C, water quench, age at 680°C for 21 hr). Their results are shown in Fig. 10.16. The stress required to produce a permanent strain of 10^{-6} is seen to be near 275 MN/m^2 (40 ksi). Yield strength values for other levels of plastic strain can be obtained directly from Fig. 10.16. Schetky (1957) reported an "elastic limit" of 395 MN/m^2 (57 ksi) for Ni–Span–C heat treated similarly to that investigated by Imgram.

As already stated, Ni–Span–C is frequently subjected to heavy cold working prior to aging to develop desired thermoelastic and hysteresis properties. Limited work conducted in the authors' laboratory indicates that this treatment

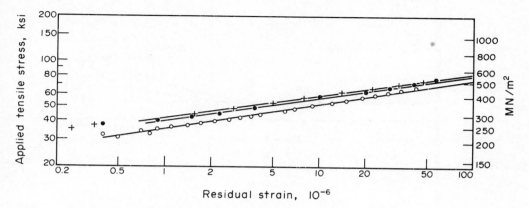

Fig. 10.16. Microyield behavior of Ni-Span-C in the
solution treated plus aged condition (Imgram
et al., 1968).

also has a marked beneficial effect on microyield behavior. Cold working 50%
followed by aging at 565° C (1050° F) produced σ_y (10^{-6}) values of about 730
MN/m^2 (106 ksi). This is more than twice the values reported by Imgram
et al. for Ni-Span-C that had not been cold worked.

With respect to microcreep, work by Imgram *et al.* (1968) on solution
treated and aged Ni-Span-C indicated that stresses of only 50 to 75% of σ_y
(10^{-6}) were sufficient to produce significant amounts of creep in 1000 hr.
These results are shown in Fig. 10.17. Recent work in the authors'
laboratory suggests that cold working 50% followed by aging at 565° C (1050° F)
improves the resistance to microcreep appreciably. At applied stresses of
434, 579, and 723 MN/m^2 (63, 84, and 105 ksi), the amount of creep detected
in 1000 hours was <2, <3, and 6 x 10^{-6}, repectively. As stated in the previous paragraph, σ_y(10^{-6}) of Ni-Span-C in this condition is about 730 MN/m^2
(106 ksi). Thus, the applied stress levels were 60, 80, and 100% of this
value.

Dimensional Stability

Imgram *et al.* (1968) also conducted a limited investigation of the
dimensional stability of Ni-Span-C in the solution treated and aged condition
and the effect of stress cycling, thermal cycling, and prestraining. Their
results are presented in Fig. 10.18. No significant change in length was
noted for the as-machined control specimen over a period of 4300 hr (6 months),
as seen in Fig. 10.18a.

Stress cycling ten consecutive times to a stress of 210 MN/m^2 (30 ksi)—
approximately 75% of σ_y (10^{-6})—caused an initial growth of about 2 to 3
x 10^{-6}, as shown in Fig. 10.18b. Thereafter, the dimensions remained constant for the next 4300 hr.

Thermal cycling five times between +100° and -75° C (+212° and -100° F) had
no apparent effect on the stability of Ni-Span-C. As shown in Fig. 10.18c,
no dimensional change was observed for a period of 4300 hr following thermal
cycling.

Small levels of plastic prestrain (25 x 10^{-6}) had no deleterious effect on

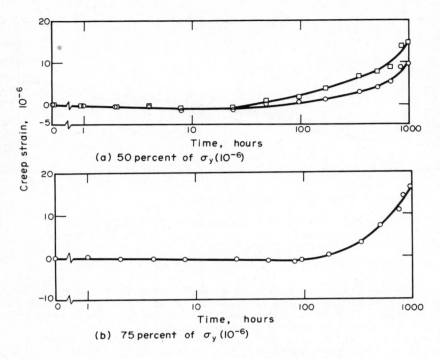

Fig. 10.17. Microcreep of Ni–Span–C in the solution
treated plus aged condition: (a) stressed at
50% of σ_y (10^{-6}); 138 MN/m^2 (20 ksi); (b)
stressed at 75% of σ_y (10^{-6}); 207 MN/m^2
(30 ksi) (Imgram et $al.$, 1968).

subsequent stability, as revealed by the data in Fig. 10.18d. However, larger
prestrains of 2% (20 000 x 10^{-6}) were followed by relatively large length
changes, continuing for long periods of time. As described in Chapter 5,
prestrains of about 2 to 5% also have a deleterious effect on σ_y (10^{-6}). It
is likely that appropriate thermal treatments following this level of prestrain
would minimize the deleterious effect. Very small levels of prestrain, on
the other hand, appear to be beneficial in that they raise σ_y(10^{-6}) significantly,
without adversely affecting dimensional stability.

10.3 CHAPTER SUMMARY

The need for materials that exhibit constant elastic modulus values over
a range of temperatures has led to the development of several iron–nickel
alloys that meet these requirements. The anomalous behavior exhibited by
these alloys is related to their ferromagnetic nature and the fact that they
experience magnetostrictive strains as they are magnetized or subjected to
stress. By proper selection of composition and processing, it is possible
to achieve thermoelastic coefficients that are slightly positive, slightly
negative, or zero over a temperature range of at least 100° C.

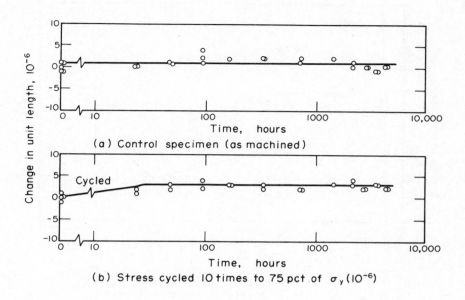

(a) Control specimen (as machined)

(b) Stress cycled 10 times to 75 pct of $\sigma_y (10^{-6})$

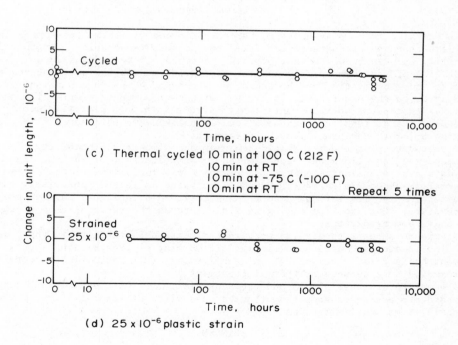

(c) Thermal cycled 10 min at 100 C (212 F)
10 min at RT
10 min at −75 C (−100 F)
10 min at RT   Repeat 5 times

(d) 25×10^{-6} plastic strain

Fig. 10.18.

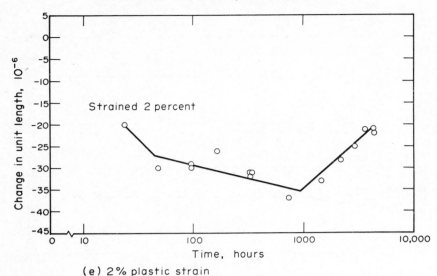

(e) 2% plastic strain

Fig. 10.18e. Effect of stress cycling, thermal cycling,
           and plastic strain on the dimensional
           stability of Ni-Span-C in the solution treated
           plus aged condition (Imgram *et al.*, 1968).

   In addition to displaying zero thermoelastic coefficients, these materials
display other useful characteristics. They possess reasonable strength and
ductility and can be worked hot or cold, machined, formed, and welded.
   Several of the commercially available constant-modulus alloys, including
Elinvar and Iso-elastic, are hardened and strengthened by cold working.
Ni-Span-C, on the other hand, is alloyed with small amounts of titanium and
aluminum that permit development of high strength through precipitation
hardening. Additional strengthening can be achieved in this material by cold
working prior to the precipitation treatment.
   Investigation of the microyield behavior and dimensional stability of
constant-modulus alloys has been limited to one material, Ni-Span-C. The
few data available indicate that σ_y (10-6) for solution treated and aged
material is about 275 MN/m^2 (40 ksi). This value is raised appreciably to
about 730 MN/m^2 (106 ksi) by cold working followed by aging at an elevated
temperature. Cold working and aging also improves the resistance to microcreep at room temperature.
   Ni-Span-C appears to display excellent dimensional stability in the
solution treated and aged condition. Cold working or prestraining several
percent has a deleterious effect on stability but this could almost certainly
be remedied by a subsequent thermal treatment.
   The nature of these alloys and the sensitivity of their behavior to composition and processing demands that they be used with full appreciation of their
potentialities and limitations.

REFERENCES

ASM. 1973. New Fe-Ni-Co alloy combats stresses, *Metal Progr.* October, 7.
Andrews, C. W. 1950. Effect of temperature on the modulus of elasticity,
   Metal Progr. July, 85.

Bozorth, R. M. 1951. *Ferromagnetism*, Van Nostrand, New York.

Cochardt, A. W. 1954. A method of measuring magnetostriction, *J. Appl. Phys.* 25, 91.

DeVoe, C. F. 1969. ULE titanium silicate for mirrors, Abstract of presentation at NASA Workshop on Optical Telescope Technology.

Fine, M. E. and Ellis, W. C. 1951. Thermal variation of Young's modulus in some Fe-Ni-Mo alloys, *Trans. AIME*, 191, 761.

Hausch, G. and Warlimont, H. 1972. Polycrystalline elastic constants of iron-nickel Invar alloys, *Z. Metallk.* 63, 547.

Imgram, A. G., Hoskins, M. E., Sovik, J. H., Maringer, R. E. and Holden, F. C. 1968. Study of microplastic properties and dimensional stability of materials, Battelle Columbus Laboratories, Air Force Materials Laboratory Report AFML-TR-67-232, Part II.

International Nickel Company. 1963. Engineering properties of Ni-Span-C Alloy 902, Huntington Alloy Products Division, Tech. Bull. T-31.

Köster, W. 1943a. Contribution to knowledge of magnitude of σ_i on basis of measurements of ΔE effect in nickel, *Z. Metallk.* 35, 57.

Köster, W. 1943b. Elasticity modulus and ΔE effect of iron-nickel alloys, *Z. Metallk.* 35, 194.

Masumoto, H., Saito, H. and Kobayashi, T. 1952. On the thermal expansion, rigidity modulus and its temperature coefficient of the alloys of cobalt, iron, and vanadium, and a new alloy 'Velinvar', *Sci. Rep. Res. Inst. Tohoku Univ. Ser. A*, 4, 255.

Masumoto, H., Saito, H. and Kono, T. 1954. Influence of nickel on the thermal expansion, rigidity modulus and its temperature coefficient of the alloys of cobalt, iron, and chromium, especially of Co-Elinvar: I. Addition of 10 and 20% nickel, *Sci. Rep. Res. Inst. Tohoku Univ. Ser. A*, 6, 529.

Masumoto, H., Saito, H. and Sugai, Y. 1955. Influence of addition of nickel on the thermal expansion, rigidity modulus and its temperature coefficient of the alloys of Co, Fe, and Cr, especially of Co-Elinvar: II. Additions of 30 and 40% Ni, *Sci. Rep. Res. Inst. Tohoku Univ. Ser. A*, 7, 533.

Masumoto, H., Saito, H., Kono, T. and Sugai, Y. 1956. Thermal expansion coefficient, rigidity modulus and its temperature coefficients of the alloys of Fe, Ni, Co, and Cr, and relations of super Invar to Stainless Invar and of Elinvar to CoElinvar, *Sci. Rep. Res. Inst. Tohoku Univ. Ser. A*, 8, 471.

Masumoto, H., Saito, H. and Goto, K. 1957a. Influence of addition of nickel on the thermal expansion, the rigidity modulus and its temperature coefficient of the alloys of Co, Fe, and V, *Sci. Rep. Res. Inst. Tohoko Univ. Ser. A*, 9, 159.

Masumoto, H., Saito, H. and Sugai, Y. 1957b. Influence of addition of copper on the characteristics of an Elinvar type alloy 'Co-Elinvar', *Sci. Rep. Res. Inst. Tohoku Univ. Ser. A*, 9, 170.

McCain, W. S. and Maringer, R. E. 1965. Mechanical and physical properties of Invar and Invar-type alloys, Defense Metals Information Center Memorandum 207.

Schetky, L. M. 1957. The properties of metals and alloys of particular interest in precision instrument construction, Massachusetts Inst. of Technology Report R-137.

Weiss, R. J. 1963. The origin of the 'Invar' effect, *Proc. Phys. Soc., London*, 82, 281.

Author Index

Subject Index